AF669316

Lean Digital Thinking is the digital age business excellence standard, and the author V. Srinivasa Rao (VSR) has explained it very well using four thinking clusters and 12 thinking principles.

VSR was part of the Tech Mahindra family and played a key role in introducing digital innovations at Tech Mahindra. The book articulates the lean digitalization journey well and explains in detail the digitalization of 12 business building blocks and a five-phase transformation approach. VSR has converted his experience and learning into easy-to-follow templates and frameworks that can be a reference guide for organizations embarking on their digital transformation journey.

—C.P. Gurnani

MD and CEO, Tech Mahindra

We are currently in the Fourth Industrial Revolution, in which technologies predominantly drive organizations. The fusion of various technologies has impacted the needs and behaviour of different businesses and academic stakeholders. It is not only the business organizations but also the educational institutions that are expected to digitalize their learning strategy, pedagogy, course offerings and overall academic operations to be relevant in the Fourth Industrial Revolution era.

In his book, the author VSR has explained very well the questions—Why digitalize? What to digitalize? How to digitalize? Furthermore, what is impressive is his explanation of digitalizing the 12 building blocks using 12 principles of unique digital business excellence standards including "lean digital thinking".

VSR has also explained the importance of "lean digitalization" and provided a five-phase approach for lean digitalization. I am confident that *Lean Digital Thinking* would enable even educational institutions to become faster digitally with optimal investments.

I recommend that academic institutions use this book to not only teach their students about the end-to-end digital transformation life cycle but also lean digitalize their institutions in a structured manner.

—Prof. M. Rammohan Rao

Former Dean and Professor Emeritus, Indian School of Business (ISB), Former Director, Indian Institute of Management (IIM), Bangalore

Organizations journeying towards digital transformation is more about culture, people and processes than it is about technology. *Lean Digital Thinking* blends a thought-provoking framework with a wealth of best practices and accessible examples. The book provides practitioners and thought leaders with a compendium reflecting VSR's distinguished career to accelerate that journey.

—Thomas Reuner

Senior Vice President, HFS Research

The world was already in the era of the Fourth Industrial Revolution when the COVID-19 pandemic began. The rapid acceleration in the adoption of digital activities and services will be the lasting legacy of the pandemic. VSR's *Lean Digital Thinking* is a timely and much-needed guide for understanding a world that is changing faster than we can comprehend. Business leaders and professionals have to apply themselves to navigate these changes in a productive manner. *Lean Digital Thinking* will help them in organizing their thoughts and actions in a way that can help sift through an overload of information. VSR has been very methodical in his approach in the book and it will serve to bring clarity of thought. The 12 principles of lean digital thinking aim to support difficult decisions that an organization needs. The why and how to digitalize is as important as when to digitalize. *Lean Digital Thinking* will be established as a useful book of reference with actionable insights for decision-makers.

Pranjal Sharma

Economic Analyst, Advisor, Author, Member, Global Expert Network, World Economic Forum, Former Executive Editor, Blomberg UTV

There are books about innovation and change and then there are books that are innovations in themselves and bring change through the very way in which they manage to organize and convey information. To change is so difficult; we all have a human tendency to resist it. To imagine an actionable pathway into ontological change—for that is what we talk about in moving to a hyperconnected world—requires a mind that has travelled all the way to not only see it but also to live it. It is given to just a few to be able to make that journey and return to the present as a teacher. That is exactly what V. Srinivasa Rao (VSR) has done. He makes a new way of thinking called lean digital thinking doable for any practitioner. This is a tremendous achievement. The fact that this way of thinking has actualized is in itself a reason to believe that a better world is possible.

—Rob van Kranenburg
Founder, IOT Council and IOTday

This book has created an unexpected opportunity to re-evaluate and accelerate digital transformation strategies and digitization initiatives. It helps organizations to become digitally ready, remain resilient against any disruptive challenges and build the new normal for the new future.

This book takes a compelling and straightforward approach to getting its message across. Digitizing a business is constantly needed for any organization to survive and succeed in the digital age. Moreover, this book provides practical insights for managing digitization with positive outcomes.

The robust framework explained in the book is very effective and efficient in helping all stakeholders influence the digital edge, solve critical digital business challenges, adapt quickly to accelerated change and leverage digital ecosystems to create new value to the customer and thrive.

The book does an excellent job explaining how to acquire a digital mindset and visualize the digital future by identifying all aspects of the life cycle of digital transformation. It benefits a large target group, including the board of directors and senior public servants. In addition, it also offers excellent added value for anyone, whether they are involved in digitization projects or not.

—Lalith De Silva
Former CEO, Sri Lanka Telecom,
Director, Sri Lankan Airlines,
Senior Advisor to Saudi Teleco

Organizations have increased the pace of digital transformation to survive and thrive in a pandemic environment. *Lean Digital Thinking*'s well-research content and its relevance to dynamic organizations' current needs create a compelling value proposition as an essential resource for leaders in all professions. The comprehensiveness of topics covered with great deal of clarity will help readers in all phases of their digital transformation journeys.

I find this book's content very comprehensive and highly relevant to current times of accelerated digital transformation. I like the way the author has structured his scholarly thought leadership and content around why, what and how to digitize. Mindset shifting is significant for digitalization and to initiate culture change. VSR has covered this topic quite well. Further, digital business blueprint, digitalization roadmap, digital solution store, COVID-19-related implications and excellence in executions are areas that are critical for success and are covered well in the book. I highly recommend the book.

—Kumar R. Parakala
President, GHD Digital,
Former Senior Consulting Partner, Regional Managing Partner and Global Leader, KPMG,
Invited Member, Forbes Technology Council

Given the noise around digitalization, the world has seen so many publications on that very subject. So why another book? A simple answer to that is *Lean Digital Thinking* is a cookbook to help getting clarity of the right things to do and the right way to do so, in the light of post-pandemic new normal. It all starts with mindset and "why" digitalization is an inevitable choice. Twelve to-be-achieved principles describe how to get ready for a digital journey, creating a better today and tomorrow. The "what" to digitalize provides an end-to-end perspective, from strategy to products to operations. VSR explains that real digitalization is much more than just automating internal processes; it starts with the business model, and he outlines an easy-to-understand roadmap to digitalization in the "how" section of his book. Very inspiring!

—Alfred Ermer
CEO, arago da vinci GmbH

"After all is said and done, there is more said than done."

This quote depicts the current understanding and application of digitalization in the world appropriately. While embracing digitalization, it is vital to complete it faster to grow the business by delivering great stakeholder benefits in the current era of the digital economy. The lean digital thinking framework provided in this book is very novel, innovative and practical, and serves as an invaluable reference to CXOs, business executives, academicians, practitioners, technologists and governments.

—Ramam Atmakuri
Executive Vice Chair, LV Prasad Eye Institute, Hyderabad, India

In the present age where everything is being perceived digital, being digital versus doing digital marks the boundary between success and failure. This book has come at the right time and portrays what needs to be done to instil the cultural mindset for "doing" digital in the new world order.

—Khushru M. Mistry
CIO and SVP, Eureka Forbes

Digitization has transformed business culture in profound ways. The sustainability of firms requires a critical assessment of existing and planned business models, strategies and products, and the author has provided a timely, eloquent and essential addition to the business executive toolkit. *Lean Digital Thinking* is a watershed moment in the future of business.

—Dr Eric Deans

CEO, Jamaica Special Economic Zone Authority

Lean manufacturing must be viewed in the context of waste reduction in the quest for driving manufacturing efficiencies. Lean digital approach constitutes a combination of lean management practices, design thinking, and advanced digital technologies and analytics. Digital transformation is thus a disruptive technology helping companies to harness real-time relevant information. The potential of digital technologies to transform the performances of companies is widely recognized. Most companies struggle to find the right approach to effectively harness digital promise. The volume by V.S. Rao reflects his in-depth knowledge and will guide the reader to adopt the right technology from a plethora of options.

—Dr P. Radhakrishnan Nair

Director, PSG Institute of Advanced studies

VSR, a long-time IT visionary, has put together in this book his decades of rich and practical experience in transforming businesses in complex settings with systematic and "lean digital thinking". It is a comprehensive and a compelling guide for organization leaders aiming to survive and stay relevant in the era of the Fourth Industrial Revolution. This timely book will help them understand and execute digital transformation strategies well.

—Dr Purnendu Sinha

Chair, Manufacturing Working Group, Institute of Engineering and Technology (IET), India

The topic(s) the book covers under the banner of digitalization are in need of an exposition. VSR observes that no one book covers the territory of digitalization comprehensively. I concur.

VSR sees the need to go beyond technology and focuses on surviving and thriving in a digital and interconnected world without being a native technology company, which he calls "digital immigrant". We need many points of view, different approaches, and maintain some '"village" mentality and cultural differences that enrich our thinking (diversity matters) in a globally interconnected world that seems to steer towards a single village at the size of the planet and its civilization.

VSR focuses on a number of organizing principles in Lean Digital Thinking that will allow us to navigate digital transformation. The chapter organization is very logical and the clustering is clear. The navigation aid is helpful—reminiscent of large textbooks—making it more widely accessible and a very usable book from a management/academe/teaching point of view.

—Prof. Iven Mareels
FTSE and Fellow of IEEE (USA), IFAC (Austria), KVAB (Belgium), EA (Australia),
Former Dean, Melbourne School of Engineering

VSR is many things—a coach, consultant, advisor and, most importantly, a practitioner. This is a practitioner's book. The need for digital transformation has been known to CXOs for a fairly long time now. In his very lucid style, VSR explains how digital organizations will differ from lean digital organizations. That's what impressed me more than anything else as this enterprise will end up as being a white elephant.

His book demystifies the challenges of digital enterprise journey by laying down simple building blocks which bring immense clarity to all, including strategists, business executives, technologists and policymakers. It is indeed a rare blend of strategic with pragmatic execution. In this book, VSR has put his wealth of experience sharing distinct insights on how lean digitalization will transform and positively disrupt the organizations.

A very timely book—at a time when all enterprises, irrespective of industry vertical, are embarking on the next orbit of a digital journey of contactless, remote delivery of services.

—Rajiv Sikka
Group CIO for Medanta multi-super specialties group of hospitals

Lean Digital Thinking comes across both as a preamble to the new world order as well as a practitioner's handbook. When VSR writes that "conventional business excellence frameworks do not work", he clearly articulates the need for a brand-new thinking cycle. The call to arms for all corporates trying to board this Noah's Ark, whom VSR calls digital immigrants, is done in a very practical and layered manner. Every digital corporate citizen will find this treatise as enriching as a cup of their favourite morning coffee.

—Ayan De
Chief Technology Officer, Exide Life

Technologies drive businesses in the Fourth Industrial Revolution. The Author, VSR, has clearly articulated how the fusion of Information Technology, Operational Technology, Communication and Collaboration Technology has created a new modern cyber system. This synthesis reflects the author's deep insight into the evolution of business and technology in the next decade.

It is interesting to learn about the troika of modern cyber system, which VSR has explained in detail, that helps organizations prepare for the digital era.

While VSR has provided a detailed framework and templates to digitalize 12 business building blocks of an organization, the essential building blocks that caught my attention are Digital Technology Strategy, Digital Technology Governance and Digital Technology Architecture. VSR has explained clearly how the digital technology foundation is very critical to digitalize the overall businesses.

We all have been experiencing uncertainty due to the COVID-19 pandemic. VSR has defined new normal that has emerged due to COVID-19 and provided many digital interventions that would help the organizations to prepare for the new normal.

—Sudhakar Varanasi
Advisor and Mentor for CXOs and start-up promoters on Transformational Initiatives

In this hyperconnected world, all aspects of human life and business are hugely impacted by technological innovations. The traditional organizations can no longer afford to ignore the huge changes that are being driven by digital technologies. There is a visible gap between those who have embraced digitalization and those who are yet to embark on the digitalization journey and this gap continues to widen. Some organizations are trying to force-fit existing business models and organizational structure to drive digital businesses. Implementation of technologies is not enough. The organizations need to go through complete transformation. VSR, through his more than two decades of industry experience, has seen this struggle from very close quarters. He is a thought leader.

In this book, *Lean Digital Thinking*, VSR has introduced the lean digital thinking framework. This will help organizations to achieve complete business transformation through the 12-12-5 model and become lean digital enterprise. The author has also devoted a complete section on the "new normal" that emerged with COVID-19 and has provided his recommendations for the digital interventions for the new normal.

A must-read for anyone wishing to master the art of digital business.

—Sanjeev Gupta
Vice President, Capgemini Engineering, GBL, India

The concept that VSR has brought forth and talked about in Lean Digital Thinking is indeed revolutionary. I have been a part of a lot of digital transformation journeys with global corporations and I can say with a sense of pride and humility that nowhere have I seen a framework or an implementation strategy as detailed, all-encompassing and succinct as this. This is an "*objet d'art*". I am sure that readers and practitioners of digital transformation will benefit immensely from this deployment framework.

—Ravi Ramaswamy
CEO, RV Consultants,
Former Senior Director and Head, Health Systems, Philips Innovation Campus, Bangalore

I used to head Digital Practice in 2012 when digital was not common. Now, in the last nine years, digitalization has moved at a very rapid pace. While the shelf life of technology is shortened and reduced further, it is vital to bring structure to digital transformation. The way VSR has described the chapters, I am very excited about the content of it. He has emphasized the cultural aspect of digital transformation. How revenue and growth can be accelerated using digital tools is still a puzzle among many business leaders, and this book presents that in a simplified way.

Also, VSR has introduced an interesting concept of "lean digital thinking", which could serve as a guide for many business, digital and tech professionals. Looking at all these aspects, I feel that getting hold of this book must be a treat for most business and digital leaders who want to enable businesses with digital horsepower.

—Anjani Kumar
CIO, Strides

VSR's new book *Lean Digital Thinking* is something that the industry has been waiting for a long time. Today, organizations have been transitioning to digital as never before. Traditional legacy organizations had been grappling with digitalization with the arrival of COVID-19, which was a game changer. With employees as well customers locked up at home, having a digitally run business no longer was a good to have. It was a question of thrive or survive. As businesses rush to change digitally, this book will serve as a ready reckoner not just giving you a list of things to do but also a number of things to avoid. VSR's 12-12-5 model for digitalization, in effect, creates more than 144 tenets of digitalization.

The name of the book, *Lean Digital Thinking*, should also give you a clue about the crux of the approach that VSR is propagating. His digitalization principles are focused on a lean digital transformation experience, not a fat one. In effect, frugal engineering occupies the centre stage of his approach. VSR also talks about the digital mindset which is crucial as organizations undertake the digital transformation journey.

In his trademark no-nonsense style, VSR takes you through each segment that would help you to:

- create new revenue streams,
- delight digital age customers,
- optimize human interventions to improve operational excellence,
- develop agility within the organization to respond to the market at speed,
- leapfrog known and unknown competition, and
- improve overall business performance.

If you are about to start reading this book, this is the best decision you could have made.

—**Shekhar Sanyal**

Director and Country Head,

The Institution of Engineering and Technology (The IET)

A contemporary read positioned at focusing on the current global "lean" and "digital" hype; the writer positions his thoughts and findings, together with his vast experience and practical knowledge, to create a new philosophy which works in post-COVID times.

This book can be referred to as a "glossary and planning guide for transforming to suit the new normal business world". The author walks the reader through a step-by-step approach, starting from inspiration, through the process of self-assessment, and onto the task of systematic innovation. The guidance offered in this book provides the reader with a map through the shrouded digital world, by aligning and adapting themselves with 12 principles and 12 business blocks which allow any corporation to assess their current position in terms of "lean digital quotient".

This is a must-read for anyone looking to stay ahead of the curve in the digital world and who holds extra utility for visionary leaders looking for a blueprint to strategically manoeuvre through the hardships of digitalization in today's world.

—Dr Hugo Wisidagama
Advisor Digital Transformation,
IPMCC Group & Associates

A must-read book in today's context of the Fourth Industrial Revolution and Digital Economy. *Lean Digital Thinking* is very unique and I am confident that it will empower organizations to acquire a digital mindset and culture required to transition from conventional businesses to digital businesses.

VSR has encapsulated his experience into this book so that both beginners and experienced professionals get the required knowledge, methods and practices for a structured lean digitalization of organizations. VSR has structured the book very well covering the "why, what and how" of lean digitalization with a 360 view of digital transformation.

VSR has emphasized the importance of building lean digital enterprises as digital enterprises are fat and costly to build and manage.

—Narottam Sharma
Group CIO, Mastek

The book *Lean Digital Thinking* written by Mr V. Srinivasa Rao, looks to be an excellent why-what-how comprehensive guide for corporates. The corporates can systematically define their digitalization vision and strategies and convert those into action. I have seen a few books on such topics, but this is a very comprehensive and focused approach that can help in aligning minds of various CXOs and arrive at fast and most appropriate realization plan. I also liked the author's recommendations useful for other stakeholders as well, such as technologists and academicians.

—Dr Subhash Rastogi
Founder, Innovative LEAN Mentoring Academy,
Director, BT & BT, Hyderabad, India

For too long, digital transformations have been associated with wrangling silos, a maze of digital methodologies, talents and datasets. Spaghetti is nice on the dinner plate but not while dealing with a mixed range of digitalization challenges, interoperability issues, migration and adoption issues. Never mind the agony of avoidable bloated budgets, extended delivery timelines, and unstructured and poorly coordinated offline and online worlds.

This is why I feel privileged to read the work *Lean Digital Thinking* by industry veteran V.S. Rao. From a solid career range overlooking many facets of digital transformation, VSR has documented an admirable holistic view of digitalization. This book may serve as a global consensus with suitable frameworks and phases to help the spectrum of talents work together optimally. With thousands of vendors offering a great product in their own regards, often whitepapers and product onboarding can be bias towards that product in isolation and it is easy for digital natives, or immigrants, to get lost in the firehose of information and still have blind spots. For this reason, early adopters through to laggards are always looking out for the unbiased astute and independent proven methods to cut through the endless sales pitches that may each have the best intentions in mind. This book serves across the spectrum of industry and hierarchy, as an authority on digitalizing business of all scales. As a digital native, I have enjoyed reading the common ground VSR enables for immigrants to natives to allow the strengths of offline and online to interplay. If you care about your career value and that of the colleagues and entities you serve, I recommend this book cover to cover.

—Scott OBrien

CEO and Co-Founder, Humense Pty Ltd

Today's organizations will need to leverage the full potential of industrial digital technologies to compete in the digital world of tomorrow. Organizations, leaders, employees, suppliers and partners need to acquire a new digital mindset. Digitalization is beyond the hyper-automation of business processes and is characterized by a fusion of technologies that impacts whole businesses and also creates new industries.

In the context of these dynamics, organizations must transform from their conventional business ecosystem to the digital business ecosystem. The author VSR has effectively translated his experience and knowledge into a book that provides a very structured approach for the lean digital transformation of organizations. Furthermore, VSR has emphasized the importance of lean digital businesses over digital businesses.

What caught my attention in the book is the concept of "lean digital thinking" and its 12 principles. Lean Digital Thinking is immensely readable, stitching crucial lessons for digitalization into contexts and scenarios that readers will find familiar. The book addresses each of the new and emerging technologies that underpin the digital transformation of modern commerce, with insightful discussion and examples. In addition, it is full of frameworks and guidance that will help any reader understand how to bring about transformational change within their organization successfully.

—Jim Morrish

Founding Partner, Transforma Insights
Former Co-Chair, Digital Transformation Working Group, Industrial Internet Consortium
Former Founder and CRO, Machina Research
Member, Global Advisory Body of the IET's Future Tech Panel

LEAN DIGITAL THINKING

Prepare for the Fourth Industrial Revolution and Digital Economy

LEAN DIGITAL THINKING

Digitalizing Businesses in a New World Order

V. SRINIVASA RAO

BLOOMSBURY
NEW DELHI • LONDON • OXFORD • NEW YORK • SYDNEY

BLOOMSBURY INDIA
Bloomsbury Publishing India Pvt. Ltd
Second Floor, LSC Building No. 4, DDA Complex, Pocket C – 6 & 7,
Vasant Kunj, New Delhi, 110070

BLOOMSBURY, BLOOMSBURY INDIA and the Diana logo
are trademarks of Bloomsbury Publishing Plc

First published in India 2021
This edition published 2021

ISBN: PB: 978-93-54350-14-6; eBook: 978-93-54350-16-0
2 4 6 8 10 9 7 5 3 1

Typeset in Adobe Garamond Pro by Manipal Technologies Limited
Printed and bound in India by Thomson Press India Ltd.

To my beloved maternal grandfather, Shri Punnaiah Kosaraju; my maternal grandmother, Smt. Ananthamma Kosaraju; and my courageous father, Shri Venkata Subba Rao Vegendla, who was a social reformer. They were real family leaders, and their support throughout my life was priceless and commendable.

Contents

Foreword

V. Srinivasa Rao, known as VSR, has worked with the Institution of Engineering and Technology (IET) spearheading the smart cities working group in India. This has led to several digital innovations like Water Twin, Quantifying City Sustainability, Smart Cities Vision 2035, Geo-Spatial Tech for Cities, and others that help in shaping the future cities. I want to congratulate him for introducing engineering innovations to contribute to India's success and growth.

With his extensive experience working in IT majors and with global customers, partners, digital specialists and professional organizations, VSR has introduced the New Digital Business Excellence Standard called "Lean Digital Thinking" in this book. I am confident that *Lean Digital Thinking* will become a significant catalyst in establishing common understanding, language and shared purpose for deploying digital technology and thinking effectively.

The impact of digital technology has brought about dramatic change in the way we live, work and play. Cyber-physical systems have become part and parcel of organizations, and technology has become a fierce driving force that is impacting the whole DNA of organizations. Hence, digitalization has become an essential agenda item for boards of directors and CEOs.

VSR has very well-articulated in this book why we need to digitalize, what to digitalize and how to digitalize, which all can easily understand. He has introduced the 12-12-5 model that helps organizations to transition from traditional businesses to deploying digital aspects in a structured manner. Twelve Business

Building Blocks, twelve Lean Digital Thinking Principles and five Lean Digitalization Phases are explained in detail.

High-tech companies have been getting into traditional businesses and, hence, traditional organizations must design and implement a digital business strategy to become relevant in today's digital age. Every digital enterprise must also be lean to optimize costs and facilitate effective communication and delivery.

I am confident that this book will aid in setting standard and understanding for organizations that want to "lean-digitalize". It should help empower business leaders with the required digital methods, tools, frameworks and prepare technology leaders to become digital innovators, strategic advisors and digital influencers.

June 2021

—Sir Robin Saxby
Founding CEO and Chairman of ARM Holdings,
Honorary Fellow of the FIET,
Fellow of the *Royal Academy of Engineering*

Preamble

We are now in the era of the Fourth Industrial Revolution, primarily driven by the fusion of different technologies. Digital natives are slowly penetrating traditional businesses and digital immigrant industries, such as manufacturing, banking, retail, travel, etc., are under tremendous pressure to catch up with digitalization and defend their forts. It is like a game of "align or perish".

VSR has worked with IT Majors and has played a crucial role in transitioning the Information Technology service organizations to Business Technology Solutions organizations, leveraging the fusion of IT, OT, CT and other technologies. He has been working on various solutions leveraging industrial Internet, consumer Internet and cyber-physical systems to prepare organizations for the Fourth Industrial Revolution and digital economy.

Today, there is no common understanding of digital or digitalization in organizations. It hinders the organizational effort to bring synergies among multiple departments, leaders and stakeholders while driving various digital initiatives.

VSR has well-articulated 12 lean digital thinking principles in this book that, I believe, will significantly help organizations, leaders and stakeholders acquire a digital mindset. He has justified why digital enterprises are fat to build, and maintain and highlighted the importance of converging lean with digital.

Like human anatomy, VSR has described the business anatomy, which includes 12 business building blocks, that must be digitalized, so that organizations can create new business opportunities, deliver unforgettable moments of experience and improve operational excellence holistically.

"Lean digital quotient", introduced in this book, helps organizations in knowing their digital strengths and weaknesses. VSR has clearly articulated the 12x12 assessment matrix and how to find an organization's lean digital quotient. I am convinced that the five phases of Inspire, Assess, Innovate, Experiment and Institutionalize to lean digitalize organizations would become a frame of reference for organizations during their digitalization journey.

I have seen how technologies evolved for many decades and how they have been continuously becoming driving forces to shape people's lives, governments, businesses and societies. Now, each organization is attempting to digitalize their businesses faster, better, steadier and cost-effectively. This requires a structured approach, and this book provides a systematic method to lean digitalize the organizations in an evolutionary manner.

COVID-19 has impacted human lives, businesses, governments and societies at large. VSR has defined a new normal in the context of COVID-19 and explained that various digital interventions would help organizations, governments and people to navigate through the new normal effectively.

I am confident that this book will become a useful reference for organizations to sail through their digital journey successfully.

June 2021

—Dr Richard Soley
Chairman and CEO, Object Management Group (OMG)

Preface

Today, digitalization has become an important initiative for nearly every organization. Most of the traditional organizations have been facing a few challenges because they are not digital natives:

1. They fail to understand the needs and wants of digital age customers.
2. They do not change their vision to align with the digital era.
3. They are not able to foresee the unknown competition emerging from different corners.
4. They force-fit existing business models and organizational structure to drive digital businesses.

There is a need for acquiring a new mindset and new thinking to bring changes in each building block of business, without which traditional organizations (also called digital immigrants) will not be ready for a digital future.

Why Do I Want to Write This Book?

We have been living in a hyperconnected world due to technology innovations like IPv6, internet, short-range and long-range networks, social, mobile, cloud and analytics technologies, distributed ledger technologies, mixed reality (MR), artificial intelligence (AI), machine learning (ML) and internet of things (IoT). **In a hyperconnected world, people to people, people to things, things to things connections are established in real time. They work together to make organizations nearly**

autonomous, in which human interventions are optimized or nullified.

Digital natives like Google, Amazon, Apple and Samsung have been leveraging the power of fusion of technologies (cyber system) and replacing digital immigrants (banking, automotive, manufacturing, music, watch, hospitality, travel industries and others). This means digital natives (technology firms) have been challenging conventional organizations, which have been there for many decades. The only choice left to traditional organizations is to become digital so that they will be successful in the era of the Fourth Industrial Revolution.

Digital immigrants have been facing severe challenges in their journey of digitalization. Digitalization is certainly not limited to implementing the latest technologies; it means a complete transformation of business DNA by leveraging these technologies. In the digital era, conventional business excellence frameworks do not work; hence, in this book, I am introducing a digital business excellence framework called **lean digital thinking** which would help digital immigrant organizations.

The Vision of My Book
My book is to become a companion, guide and advisor to

1. business executives,
2. government officers,
3. technology executives and
4. professors and students of academic institutions.

This will enable them to achieve objectives in their digital journey at an accelerated pace.

Objectives of My Book

1. Build a digital mindset with **lean digital thinking** to identify digital business innovations.
2. Establish a common understanding of what is digitalization and digital business.
3. Establish a common digital language and a shared digital purpose.
4. Bring clarity on how digital technologies impact the entire business.
5. Answer the most challenging questions:
   a. Why digitalize?
   b. What to digitalize?
   c. How to digitalize?
6. Find the lean digital quotient (LDQ) to find digital strengths and weaknesses of organizations.
7. Provide a five-phase digitalization roadmap that ensures a better return on investment (ROI).
8. Introduce new business models, products, services, processes, digital workplaces and operating models driven by digital technologies.
9. Provide insights on how to leverage digitalization to address COVID-19 challenges.

Why Do You Need to Read This Book?

In the digital immigrant organizations, existing business models, strategies, products, services, processes and culture have been failing to meet the needs of the digital economy and the Fourth Industrial Revolution. Hence, such organizations' existence itself is being challenged. Many organizations which were leaders in the past (Nokia, Kodak, Borders and others) do not exist today as they failed to embrace the changes happening in the market.

The fusion of various technologies like operational technology, communication and collaboration technology and information technology has created a powerful cyber system. The modern cyber system brings supercomputing power, super storage and algorithmic virtual brains that help organizations to become nearly autonomous. Hence, it significantly optimizes human interventions in business operations.

The modern cyber system becomes a driving force which helps build digital businesses. Unfortunately, except for a few, most of the traditional organizations have been struggling to prepare for the digital future by leveraging the cyber system. You need to read this book to understand how, by embracing the cyber system using **lean digital thinking**, organizations can

1. create new revenue streams,
2. delight digital age customers,
3. optimize human interventions to improve operational excellence,
4. develop agility within the organization to respond to the market at speed,
5. leapfrog known and unknown competition and
6. improve overall business performance.

Most organizations have been building digital businesses which are fat and expensive to develop and manage. By reading this book, you will understand the difference between a digital enterprise and a lean digital enterprise. The future is in **Lean Digital Enterprises** as digital enterprises will end up being white elephants.

Why Is the Content in This Book Invaluable?

There are a few great books on digital transformation or digital businesses written by very renowned authors today. Most of these books cover more about technologies and less on how technologies impact the whole DNA of businesses.

It does not mean that there are no good books on digital businesses. But in these books, there are a few unmet needs. Some authors have covered only digital strategy—including data, innovation, customer, competition and value proposition. However, they missed out the digitalization of other important digital business building blocks.

In a few other books, authors have also addressed how digital capabilities and leadership capabilities of an organization are to be nurtured. But they have missed the most critical aspect of acquiring a new thinking/mindset to build these capabilities. With traditional thinking, organizations fail to digitalize their businesses.

In this book, I have introduced a brand new way of thinking called **lean digital thinking** that includes 12 principles. **Lean digital thinking** prepares the human brain for the digital journey that yields better business results.

I have covered in detail how the 12 principles of **lean digital thinking** impact each of the 12 business building blocks of an organization, creating a 144-dimensional digitalization matrix; something that no author has touched upon till now.

The readers will benefit from the book, as mentioned below:

1. Business executives will be able to easily understand, without jargon, how various digital technologies impact businesses.
2. Technology executives will get an excellent knowledge on how they could become digital innovators or internal consultants, and design digital business models, strategies and products.
3. It is a practitioner's book and readers will find step-by-step guidance if they want to digitalize their businesses.
4. The topics are arranged in such a way that different readers can read their choice of topics rather than going through all the topics. But a few topics are a must-read for all readers.

5. There are a few incredibly unique topics, like acquiring a digital mindset that covers 12 **lean digital thinking** principles; digitalizing 12 business building blocks; COVID-19 as the new normal, which bring out new energy and new digital innovations.
6. This book has the potential to become a textbook for professors and students of academic institutions.

How Is This Book Organized?

Chapter 1	Chapter 2	Chapter 3	Chapter 4
How new technologies and contactless lifestyle have been putting CEOs under a lot of pressure and dilemma	The emergence of the modern cyber system due to various technologies, fusion describes the tectonic impact of the modern cyber system and the need for digitalization of organizations	The troika of the modern cyber system—digital economy, the Fourth Industrial Revolution, hyperconnectivity—the new trigger point for digitalization of organizations	Digitalization opportunities to meet the needs of digital age customers by leveraging the physical and digital worlds

Chapter 5	Chapter 6	Chapter 7	Chapter 8
Acquire a new mindset with **lean digital thinking** that includes four **lean digital thinking** clusters and twelve principles	Visualize the digital future by leveraging three **lean digital thinking** principles	Achieve digital execution excellence by leveraging six **lean digital thinking** principles	Embrace social responsibility by leveraging one **lean digital thinking** principle

Chapter 9	Chapter 10	Chapter 11	Chapter 12
Design unforgettable moments of experience by leveraging two **lean digital thinking** principles	Digitalize the business future—business vision, business model, business strategy and business outcomes	Digitalize technology foundation—technology strategy, technology governance and technology architecture	Digitalize offerings—products and services

Chapter 13	Chapter 14	Chapter 15	Chapter 16
Digitalize business operations—workplace, processes and operating model	Digitalization roadmap covering five phases of the digital journey: a) Inspire, b) Assess, c) Innovate, d) Experiment and e) Institutionalize	Digital business solution; use case examples	COVID-19 pandemic and the new normal— business new normal, governance new normal, society new normal and lifestyle new normal

Chapter 17	Chapter 18		
Digital interventions in business, governance, society and lifestyle in the new normal emerged due to COVID-19	Employee health and safety during and after, and how digital interventions improve the health and safety of employees		

Reading Recommendations

Target Market

- Global Market
- All Industries

Target Audience

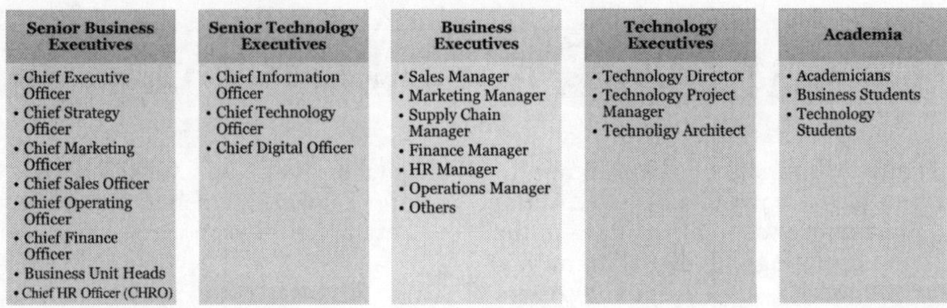

Source: Author

Figure 1: Target Readers for the Book

I advise the readers to read all the chapters. Reading all the chapters will help you get a 360-degree view of digitalizing an organization. In case you have minimal time, I give the following recommendations for your reading.

Reader Category	Recommended Chapters
All readers	**Read the first nine chapters (Chapter 1 to 9)** to understand why you must choose to digitalize and how to acquire a digital mindset. These nine chapters will help you get enough knowledge to kick-start your digital journey. You may read all the other chapters as per your requirement and priority.

Readers who are interested in visualizing digital business future	**Chapter 10** Digitalization Cluster 1: Digitalize Business Blueprint **Chapter 11** Digitalization Cluster 2: Digitalize Technology Foundation **Chapter 14** Digitalization Roadmap: Five-Phase Lean Digital Journey **Chapter 16** COVID-19: The New Normal
Readers who are interested in establishing a digital technology foundation that supports digital businesses	**Chapter 11** Digitalization Cluster 2: Digitalize Technology Foundation
Product managers and solution architects	**Chapter 12** Digitalization Cluster 3: Digitalize Offerings **Chapter 16** COVID-19: The New Normal
Business process managers, operations management executives, finance executives, HR managers, supply chain managers, sales, marketing executives and others	**Chapter 12** Digitalization Cluster 3: Digitalize Offerings **Chapter 13** Digitalization Cluster 4: Digitalize Business Operation **Chapter 16** COVID-19: The New Normal **Chapter 17** Digital Interventions in the New Normal That Emerged with COVID-19 **Chapter 18** COVID-19: Employee Safety and Health

Executives who will own and drive internal digital programs and initiatives	**Chapter 10** Digitalization Cluster 1: Digitalize Business Blueprint **Chapter 11** Digitalization Cluster 2: Digitalize Technology Foundation **Chapter 14** Digitalization Roadmap: Five-Phase Lean Digital Journey
Academicians, business students (BBA/MBA)	**All Chapters:** A textbook for management institutions that will prepare the students for the Fourth Industrial Revolution. It will also empower the professors to design innovative digital business models, products, business processes and business operating models
Professors and technology/ engineering students	**Chapters 1 to 9;** **Chapter 11** Digitalization Cluster 2: Digitalize Technology Foundation

Table 1: Reading Recommendations

Acknowledgements

> A person who wants you to be always successful and gives you motivation, confidence and courage to bounce back in tough times is the true well-wisher.
>
> **—VSR**

This book is the outcome of the efforts, perseverance and encouragement of my well-wishers and friends. Social media, for me, is a teacher, mentor and guide. I want to thank the speakers and writers who have shared their knowledge and wisdom on YouTube, SlideShare, Twitter and blogs on digital technologies and businesses. I derived great inspiration from these social media knowledge champions.

I wish to thank two of my friends and well-wishers who continuously motivated me while writing this book. **Vijayanand Kondapaneni**, the former executive vice president of Zee Entertainment Ent. Ltd challenged me and brought the best out of me while writing this book. My classmate and dear friend, **Yugandhar Cindepalle**, chief technologist at Booz Allen Hamilton, has provided excellent inputs in his personal capacity on the chapter on digital technology architecture.

Part 6, Chapter 15 of my book includes a few innovative digital business solutions. I wholeheartedly want to thank the senior leaders from various renowned organizations, who provided detailed information about their solutions, which would be useful for the readers to understand real-life digital business solutions.

1. **Adarsh Kumar**, Co-founder and CEO, TagBox Solutions
2. **Amrita Chowdhury**, director, Gaia Smart Cities
3. **Ashish Jain**, principal designer, LV Prasad Eye Institute Centre for Innovation
4. **Juhi Bhaskar**, technology solutions engineer, LV Prasad Eye Institute Centre for Innovation
5. **Mohak Sood**, head of Strategic Growth, Aeries Communications
6. **Raghu Gullapalli**, chief evangelist, LV Prasad Eye Institute Centre for Innovation
7. **Ramam Atmakuri**, vice chairman, LV Prasad Eye Institute
8. **Ravinder Paul Singh**, president and CEO, Merkado RHA
9. **Dr Rishi Mohan Bhatnagar**, president, Aeris Communications
10. **Sahil Gupta**, director of Growth and Ecosystem, Techforce.ai
11. **Saumitra Singh**, CPO, TagBox Solutions
12. **Sriram Papani**, CEO, Techforce.ai

My sincere thanks to **Amith Mikkilineni**, technical delivery director of Intelligent Process Automation for Cognizant; **Aravind Chembeti**, senior technical manager at Innominds; **Sailesh Nadendla,** ERP Delivery head, Tech Mahindra; **Sathya Venkatesh Polepalle**, head of PDIT Delivery, Tata Consultancy Services and **Akshaya Utla**, entrepreneur and freelance consultant, for providing useful inputs on the first draft version of my book. I want to convey my gratitude to **Nikhilesh Vegendla** for helping me collate the list of tables and figures for my book.

Without the tremendous motivation, positivity and support from my mother **Krishna Kumari Vegendla** and the wholehearted support of my better half **Shilpa Vegendla**, I could not have completed this book.

My special thanks to **AS Murthy** (ASM), former CEO of Satyam Computers, **Kumar Parakala,** president of GHD Digital and **Sunil David,** regional director (IoT) at AT&T, for India and

ASEAN for their valuable inputs. I thank BT&BT for allowing me to use the **lean digital thinking** registered trademark, while it was conceptualized by me as part of BT&BT.

I humbly convey my thanks to the Bloomsbury team, **Praveen Tiwari, Nitin Valecha** and **Shreya Chakraborti** for their excellent support. I express my gratitude to **Elizabeth James**, who reviewed the manuscript and gave inputs to improve the quality of the book.

V. Srinivasa Rao (VSR)

October 2020

PART 1

Why Digitalize?

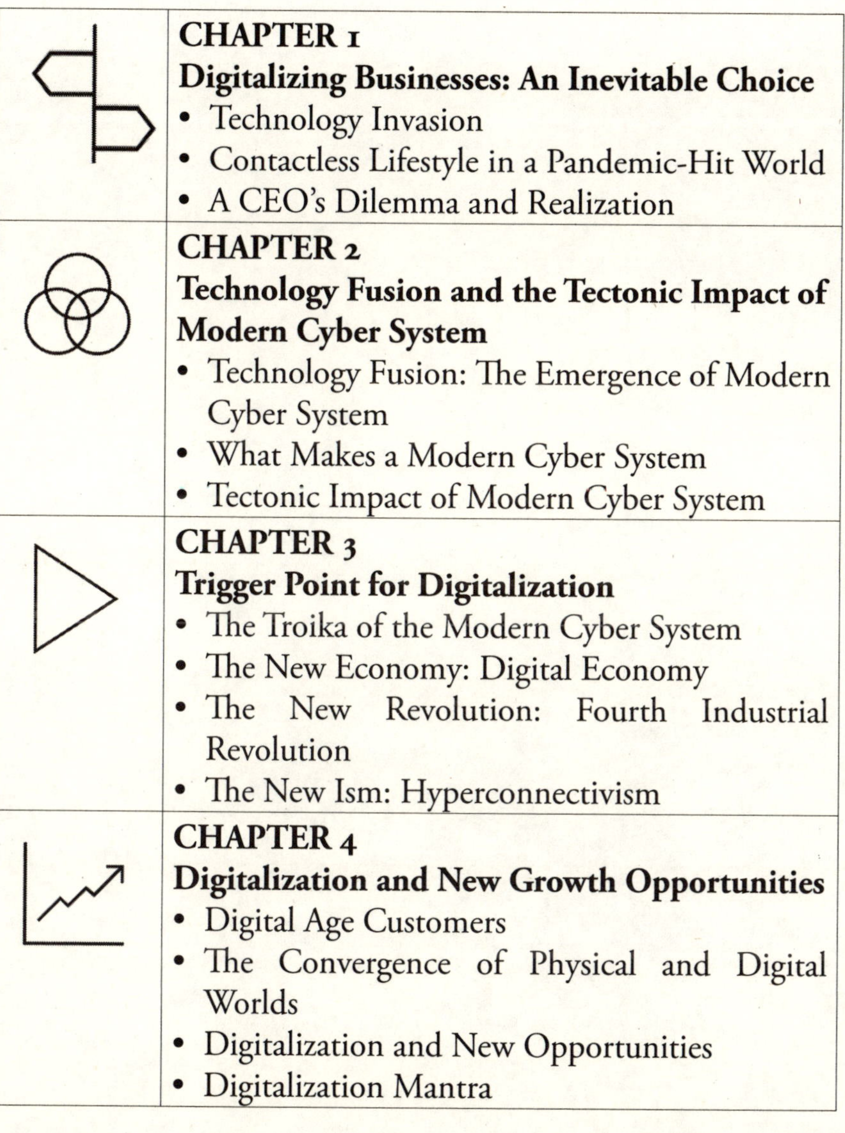

CHAPTER 1
Digitalizing Businesses: An Inevitable Choice
- Technology Invasion
- Contactless Lifestyle in a Pandemic-Hit World
- A CEO's Dilemma and Realization

CHAPTER 2
Technology Fusion and the Tectonic Impact of Modern Cyber System
- Technology Fusion: The Emergence of Modern Cyber System
- What Makes a Modern Cyber System
- Tectonic Impact of Modern Cyber System

CHAPTER 3
Trigger Point for Digitalization
- The Troika of the Modern Cyber System
- The New Economy: Digital Economy
- The New Revolution: Fourth Industrial Revolution
- The New Ism: Hyperconnectivism

CHAPTER 4
Digitalization and New Growth Opportunities
- Digital Age Customers
- The Convergence of Physical and Digital Worlds
- Digitalization and New Opportunities
- Digitalization Mantra

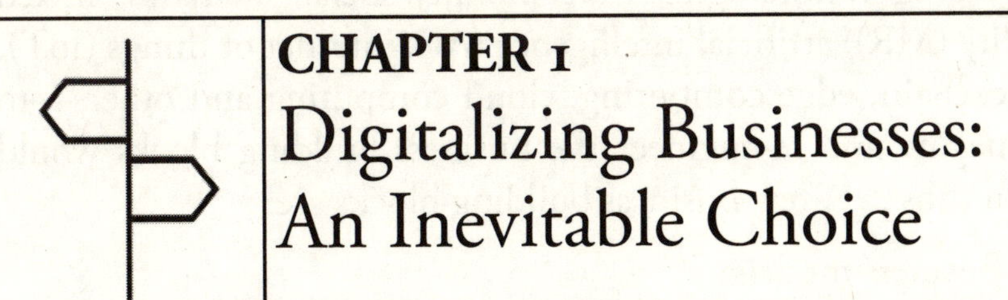

CHAPTER 1
Digitalizing Businesses: An Inevitable Choice

Technology Invasion

Electricity, internet, supercomputing power, mobile devices and social media are available today at an affordable price. Every 60 seconds over the internet, millions of messages are sent, videos are viewed and thousands of online transactions are done. This is a major disruption and has been significantly impacting the customer's needs, wants and purchasing behaviour.

Physical things are becoming intelligent due to the availability of sensors at a lower price and innovations in microprocessors as well as short-range and long-range communication networks. These smart things can sense, think, communicate and predict like humans to some extent. Hence, these smart things have been performing many tasks that have been done by human beings for many decades.

This means technological innovations have been impacting customer needs and wants. Also, it has been changing the basic configuration of products, physical objects, services, operating models and the work environment of organizations to optimize human interventions, which had not been possible in the past.

Technologies in the past were used to manage information, machines, physical objects, equipment and mostly to improve operational excellence. But now, the board of directors, CEOs, chief experience officers (CXOs) and business unit heads have realized that technologies have been impacting the whole business DNA of the organization beyond operations.

Those organizations which fail to introduce newness by leveraging technologies—like mobile, social, analytics, mixed reality (MR), artificial intelligence (AI), internet of things (IoT), blockchain, edge computing, cloud computing and others—to reimagine and re-engineer the business building blocks would soon sunset. A few business building blocks are:

1. Business models
2. Products
3. Services
4. Processes
5. Operating models
6. Workplace

Board of directors, CEOs, CXOs and business executives are expected to be proficient with digital technology to understand its impact on their business.

Technology is no more an enabler but the driving force of the business. And technology executives like chief information officers (CIOs), chief technology officers (CTOs) are expected to be part of the board of directors to steer the companies in the era of the Fourth Industrial Revolution.

Organizations and their businesses must ALIGN with technological innovations, else they will successfully EXIT from the business world.

Contactless Lifestyle in a Pandemic-Hit World

COVID-19 has unimaginably disrupted the whole world. The virus was detected in Wuhan, China around November 2019. COVID-19 virus has been spreading across the globe at an extremely high speed and millions of people have been infected.

In the years 1720 (plague), 1820 (cholera) and 1920 (Spanish flu), there were global pandemics which took the lives of millions of people. After a hundred years from 1920, this generation is facing the COVID-19 pandemic. There is uncertainty, ambiguity and fear in the whole world as there are many waves of COVID-19. However, due to vaccine availability, the severity of the pandemic may get reduced. The vaccination of the major population in the world may take at least 18–24 months. Having said that, people are expected to wear masks, maintain physical distance and continuously wash or sanitize their hands.

Governments have declared lockdown forcing billions of people to stay at home. The fear of the virus and the lockdowns continue to have a severe impact on the economy, lifestyle, society and businesses.

Retail, tourism, travel, movies, manufacturing, banking and transportation industries took severe hits due to COVID-19. Many governments announced bailout packages. Unorganized workforce got displaced entirely and are in deep trouble. The earning power of individuals has gone down tremendously.

The situation during a pandemic is that people fear to meet others. Even when essentials like rice, food, medicines are delivered, there is a fear that delivery personnel may be carriers of the contagious virus. People are avoiding contact with exposed surfaces and objects to prevent infection. This means there is a new **contactless** lifestyle emerging:

1. There are virtual meetings using various digital tools, like Cisco Webex, GoToMeeting, Google Hangouts and others, ensuring less or no face-to-face meetings to the extent possible.
2. No travel to offices and working from home have become the norm in knowledge-intensive industries and functions (IT and others).
3. People want to see movies at home; Amazon Prime and Netflix have become default movie theatres.

4. The number of patients going to hospitals has come down; telemedicine has been picking up due to the fear of the pandemic.
5. Delivery of consumer goods using crewless vehicles (drones) at selected distribution points has also seen a rise.

The pandemic-influenced world is being driven by the following:

1. Contactless lifestyle
2. No travel
3. No face-to-face meetings
4. Working from home
5. Others

Hence, digital technologies—mobile, social, analytics, cloud, augmented reality (AR), virtual reality (VR) and AI—have been playing a vital role in shaping up the future of the world. A few lifestyle changes and industry practices being used during the COVID-19 period will become the new norms, in which digital technologies will play a crucial role.

A few industries, like manufacturing, construction and others, require the workforce at the physical sites. Digital technologies play a key role in these industries. For example, at the workplace, each worker will be distributed a smart cap which alerts them if the physical distance between the workers is less than six feet. Smart thermal scanners at offices will automatically measure the body temperature and send guidelines to the employee if the temperature is beyond the allowed limit (high temperature is one of the indicators of COVID-19).

To summarize, digital technologies play a significant role in helping us survive and enjoy even in a pandemic-hit world.

A CEO's Dilemma and Realization

On 25 November 2019, at 9:00 AM, there was pin-drop silence in the Sunrise Retail boardroom. Chairman Nandu asked Sushant, the CEO, in an impatient voice, why there was a consistent decline in their retail business for the last six quarters. Nandu and the other board members expressed severe concerns about Sushant's—a star performer for the past 10 years—failure to turn around the company.

Nextgen Retail, a new-age retail company, had grabbed a significant share of the retail business. Nextgen Retail uses mobile, social media, consumer analytics and consumer IoT (CIoT) technologies to provide exceptional moments of experience to customers.

When a customer enters the retail store, they use Nextgen Retail mobile app through which they get the details of discounts on various products within the store. Nextgen Retail does not have any point of sale (POS) counters for the customers to pay for the items.

Using the mobile app, they can directly scan the QR code on each product, and then pay the money using the digital wallet within the app. The buying experience which consumers get within retail stores using digital technologies is fantastic. In addition to this, Nextgen Retail best leverages its digital commerce platform with which consumers can experience the products using virtual reality and augmented reality.

On 30 December 2019, at 7:00 AM during a breakfast meeting with his core team, Sushant discussed their declining revenues and loss of customers. Sunrise Retail must embrace digital technologies to transform their retail business.

Chief Marketing Officer (CMO) Gaurav passionately proposed that they begin a few Twitter and Facebook campaigns and develop a mobile app to brand their retail offerings and products.

Chief Strategy Officer (CSO) Keshav suggested modernizing retail stores with great interiors.

Sushant felt something was missing in their efforts to prepare their organization for the future. Sushant and his team had spent millions of dollars to modernize the retail store. They used digital marketing agencies to promote their organization; still, they did not see any business growth!

Sushant shared his serious concerns with his leadership team that they were losing ground to the new competition which was digitally empowered. In the previous year, they had simply copied what other retail shops had been doing. They had spent millions of dollars on many digital initiatives which their competition had implemented as well. However, they had failed miserably. They had lost their soul, their pride while their employees were feeling insecure and investors losing hope.

Sushant asked them a fundamental question: "What is the meaning of **digital** to you? I want to hear from each one of you."

1. CMO Gaurav immediately responded that digital meant social media, digital campaigns and millions of followers on social media.
2. CIO Rajesh replied that digital meant mobile, IoT, AI, MR, mobile, cloud, edge computing, analytics, blockchain, SaaS (Software as a Service) applications, data centre and others. (Nobody in the room understood what the CIO was talking about.)
3. Senior Vice President (SVP) Digital Commerce Aakash responded that digital meant digital wallet and buying of products using the internet.
4. SVP Retail Stores Kalyan admitted that he did not have any clue about digital and had been struggling to deal with store layout and shelf filling issues.
5. SVP Supply Chain Sarkar added that he did not have much understanding of digital either. He wondered why Sushant had been so obsessed with digital and acting funny these days!
6. CMO Gaurav again passionately told Sushant that they had over one million followers on Facebook, 7,00,000 followers on Twitter and thousands of subscribers on YouTube. Sushant asked the CMO how many of these followers and subscribers were customers? He did not have an answer to that.

Sushant, addressing his team in a strict tone, said, "Guys, now I understand why are we losing ground even after having extensive retail business experience? Based on your responses to my question, I am clear that we do not have a common understanding of digital and most of us have not even attempted to understand the impact of digital technologies on our business."

After taking a long breath, Sushant continued his speech and told his team:

1. We do not know about designing new business models using digital technologies.
2. We do not know about opening new channels to reach our customers.
3. We have not tried to innovate any new services leveraging digital technologies.
4. We have not used digital technologies to improve our customer engagement and experience.
5. We have not leveraged digital technologies to reduce human interventions in our business processes (in-store process, merchandising, HR, pricing and promotion management, category management, forecasting and replenishment, etc.). Digital is beyond **marketing** (digital marketing) and we completely missed this point.
6. We are not at all clear on how we use digital technologies to improve supply chain efficiency.
7. Most importantly we have not unlocked the full potential of digital technologies to predict the business future, to create new opportunities and to improve business performance.

Sushant also highlighted the COVID-19 pandemic and how consumer behaviour and buying habits have been changing. He cautioned that their current supply chain model and physical store layouts will not work. New delivery channels, commerce platforms, delivery at the doorstep with drones may emerge. Hence, Sunrise Retail needed to embrace digital technologies throughout its business value chain.

Sushant concluded his meeting by saying: "We need a **common language**, **common purpose** and **common understanding** of digital to prepare for a better future."

Sushant smiled at his team and said that he had found the solution to their business problem. He asked his team to get ready for the next big digital battle with their competitors and to acquire lifelong customers.

Sushant looked at the CMO and the CIO and told them to come out of their comfort zones. He asked them to be strategic partners in devising business innovations, new business opportunities and new revenue streams. He requested them not to limit their thinking to the cloud, software applications and digital campaigns which were not digital in the true sense of the word.

Sushant cautioned that Sunrise Retail leadership team's thinking must change. Driving digital business (software-defined business) is different from driving the current traditional business.

He emphasized the importance of leveraging the fusion of technologies to adequately prepare their business for the Fourth Industrial Revolution and the digital economy in a pandemic-hit world.

He also warned his team to ALIGN or EXIT.

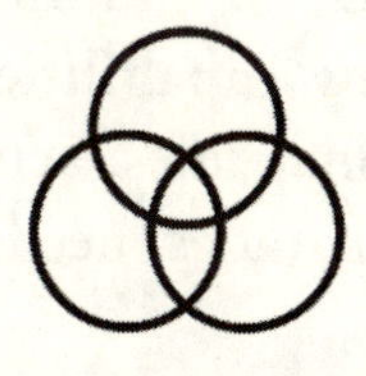

CHAPTER 2
Technology Fusion and the Tectonic Impact of Modern Cyber System

Technology Fusion: The Emergence of Modern Cyber System

In the past, technology was only an enabler and, most of the times, was considered to be an overhead cost, not an investment. Most organizations implement technology to improve productivity, operational efficiency and speed of the automation of processes.

Technology has been there for many decades and is nothing new. However, all those different kinds of technologies in the past were proprietary and disconnected. Hence, the value delivered and the impact created by these disjointed technologies on organizations was minimal and localized. Today, due to significant advancement in the technologies, internet and standardized communication protocols, technology fusion is happening.

The **fusion** of communication and collaboration technology, operational technology, process automation technology, user interface technology and information technology has created the modern **cyber system**, which is the genesis of the digital economy.

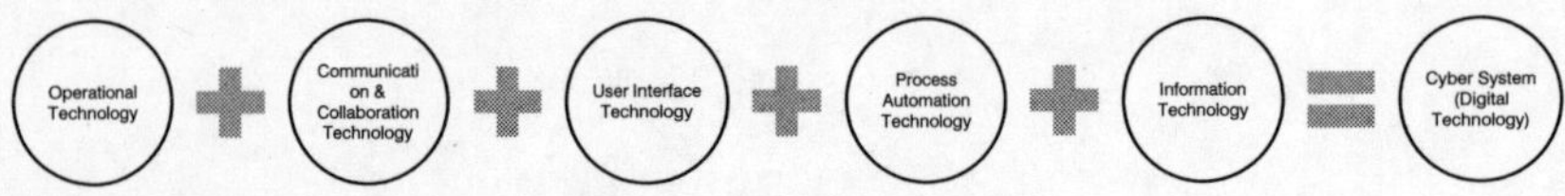

Source: Author

Figure 2.1: Modern Cyber System

This modern cyber system with the confluence of multiple technologies and internet as the glue brings **new capabilities** and has been significantly changing the traditional rules, acts, policies, tools, processes, methods, techniques, attitudes, needs, expectations and aspirations.

S. No.	New Capability	Technology That Provides This Capability
1	Anywhere, anytime instant communication and collaboration among distributed people (real time)	Communication and collaboration technology
2	People perform their tasks and get information from anywhere, anytime using any device (mobile devices, car console, refrigerator door console, tablet, laptop, wearable devices) through body movement like gesture, touch, mood, etc.	User interface technology
3	People and intelligent things communicate in real time and perform the work in a collaborative manner	Process automation technology, operational technology, communication and collaboration technology

4	Non-existent virtual world creation by simulating the physical world (objects and locations) or augmenting the physical world to provide an immersive experience to people	User interface technology
5	Hyperautomation of process workflows and transactions (digital workflows) with less/no human intervention	Process automation technology
6	Intelligent things like machines, cars, equipment and other things come alive and can sense, think, communicate and perform work themselves (autonomous) without human interventions	Operational technology, communication and collaboration technology
7	Elastic super storage to store high volume (exabytes/zettabytes), variety (text, audio, video, XML, transactions) and velocity of data (terabytes of data in minutes)	Information technology
8	Elastic supercomputing power to process humongous data with intelligent algorithms (AI-/ML-/DL-enabled virtual brains) and complex transactions to predict the future and to prescribe the actionable insight	Information technology

Source: Author

Table 2.1: Modern Cyber System—New Capabilities

Do you know how these capabilities significantly and ruthlessly impact your organization's vision, business models, strategy, products, services, operating model, people, culture and processes? If your organization does not acquire these capabilities and bring about the required changes, it may not take much time for your organization to face serious growth challenges.

Real-Life Example of Modern Cyber System

The modern cyber system is so powerful that it disrupts the whole organization's ecosystem significantly. Take a look at Table 2.2 to see how it changes the entire experience of a tourist, visiting a tourist location. This experience is completely new.

S.No.	Scenario	Technology
1	A tourist who wants to visit the Borra Caves, before booking the tickets, makes use of a mobile device to get an immersive experience of the caves. This is made possible by VR.	User interface technology
2	The tourist wants more information about the tourist place and nearby attractions. So he/she makes a video or audio call from a mobile device or desktop to speak with the staff at the location to get more details. The tourist takes part in social media conversations and gets better insight into the tourist location based on past experiences shared by other tourists.	Communication and collaboration technology

3	The tourist searches online for the best accommodation near tourist places, including homestay options. With matchmaking algorithms, the tourist gets the best possible accommodation of his/her choice.	Process automation technology, information technology
4	Before reaching the accommodation venue, using a mobile device, the tourist sets the temperature of his/her room, selects the choice of music to be played and lighting colour in the room.	Process automation technology
5	Once the tourist reaches the hotel or homestay accommodation, he/she goes to the room directly and opens the NFC door lock using an NFC-enabled mobile phone and code.	Operational technology
6	When the tourist gets into the room and lies down on the bed, which is embedded with sensors and software, the lights are switched on, the AC begins cooling at the set temperature and the chosen music starts playing.	Operational technology

7	The tourist is at a restaurant and gets the menu card in the local language Telugu but wants to see the same in his/her mother language, say German. Using a mobile phone and augmented reality, the tourist easily reads the Telugu menu in German.	User interface technology
8	At the tourist location, the tourist uses the mobile device with augmented reality to see all the information superimposed on the statues.	User interface technology
9	During the trip, the tourist comes down with fever and cold and goes to the nearby hospital. The doctor extracts his/her medical history along with DNA details and offers personalized medicine based on an algorithm.	Process automation technology, information technology
10	The tourism department captures the data about the tourist at every touchpoint (hotel and tourist locations) and applies AI/ML to regularly provide personalized recommendations on how to best use the holiday time.	Information technology

11	The tourist books slots at a car parking place at identified tourist locations well in advance using a mobile device.	Operational technology
12	Using the mobile device, the tourist creates an album of his/her choice by combing all the photographs, videos and audio taken during the holiday and sends the album to friends and family in real time. The tourist also shares his/her experiences with family and friends over the mobile device in **real time**.	Communication and collaboration technology

Source: Author

Table 2.2: Real-Life Example of Modern Cyber System

Each technology has its significance and impact, and when the convergence of all these technologies happens, the impact is manifold.

If you do not care about the modern cyber system, you do not have a future in business. Hence, there is only one choice: ALIGN or EXIT.

What Makes a Modern Cyber System

Operational Technology

Operational technology is nothing but integrated hardware and software that automatically monitors and controls **physical things** like machines, cars, valves, equipment, cameras, medical devices, ATMs, chairs, bed sheets, trucks, retail bins and others. Operational technology optimizes human interventions.

Sensors

Operational technology uses sensors that sense data from physical things, devices or environment. There are different type of sensors like temperature sensors, proximity sensors, pressure sensors, water quality sensors, chemical sensors, gas sensors, smoke sensors, level sensors, image sensors, accelerometers, gyroscopes, humidity sensors, leak/moisture sensors, window door open/close sensors, video doorbell sensors, thermostats, motion sensors and others.

Microchip and Embedded Software

Sensor data is processed by an embedded software within microcomputers (microchips) and sends data or action commands to the gateway devices, things or internet. Embedded software can perform simple operations like switching lights on/off using a remote control or complex ones like operating numerous electronic components running in an aircraft or car.

With the availability of standardized communication protocols and internet, things are now able to exchange data and communicate with each other even when they are distributed across different locations. More about communication and collaboration technologies will be discussed later. For example:

1. Parking sensor in a parking meter reads the data, microprocessor processes the data and finds whether the parking place is vacant or not.
2. The CO2 sensor within a room reads the carbon dioxide level and the microprocessor determines whether the CO2 levels are within the safety threshold.
3. A machine reads the instructions on the radio-frequency identification (RFID) tag attached to a bottle and the microprocessor instructs the machine to automatically fill the right liquid as per the instructions inscribed on the RFID tag.

Communication and Collaboration Technology

Communication and collaboration technology has erased boundaries, enabling people and things to communicate and collaborate anywhere and anytime. Some 100 years back, any communication from Europe to India used to take months as the letter came through ships! Today, it is a matter of a few seconds irrespective of the location of the sender or the receiver. It is not just human beings, due to the convergence of operational technology and communication technology, but **things** (car to car or machine to machine) are also communicating with each other in real time.

IPv6

IPv6 provides 343 trillion trillion trillion IP addresses. Hence, trillions of things and devices can be connected to the internet (any device or thing to be connected to the internet must have an IP address). With IPv6 literally, anything can connect to the internet.

Short-Range Transportation Wireless Networks

These enable communication between physical things and devices across short distances (machine to machine, mannequin to mobile phone) using short-range transportation wireless networks like Bluetooth ZigBee, 6LoWPAN, Z-Wave, Wi-Fi, NFC (near-field communication) and LoRaWAN.

A gateway is used to connect different things that use different network protocols and aggregate the sensor's data from different things before sending the data to a broadband network or another thing.

5G Broadband Networks

Long-distance communication between physical things, devices and people is facilitated by 5G broadband networks. These enable HD quality, ultra-high bandwidth, effective interactive multimedia, hyperfast streaming of data at a rate of >1 Gbps. High speed and low/no latency video call, audio call and telepresence have become very easy to make using 5G.

Social Media/Collaboration Technology

This is the most powerful technology today which has enabled people to carry out two-way collaboration and conversations in real time, increasing the proximity among people who are far away from each other.

For example, health sensors inside diapers read blood pressure and the temperature of a baby. This data is processed by a microchip and sent through the 5G wireless network to the mother's mobile phone. Now, the mother can see the health status of the baby while working from her office. **This is the power of the fusion of operational and communication technologies.**

User Interface Technology

While keyboard and touch-based interface with computers, mobile phones, tablets, machines and devices have been there for quite some time, many advancements are happening in this space. Today, users can easily interface with computers, mobile phones, tablets, console screens (car/refrigerator), wearable devices, machines and other devices using any one of the following options.

Voice-Based User Interface

Voice-based user interface has been maturing day by day. People can operate computers, mobile phones, tablets, machines with their voice without typing a single word. Chatbots, virtual assistants, mobile apps and many more can be operated with voice.

Many organizations also have been building custom built voice interface applications and systems like Google Voice, Google Home, Apple's Siri, Microsoft's Cortana, Amazon's Alexa. For example, Amazon Echo's voice interface system is integrated with SAP Business and by instructing Amazon Alexa, it brings out the required information, dashboards or reports.

Gesture-Based User Interface

Touchless human interface or gesture-based human-computer interface allows human beings to use gestures like scrolling, pinching, tapping, swiping to perform their tasks without touching. For example, with a gesture-based interface at home, a person can operate a TV, turn on/off the AC, lights and other devices. This technology is still emerging and will take some more time to reach maturity.

Virtual Reality

Due to various reasons, we may not get an opportunity to experience real-world physical objects and environment. Virtual reality (VR) enables the computer to create or simulate three-dimensional virtual objects and virtual environment to provide an opportunity for people to get an immersive experience of the real world.

For example, before the construction of an office, an organization can create a three-dimensional virtual office using VR and allow their employees and clients to walk through the conference room, cafeteria, gym, cubicles, cabins, huddle rooms and aisles before finalizing the design. They will also get to experience things like colour and tile texture. VR helps the organizations to improve and change the office interiors and facilities before the construction of the actual office.

Augmented Reality

While VR creates a virtual world, augmented reality (AR) enhances the physical world by superimposing information, images, videos, addresses, location coordinates and recommendations on real-world objects and environment. For example, a person walking in a retail store can see product details like price and product specifications superimposed on the product through a mobile device or glasses, helping the customer take a right purchase decision.

Process Automation Technology

In any organization, there are both industrial processes and business processes. Processes were automated to improve quality, productivity and efficiency to some extent with less human interventions. In the last decade, the availability of many breakthroughs in technology has been enabling the creation of hyperautomated processes with optimized human interventions.

Robotic Process Automation

Instead of a human being, a software robot (unlike physical robots used in industrial process automation) mimics human beings and interacts with business applications. This is called robotic process automation (RPA). While RPA is remarkably interesting, it must mature to be applied to all business processes. At present, it is best suitable for business processes, which are routine, repeatable and driven by business rules. It has voluminous transactions with high manual data entry into software applications.

At present, recruitment, invoices, financials, software installations, vendor onboarding, reports creation processes are best suitable for RPA. If decision making requires human intervention at regular intervals, then RPA may not be the best fit today. However, with the intervention of artificial intelligence (AI), machine learning (ML) and blockchain smart contracts, RPA may reach the next level of maturity in the coming years

For example, an RPA software robot automatically opens the invoice document in excel format, scans all the relevant data, verifies the data integrity and correctness as per the configuration. Later, it automatically enters the scanned data from the excel sheet into the financial transaction software system. For any data corrections in the invoice, the robot automatically sends the list of mistakes to the invoice requester. Once the requester corrects the invoice data, the robot automatically approves the invoice and sends a mail to the customer.

Industrial Process Automation

Industrial processes are different from business processes. Industrial processes usually bring about physical and chemical changes in the materials and substances used.

Physical change means a change in the form and shape of the material through cutting, polishing, folding and other techniques. Chemical change means there will be a rearrangement of atoms within the substance to create a new substance from the original one. Computer programs, intelligent machines with controllers and robots reduce/optimize human interventions by automating these industrial processes.

Computer-aided designs and computer-controlled smart machines with robotics have enabled the manufacturing of products like cars, washing machines and others with less human intervention.

Manufacturing based on physical changes in the material has been happening through cutting, polishing, folding and other techniques (this means removal) till now. However, with the invention of 3D printing, the whole manufacturing is becoming additive. This means intelligent machines can now manufacture a component or a product automatically with computer programs by printing the component or product using a 3D printing machine and with additive materials.

Operational technology, 3D printing with AI, ML and robotics aim at bringing autonomous industrial processes to the forefront. This digital transformation is called Industry 4.0.

Information Technology

Data is the new oil or the new gold mine. Operational technology, communication and collaboration technology, user interface technology and process automation technology generate structured, semi-structured and unstructured data at high velocity and volumes.

The critical challenge is how to store, retrieve and process the large data to gain real-time hindsight, insight and foresight with the convergence of data from different sources in different structures.

In the past, only around 15% of the structured data (transactions data) was used to get information or insight. But today with the inventions of new data technologies, structured, semi-structured and unstructured data consolidation gives unimaginable insight into an organization. Information technology is so powerful and mines data to answer the following questions about an organization in various dimensions:

1. What happened? (facts and history)
2. Why did it happen? (diagnosis of facts)
3. What might happen? (predicting the future)
4. What should we do when it happens? (prescribed actions to manage the predicted future)

Cloud Computing

Computing power, storage, network bandwidth are essential components of every organization today. Unfortunately, these resources are costly but cloud computing has made it possible for organizations to rent elastic supercomputing power, superelastic storage and elastic network bandwidth at affordable prices. Without cloud computing, it would be difficult to leverage the power of technologies like AI, ML and deep learning (DL), which need high computing power.

While Public Clouds, Private Clouds and Hybrid Clouds (Public + Private clouds) have significantly helped organizations in their digital journey, in parallel, they also created many problems.

In the era of the Fourth Industrial Revolution and digital economy, there is a surge in the enterprise workloads that includes:

1. Analytics workloads
2. Batch workloads
3. IOT/smart things workloads
4. High-performance workloads
5. Big data workloads

These workloads require massive computing power, storage, network bandwidth and security.

Organizations do not want to lock-in with one or two cloud providers. Based on their needs and suitability, organizations have been using multiple clouds at a time. For example, for CRM, salesforce, for cloud infra AWS and Microsoft Azure, and storage and database using private clouds. According to Forbes Insight research, companies have an average of 17 major cloud providers.

It is essential to migrate software applications from one cloud to another cloud easily. Also, instances created in multiple clouds must be optimally utilized without wastage.

The fragmented clouds impact the efficiency and cost-effectiveness of organizations' operations. Hence, multiple cloud management platforms like BMC, App Formix, Embotics, Zenko, IBM Cloud Manager have to be used.

Edge Computing

Unlike cloud computing, in which computing is done at a centralized location, in edge computing, it is done at or near the source of data. Edge computing brings distributed processing power, storage and network. This reduces latency and improves the speed of response. This helps IoT-enabled self-driving cars and machines in shop floors to function without consuming too much internet bandwidth. Edge computing also improves the privacy and security of sensitive data.

Quantum Computing

Quantum physics is different from Newtonian physics (also called classical physics). For many decades computers have been built and operated based on Newtonian physics, which deals with atoms (particle). When we get into the subatomic level (particle and wave), Newtonian physics does not work; therefore, quantum physics is required.

In classical computers, computing bit has only two possible states, either zero or one. In quantum computers, the computing bit is called a qubit. It has multiple states of zero, one and combinations of zero and one at the same time. Hence, computational power is exponentially high. Due to this unimaginable computational power, we could solve the problems which we can't even imagine solving today. In the past, we did not have enough computing power to simulate the complex mathematical algorithms. Today, we have supercomputing power available at affordable prices. Hence, with AI, we can solve problems which we were not able to solve earlier. Similarly, quantum computing will completely revolutionize the world and enable us to solve overly complex issues of business and society. It may take 10 years from now for quantum computing to reach its maturity level.

Big Data Technology

Today's enterprises are generating different types of data:

1. **Social data:** Data produced by consumers using various public social networks like Twitter and Facebook. Some enterprises are also building private social networks considering privacy and security, which also creates high-volume data.
2. **Transactional data:** Data generated by transactional systems like enterprise resource planning (ERP), customer relationship management (CRM), supply chain management (SCM), supplier relationship management (SRM) and product lifecycle management (PLM).
3. **Things-generated data:** Data collected from machines, equipment, wearable devices, smart meters and call records from help desks.

In general, big data means 3 Vs:

1. **Volume:** Organizations generate data in the size range of gigabytes (1024 MB), terabytes (1024 GB), petabytes (1024 TB), exabytes (1024 PB) and, in the future, zettabytes (1024 EB) and yottabytes (1024 ZB).
2. **Velocity:** Machines, equipment, devices and social media streams, enterprises generate data at very high speed (per minute, per hour, per day).
3. **Variety:** Transaction systems produce structured data, things generate semi-structured data and social media produces unstructured data.

In memory appliances, NoSQL databases (supports unstructured and semi-structured data), MapReduce, Hadoop, Hive, Pig, Apache Spark, Flink, Nifi, Kafka, Samza are a few examples of big data technologies which process high-volume, high-velocity and variety of data which was not possible with traditional relational database technologies.

Big data technologies with cloud computing are enabling organizations to create new inventions and innovations, which otherwise was not possible in the past due to low computing, storage power and limited data management technologies.

Analytics Technology

Analytic technology uses the data generated from different sources to provide hindsight, insight and foresight into businesses in different dimensions. Various techniques—like data mining, statistical analysis, quantitative analysis and predictive modelling—is used on the **data** to do so. The different types of analytic technologies are:

1. **Descriptive analytics:** Descriptive analytics answers the questions "what happened?" and "why did it happen?" by mining and analyzing data.

2. **Predictive analytics:** Predictive analytics uses statistical, ML techniques and data mining techniques to answer the question—"What might happen in the future?"
3. **Prescriptive analytics:** This focuses on finding the best possible action for a given situation among various choices, based on given known parameters—"What should we do when it is going to happen?" Prescriptive analytics uses optimization, simulation, game theory and decision analysis methods.
4. **Autonomous analytics:** Autonomous analytics involves implementing the actionable insights derived from prescriptive analytics. For example, autonomous cars and autonomous machines on shop floors.

Artificial Intelligence Technology

Artificial intelligence (AI) means a simulation of intelligent behaviour using mathematics, sociology, philosophy, computer science, psychology, neuroscience, biology and other disciplines. AI augments human intelligence in the following areas so that human beings can perform their work faster, better and steadily:

1. Planning
2. Learning
3. Problem-solving
4. Reasoning
5. Speech recognition
6. Decision-making

Organizations have been generating a lot of data from different systems. This data can be structured or unstructured in the form of text, audios, videos, images or numbers. But this data has not been adequately leveraged to improve business performance.

ML is the part of AI with which machines are trained with the help of a variety of data sets, received from different sources (things, social media, transactional systems, human body and others). The machine continuously learns based on the data sets, identifies patterns and creates predictive models in the form of algorithms. These algorithms act like virtual brains and perform a few human actions without human interventions. These algorithms are improved continuously based on new data sets. These are used in different situations to reach approximate or definite conclusions and recommendations in the context of that situation.

DL is also part of AI and is the next level of ML. The machines are trained just like in ML with the help of a variety of data sets, but DL applies the principles of the human brain's neural networks on the received data sets. After rigorous training and learning, it creates algorithms which gain nearly human intelligence to perform work like human beings.

A few examples of AI are given below:

1. Deep Blue (chess computer) and Google's AlphaGo (Go computer) are the best examples of AI-based machines, which have reached the level at which they can beat human beings in chess and Go, respectively.
2. Google's autonomous car uses AI to drive the vehicle without any driver.
3. Visual inspections of cars in automotive plants are done automatically.
4. AI-driven drones are effectively used to pick products automatically from the inventory and deliver them to the customer's doorstep.
5. Drones are also used for security surveillance and progress monitoring of construction in real time without human interventions.
6. Microsoft has been researching climate change by sending aerial robots into the air.
7. Cancer diagnosis and other health issues are found very effectively using AI to take timely preventive and corrective actions.
8. AI-enabled machines are making the shop floors autonomous and facilitating production with negligible human intervention.

Blockchain Technology

While performing transactions, we depend on trusted intermediaries to whom we pay a considerable fee. A few examples are:

1. Banks play the role of the intermediary when transferring money from one country to another country. Banks verify the authenticity of the sender and receiver as well as the availability of cash.
2. While registering land, the registrar plays an intermediary role between the buyer and the seller. The registrar certifies that the property is legal and that there are no disputes.
3. Auditors play an intermediary role between companies and the government. The auditor verifies accounts and certifies their correctness.
4. The procurement head acts as an intermediary between the business unit and the supplier. The procurement head approves the purchase order as per organization policies.

Intermediaries use policies, contract clauses and business rules while performing their tasks. At present, software applications are being used by these intermediaries to perform their work. The loophole in these software solutions is that a super system administrator can access the software code and database, and can tamper with the same, so that wrong policies and rules will be applied before approval by an intermediary.

Now imagine a situation in which we remove the intermediaries and do not allow the super system administrator to tamper with the data and transactions using smart contracts running on distributed nodes.

Blockchain distributed ledger technology makes this possible. It plays a very vital role in minimizing the need for intermediaries through distributed nodes (computers) and clever software code. Figure 2.2 depicts a simplified version of how blockchain technology works.

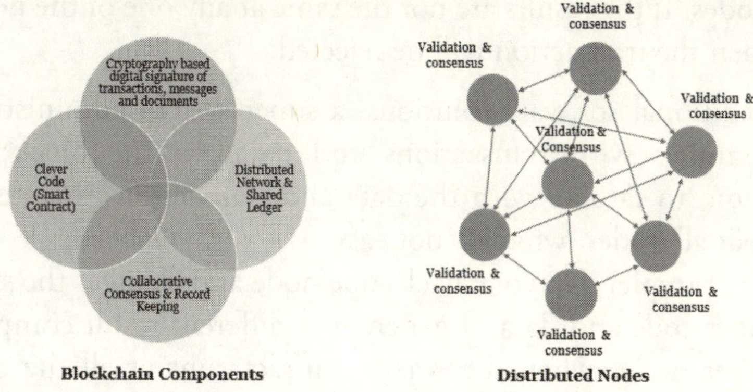

Source: Author

Figure 2.2: Blockchain Components and Distributed Nodes

This is how blockchain technology is set up in an organization (refer to Figure 2.2 for a simplified version of how blockchain works):

1. Distributed nodes (computers) network is established as per the criteria defined by the organization. The smart contract code is uploaded at each node.
2. A smart contract is a pre-written logic (computer code) written as per organization policies, rules, agreement clauses and others. It is stored and replicated on a distributed storage platform (e.g., a blockchain). When it is executed/run by a network of computers, it results in data updates in a ledger at each node (payments, approvals, etc.).
3. After the distributed ledger network and smart contracts are set up, sender digitally signs the transactions, messages and documents using cryptographic algorithms with private and public key combination.
4. Nodes automatically run the **same smart contract** (clever code) on all distributed nodes in the network to apply the required policies and rules, which produce the results.
5. Ideally, the results generated at all the distributed nodes must be the same. The smart contract does that validation at all

nodes. If the results are not the same at any one of the nodes, then the transaction will be rejected.

In traditional software solutions, a super system administrator can tamper with transactions and data. In the blockchain solution, to tamper with the data and transactions, it must be done at all nodes, which is not easy.

For example, if anyone hacks one node and changes the smart contract code and data, it generates a different result compared to other nodes when the smart contract runs at all the other nodes. Therefore, the transaction is rejected.

To further strengthen security, the blockchain system creates blocks of predetermined size by packaging a predefined number of validated transactions. Hence, there is a strong interrelationship between all the blocks and this is known as the blockchain. Any unwarranted change in the transaction in a block will make the whole blockchain invalid. Organizations can choose from public blockchain, private blockchain and permissioned blockchain according to their requirements.

Hence, blockchain technology delivers the following value by making organizations nearly autonomous:

1. Build and provide access to **tamper-proof** and **trustworthy** data and transactions
2. Increase **transparency** in transactions
3. Reduce the creation of **physical documents** while completing the transactions
4. Reduce the **effort and cost** of operations while offering better customer services
5. Reduce **reconciliation/audit** effort
6. Reduce **intermediary efforts** and utilize employees' energy for high-end work

Many blockchain protocols have emerged as mentioned below:

Unified Master Access (UMA): UMA protocol helps in establishing Decentralized Finance (DeFi) Systems. At present, finance derivatives like Futures and Options are available only for institutions. Now with DeFI, peer-to-peer insurance companies, money lending companies, banking, all is possible with customizable financial smart contracts. UMA will seriously impact the current Centralized Finance (CeFi) and allow individuals to perform transactions between them without intermediary financial institutions.

Injective: It is a fully decentralized blockchain-based derivatives exchange that solves high transaction fee, liquidity and slow speed. Injective protocol exchange helps individuals to create their own derivatives market and trading pairs.

Ocean: Ocean protocol unlocks the value of data. Today, data about individuals and organizations are being collated by a few organizations like Google, Amazon, Microsoft, and that data is being used to design and deliver value to various stakeholders. In general, the data owner does not get any value/reward from these companies for using their data. Blockchain-based Ocean protocol is established to address this problem. The data owners could securely save the data in the blockchain and provide this data to data consumers like data scientists and mathematicians.

With Ocean protocol, thousands of data marketplaces can be created by safeguarding the data owners' data privacy. Data consumers get exceptional value as they get much data from data marketplaces to train their ML algorithms. Data owners also get 'Ocean Tokens' for the consumption of their data.

Enjin: This is blockchain-based gaming platform that creates tokens to be used across multiple video games. Enjin is, thus, the first gaming cryptocurrency.

Many other blockchain projects are being executed to support tokenomics, such as Decentraland - Mana, Cardano - ADA, Algorand, Siacoin, TRON, Lumen, Cosmos, Basic Attention Token, OMG Network, Theta, etc.

Geospatial Technology

Geospatial technologies include Global Positioning System (GPS), geographical information system (GIS), light detection and ranging (LiDAR) technologies. Geospatial technologies are being used in various industry functions like supply chain management, insurance, retail, healthcare and others.

Geospatial data helps in identifying accident locations, network outage locations, finding the optimal route for trucks, determining the properties which have not paid property tax to municipalities, etc. Banks and retail industries have been using Geospatial technologies to introduce location-based discounts and services. LiDAR technologies help avoid accidents by estimating the distance between two vehicles using light beams.

Tectonic Impact of Modern Cyber System

The cyber system, today, has become the driving force of governments, businesses, people and society. Hence, each organization and individual must assess its impact and prepare for an uncertain, complex and unknown future—**the digital future**.

This new cyber system has had a **tectonic impact;** it has made past global leaders extinct from the business scene as they failed to prepare for the digital future by leveraging unimaginable new capabilities of the cyber system.

1. Nokia and Blackberry which were the best mobile phone providers are no more in business.
2. Amazon has made Borders bookstores go out of business.
3. Netflix has replaced Blockbuster Video.
4. Airbnb has been challenging the traditional hotels, hotel chains in an asymmetric competition.
5. Uber has disrupted the whole taxi/cab ecosystem and created many new opportunities.

6. Smartphones have replaced cameras, navigators, watches, Walkman, calculators, calendars, alarms, notebook and many other objects.
7. Social media and the internet have enabled consumers to become journalists and content writers, challenging conventional media houses and newspapers.
8. The *Wall Street Journal* stopped its print edition due to losses in Europe and Asia and requested readers to switch to their digital editions.

Hence, digitalization of businesses by leveraging the modern cyber system (technology fusion) is the only option for organizations.

CHAPTER 3 Trigger Point for Digitalization

The Troika of the Modern Cyber System

The cyber system has the potential to change all the systems in the world. Organizations must understand the significant changes and impact of the **troika of the modern cyber system** for them to survive and excel in the future.

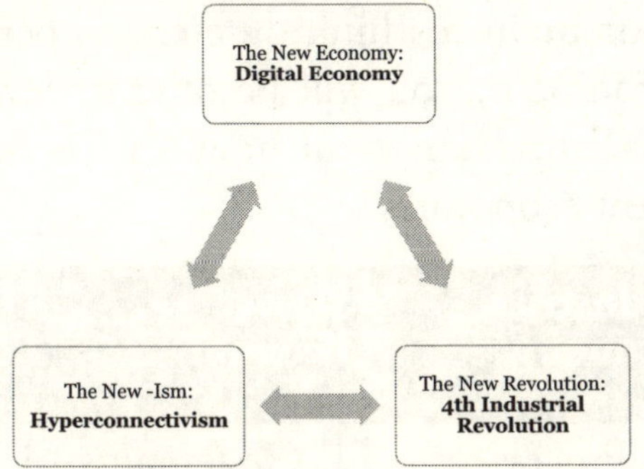

Source: Author

Figure 3.1: The Troika of Modern Cyber System

Unlocking the potential of the cyber system makes organizations truly **software-defined** to meet the needs of digital age customers and optimize human interventions. This allows humans to focus on other unimaginable digital inventions and innovations.

The technology constituents of cyber system will change as and when new technological innovations emerge. Hence, the continuous upgradation of the cyber system is essential.

The New Economy: Digital Economy

While we have seen tremendous growth of the world in many facets over more than 150 years, there are still many social, economic, cultural and environmental problems, which business organizations can solve and build profitable businesses. Making successful organizations by creating a tremendous social, economic and ecological impact is the next wave of business revolution.

We have successfully evolved from an agriculture economy to an industrial economy to a service economy to a digital economy. Today, the digital economy is driven by the power of the **cyber system**.

Social, environmental, political, cultural and economic issues are contextual to the period of that economy. Unless you train and prepare your brain to think different as per the changing economic environment, you will become irrelevant. Table 3.1 provides an excellent example of how tea has been made and served in different economies.

Agri-cultural Economy	Industrial Economy	Service Economy	Digital Economy
1. Harvest tea leaves from the farm 2. Make tea powder at home 3. Prepare tea at home	1. Process tea leaves in the manufacturing plant 2. Sell ready-made tea powder packets in shops 3. Prepare tea at home using the packets	1. Restaurants in each street 2. Ready-made tea available for consumption in restaurants	1. Digital commerce platforms sell tea leaves 2. Place an order on the internet to get tea leaves 3. Digital commerce platform to order tea that will be delivered at the doorstep

Source: Author

Table 3.1: Evolution of Tea Service in Different Economies

The Driving Forces Behind Different Economies

	Agricultural Economy (18th Century)	Industrial Economy (19th Century)	Service Economy (20th Century)	Digital Economy (21st Century)
Powered by	Man and animals	Steam engines and electricity	Internet and web, electricity, electronics and IT	Cyber system
For Delivering	Commodities and goods	Products and services	Services and products	Unforgettable moments of experience with digital age products and services
Human-Machine Interaction	Manual	Mechanization (tools and machines)	Automation	Nearly autonomous

Source: Author

Table 3.2: The Driving Forces Behind Different Economies

Agriculture Economy

In an agricultural economy, wealth was generated mainly with the help of CULTIVATION.

In this economy, products, services, job opportunities, exports and imports significantly revolved around farming, agriculture and animals. Food production was a critical enabler for economic growth. A few examples are coffee beans, grapes, rice, wheat, bananas, sugar, crops and livestock.

Industrial Economy

The inflection point—the shift from an agricultural economy to an industrial economy—was marked by the PRODUCTION of goods and products using science, engineering and technology on an unprecedented scale.

In this economy, products, services, job opportunities, exports, and imports significantly revolved around mechanized manufacturing and production industries. The greatest inventions during this period were the steam engine, steamship, machines in factories, electric motor, electric generator, light bulb, telephone and telegraph. New transportation (rail, road and sea) and communication (telegraph/telephone) channels helped in taking products and related services beyond local regions. The production of refrigerators, washing machines and internal combustion engine-based vehicles started during this period.

Service Economy

The inflection point—the shift from an industrial economy to a service economy—was marked by the provision of INNOVATIVE SERVICES to customers by leveraging knowledge, information and ideas.

In service economy, products, services, job opportunities, exports and imports significantly revolved around service industries like banking, hospitality, retail, healthcare, telecom, e-commerce, restaurants, financial services, logistics, maintenance, etc. Multi-speciality hospitals, 3-star or 5-star hotels, software companies, global scale financial institutions, retail chains, automotive dealers, airlines, pharmacies and e-commerce sites were established during this period. Offering products as service also became a norm.

Digital Economy

Digital economy arrived with software-defined business models, products, services and platforms by leveraging the modern cyber system.

The digital economy is driven by the **modern cyber system** that emerges with the fusion of multiple technologies.

CYBER SYSTEM = [mobile + social + analytics + artificial intelligence + machine learning + deep learning + blockchain + 3D printing + Internet of things + augmented reality + virtual reality + cloud computing + edge computing + sensor networks + 5G networks and others]

The cyber system plays a vital role in creating new jobs, industries, wealth opportunities, products and services that contribute to economic growth. In the coming years, there will be no industry left untouched by the impact of the cyber system.

In the past, different disconnected technologies did play an **enabler** role in the growth of agriculture, industry and service economies. However, in the digital economy, the cyber system has become a **key driver**, not just an enabler.

In Figure 3.2, one can see how in each economy, there is the presence of activities relating to other economies too. For example, in the agricultural economy, the GDP was highly dependent on agriculture. It does not mean that there were no activities relating to industrial economy and services economy. But the GDP impact of these activities was minimal.

But the beauty of the digital economy is that with the power of the cyber system, the agricultural, industrial and service industries can equally contribute to the GDP growth. There are considerable possibilities of increasing the world's GDP exponentially with the help of the cyber system.

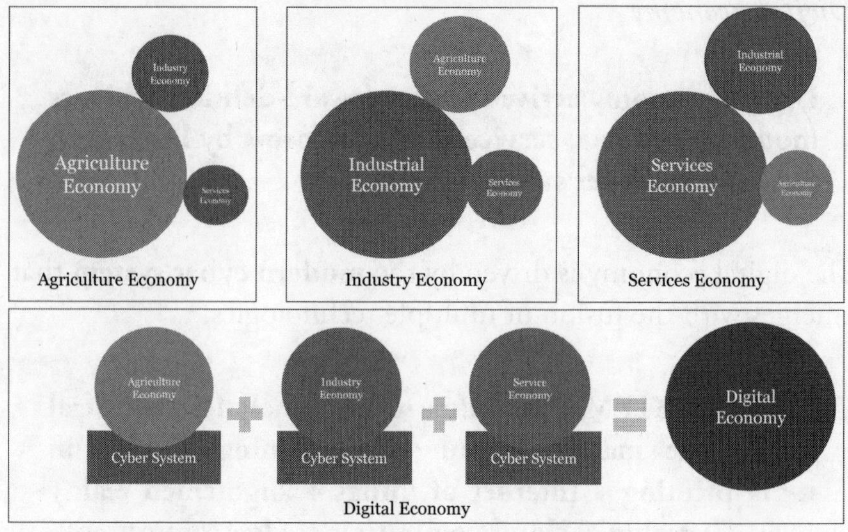

Source: Author

Figure 3.2: Evolution of Economies

The New Revolution: Fourth Industrial Revolution

Before the Fourth Industrial Revolution, the cyber system was not a driving force but just an enabler or a support system for other systems. In the past, it was mainly used for improving system performance and not for total system transformation.

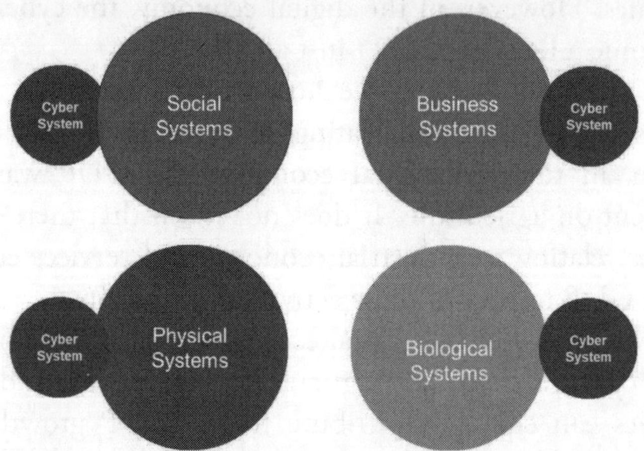

Source: Author

Figure 3.3: Cyber System as an Enabler for Other Systems

In the era of the Fourth Industrial Revolution, the **cyber system** has become a central driving force and has been reinventing and re-engineering all the other systems.

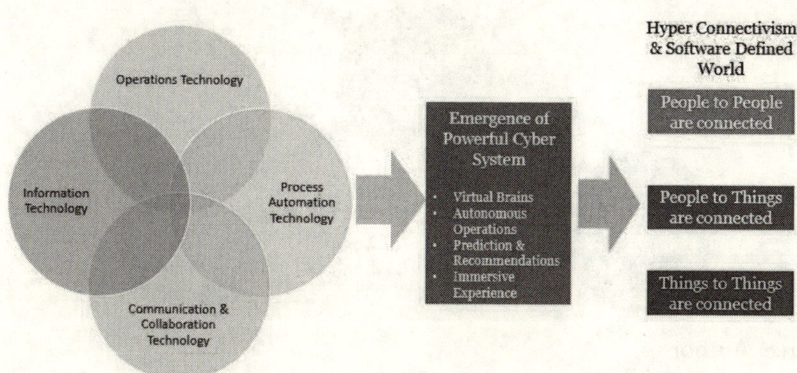

Source: Author

Figure 3.4: Modern Cyber System as a Driving Force for Other Systems

It is imperative to reimagine, redesign and re-engineer the physical, social, business and biological systems. New inventions, innovations and opportunities will emerge out of this revolution. All industries, governments, society and individuals must be ready to brace themselves for a significant **change**.

	Physical Systems	Social Systems	Business Systems	Biological Systems
Cyber System	Cyber-physical systems	Cyber-social systems	Cyber-business systems	Cyber-biological systems
Industries	Manufacturing, construction, energy, aerospace, automotive, high-tech, government and public sector	Government and NGOs (public health, public education and governance)	All industries and government	Life sciences, healthcare industries and government

Source: Author

Table 3.3: Modern Cyber System and Its Impact on Other Systems

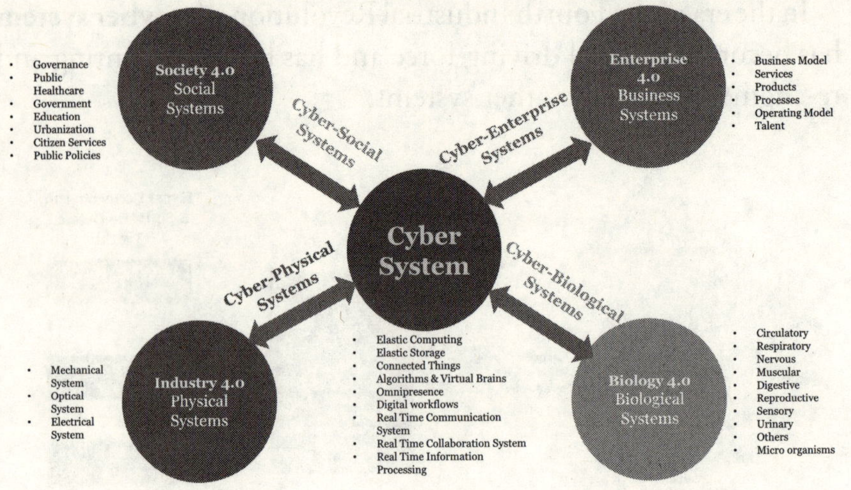

Source: Author

Figure 3.5: The 4.0 Systems

Society 4.0

When society is powered and driven by the cyber system (cyber-social systems), it is called Society 4.0.

There are so many problems in society like poverty, hunger, poor health, low-quality education, gender inequality, unequal justice, impure water, poor sanitation, costly non-renewable energy production, unemployment, low per-capita income, poor infrastructure, urbanization challenges, elimination of aquatic species because of pollution in oceans/rivers, deforestation, farmer suicides, corruption, outdated public policies, poor governance and many more.

Society is made up of different kinds of people. For example, based on generation, there are four categories: Gen Z (1995–2015), Gen Y (1980–1994), Gen X (1965–1979) and baby boomers (1946–1964). In today's hyperconnected world, the problems, needs and wants of people from each of these segments have changed significantly.

The cyber system has the potential to resolve all these problems more effectively.

Governments must re-engineer their services and governance to resolve the problems mentioned earlier and to meet the needs of the people from different segments of society. With internet in nearly every household, people will expect to get services and benefits of the government schemes at their doorstep. At the same time, even crime is becoming highly technology-driven.

Hence, we need to establish Society 4.0, which is powered by the cyber system.

In Society 4.0, the cyber system provides an opportunity to

1. design unimaginable new public policies,
2. establish digital (smart) governance,
3. introduce omnipresent citizen services (anywhere, anytime, any device) and
4. build smart cities and villages.

The cyber system generates a new breed of leaders like **techno-social reformers, techno-political leaders and techno-economy reformers, who will drive Society 4.0.** For example:

1. Governments can define a new policy to make generic medicines available from "Any Time Medicine Machines" (ATMMs) established at different locations. When a patient/person inserts a smart card into the ATMM, it would retrieve health records along with medical prescriptions. The patient can select the required medicines as per the prescription and pay the money.
2. Governments are now establishing **fibre to home** connectivity with which citizens can get a telephone, Internet Protocol Television (IPTV) and internet at home. Through IPTV, citizens can avail government services without visiting government offices. For example, the government of Andhra

Pradesh has been offering **free** education to students at home using IPTV.

3. Governments have been providing citizen services anywhere and anytime, using any device (mobile, tablet, desktop, wearable device). Countries like Estonia, UAE have leveraged the cyber system and transformed their governance to increase citizen happiness. With the establishment of telemedicine at primary health centres (PHCs), specialists from cities/towns can advise the patients about the right medication and treatment.
4. India has started a mega initiative called smart cities under which 100 cities will be identified to make them smart using the cyber system. This is mainly to address urbanization challenges. In the 2018 budget, India allocated US$30 billion for smart cities.

Enterprise 4.0

> **When business is powered and driven by the cyber system (cyber business systems), it is called Enterprise 4.0.**

The cyber system has the potential to create new businesses, business models, products, services, processes and operating models in the business. Digital natives like Amazon, Apple, Airbnb and Uber are born digital. However, many organizations, which are still driving conventional businesses, have been trying to jump on the digital bandwagon.

Banks have been offering non-banking services by leveraging the cyber system. Banks have started enrolling real estate developers and car dealers into their loan management digital platform. Millions of customers have been brought onto this digital platform. With the help of a matchmaking algorithm, any bank customer can search for a car or a house as per their preferences.

The digital loan platform provides precise details by mapping customer preferences. Also, the customer can directly communicate with the real estate developer or car dealer. Once the customer selects their preferred house or car, the bank automatically processes the loan application and issues the loan to the customer. Banks have implemented a new business model to improve their loan business.

At some retail stores, beacon-embedded mannequins send signals to the smartphone of the customer in the store and provide information on clothes and accessories the mannequin is wearing, their price and where to find the items within the store—redefining customer engagement.

Digitalized supply chain process redefines logistics and transportation. The **farm-to-fork** solution allows the customer to track fruits from the farmland to the warehouse to the delivery vehicle to the restaurant using the internet of things (IoT), blockchain and GPS.

The cyber system has the potential to change the whole business DNA. The Enterprise 4.0 journey has begun.

Biology 4.0

When biology and the cyber system converge (cyber-biological systems), it is called Biology 4.0.

In the era of Biology 4.0, we digitalize the human body with nanosensors, microprocessors, embedded software and communication networks. This means the injection of microcomputers/nanocomputers into the human body (invasive) or wearing smart devices (non-invasive) that provide excellent insight into the functioning of the human body in real time.

When we digitalize the body of two people, they can easily exchange data by shaking their hands. Radio-frequency identification (RFID) tags are now implanted inside the hand and using NFC or any other short-range communication network, a person can open the door lock or car lock just by tapping their fingers on the digital locks, without any remote-controlled device.

Digital pacemakers monitor and regulate heartbeats. It continuously transmits the heart data to the patient's mobile and allows the clinician to supervise remotely through a mobile or desktop. The data transmission frequency can be set by the doctor based on the health of the heart. Based on vital data of the heart, the mobile app also suggests the right kind of physical activity.

Brain-computer interface is the next big thing. The brain sends signals, responses or commands to computers, machines, intelligent things to interpret and act. For example, a person with nerve damage in the arm is unable to move his/her arm as it fails to receive brain signals. In this scenario, an electrode implanted into the arm directly receives brain signals allowing the person to move his arm. Prosthetic legs also receive brain signals directly allowing the person using them to walk.

Human body powered by the cyber system empowers human beings with additional capabilities and helps overcome many disabilities. The era of Biology 4.0 has begun.

Industry 4.0

When industries are powered and driven by the cyber system (cyber-physical systems), it is called Industry 4.0.

In industries, there are many physical processes which convert raw material into finished products. Different types of machines, tools and equipment are used to execute processes like hardening, tempering, annealing, through hardening, surface hardening, metalworking, machining, casting, joining, shearing and forming, surface finishing and others.

Traditionally the machines, equipment and tools used in industries to execute physical processes are powered by:

1. steam engines (Industry 1.0),
2. electricity (Industry 2.0) and
3. electronics and information technology (Industry 3.0).

In the era of **Industry 4.0**, the cyber system automates physical processes. They are run by intelligent machines, equipment and tools powered by IoT, robotics, artificial intelligence (AI), machine learning (ML), cloud computing, edge computing and others.

For example, in smart bottle factories, machines and products (bottle) communicate with each other as follows:

1. **Robotic Machine 1:** Bottles move on a conveyor and Machine 1 automatically reads the instructions on the RFID tag of an empty bottle and fills it with the relevant liquid as per the instructions. The filled bottle now moves to the next machine on the conveyor.
2. **Robotic Machine 2:** Machine 2 automatically puts the relevant cap on top of the filled bottle, then it moves to Machine 3 on the conveyor.
3. **Robotic Machine 3:** At this point, the appropriate label is pasted on the bottle, after which bottle moves on the conveyor for quality inspection.

In the above example, machines are made intelligent with the help of sensors, microcomputers and short-range network communications. When the bottle moves on the conveyor and comes to Machine 1, it reads the instructions on the RFID tag on the bottle and executes appropriate algorithms (virtual brain) using a microcomputer (edge computing) to pass the command and fill the bottle with the proper liquid. The other machines work similarly. Human interventions are eliminated and machines perform all the tasks.

The cyber system enables raw materials, products and machines to communicate with each other to perform the relevant functions without human interventions.

The era of Industry 4.0 has begun. The Fourth Industrial Revolution powered by the cyber system is driven by a new ideology called HYPERCONNECTIVISM. This new ideology has tremendous power to change the world.

The New Ism: Hyperconnectivism

Due to the cyber system, an unimaginable thing is happening in the world today. We are giving life to **things** like machines, plants, equipment, chairs, cars, bedsheets and many others. These things can now speak, sense, think, predict, communicate and work like human beings. This means, people and things work as a team and compete like rivals.

People to people communication has reached greater heights and it is possible to communicate and collaborate in real time using video, audio and text leveraging different channels. A new ism called **hyperconnectivism** is emerging, in which **people to people**, **people to things**, **things to things** are connected in real time.

Hence, we need to reimagine and redefine our businesses, organizations and society. Organizations must not limit themselves to human beings and consider **things** as important resources as well. For example, if an organization has 25,000 human employees, it may have another 25,000 "thing" employees (bots, robots, intelligent vehicles, intelligent machines and many more). This means the organization now must manage 50,000 resources.

With hyperconnectivity, new business models, products, services, policies, processes, operating models and work environments have already started emerging. The organizations that do not understand the implications of a **hyperconnected world** and its impact on their businesses may become extinct in the next 5–10 years.

A new journey has just begun with hyperconnectivism.

Six Ideals of Hyperconnectivism

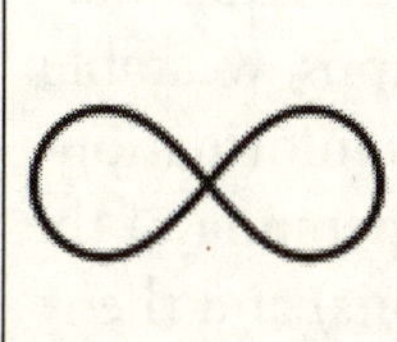	1. **The organization is boundaryless:** Globally distributed teams, suppliers and partners effectively work with instant communication and collaboration. Crowd learning, crowd innovation and crowd execution with extended and globally distributed specialists outside the organization help the organizations to prepare for a better digital future in a timely manner. **This is called the extended business ecosystem.**
	2. **Nearly autonomous execution:** People and things work together in an organization. Intelligent things like machines, equipment, robots, chatbots and virtual assistants perform their tasks in collaboration with people. This means human interventions for recurring tasks will be optimized significantly with the help of intelligent things.
	3. **The organization is omnipresent**: Employees, customers, suppliers and partners complete transactions, get information and can communicate anywhere and anytime using any device (mobile phone, mobile tablet, wearable device, console, laptop, etc.) that makes the organization truly omnipresent.

	4. **Immersive experience for emotional engagement of people**: People get into a virtual environment and experience the physical world virtually. Virtual cars, trees, plants, statues, mountains, etc., make people feel as if they were performing tasks in the physical world. In a hyperconnected world, there is an information aura around each person which is collated from social media, transactional systems, call records, mobile apps, wearable devices and many others. This information aura helps to understand the **persona** of the person (likes, dislikes and habits) and then personalize services to provide unforgettable moments of experience.
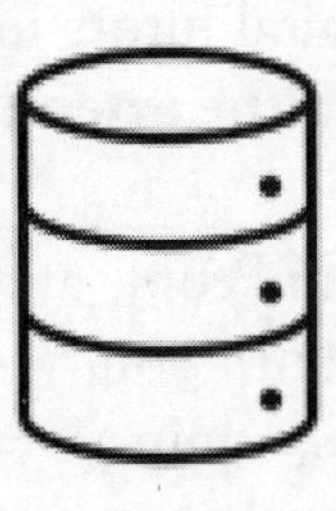	5. **The organization generates a humongous amount of data**: Boundaryless and connected organizations generate high volume, high velocity and variety of data in real time. For example, a jet engine generates 10 TB data in 30 minutes. This data is used to provide predictions and actionable insights by leveraging elastic super storage and elastic supercomputing power.
	6. **The organization deals with a new generation of people**: People are connected to the internet in real time, irrespective of age. They have low tolerance levels and expect to complete their tasks instantly. Behaviour, attitude and aspirations of employees, customers, partners and suppliers are quite different today.

Hyperconnectivism helps in building the following:

1. Software-defined societies (digital societies)
2. Software-defined businesses (digital businesses)
3. Software-defined human bodies (digital humans—cyborgs)
4. Software-defined industries (digital industries)

Organizations with their conventional thinking of driving businesses fail to implement hyperconnectivism to prepare for the Fourth Industrial Revolution. Embracing brand-new thinking is imperative to become future proof and future-ready.

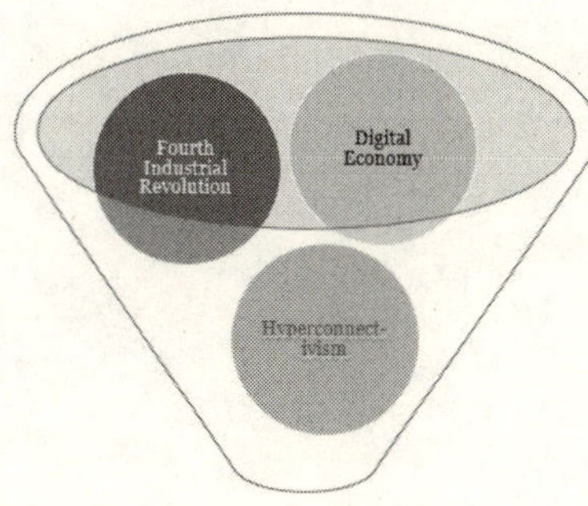

Source: Author

Figure 3.6: Troika of Modern Cyber System—Trigger Point for Digitalization

CHAPTER 4
Digitalization and New Growth Opportunities

Digital Age Customers

Whichever economy or revolution you are in, your customer must be the focal point of your organization's growth and sustenance. Due to the **cyber system**, which comprises the internet, mobile devices, social media and other technologies, the characteristics of customers have dramatically changed in the digital economy.

1. The digital age customers are tech-savvy and use multiple devices like laptop, mobile phone, tablet, wearable devices, etc.
2. They want to complete their work anywhere and anytime using the device of their choice.
3. They expect personal attention and superior experience at every interaction point.
4. They look for continuous newness (new features, new products and new services) within a shorter time span.
5. They make purchase decisions based on internet searches and social media comments.
6. They have highly collaborative social networks and are well informed about products and services (better than the enterprise sales team).

7. They have a low tolerance limit, are easily dissatisfied and shift service providers very fast.
8. They voice their opinions on social media very openly and can elevate or damage the organization's brand in a few hours.
9. Contactless lifestyle (digital) has become the norm in our pandemic-hit world.

The Convergence of Physical and Digital Worlds

The cyber system leads to the convergence of physical and digital worlds. The physical world includes cars, houses, machines, equipment, flats, people, animals, roads, poles, trees, plants, vehicles, trains, flights and many more. The digital world includes intelligent things, virtual environment, communication networks and others.

When the physical and digital worlds converge, the physical world gets augmented with the additional features and intelligence. For example:

1. The convergence of trains (physical world) and sensors, networks and other technologies (digital world) makes it possible to predict when a train might derail.
2. A machine (physical world) on converging with sensors, networks, internet and other technologies (digital world) becomes intelligent and performs tasks without any human intervention.
3. Trucks carrying fruits (physical world) when equipped with sensors, communication networks and information technology (digital world) sends regular alerts on the condition of the fruits. So one can keep a track of whether they are being maintained at the required temperature or have been spoiled and even the location of the truck.

The impact of the convergence of the physical and digital worlds are as follows:

1. It makes devices, machines and things smarter and reduces human interventions.
2. It enables remote monitoring and controlling of assets, equipment, appliances, devices and products.
3. It facilitates predictive and preventive maintenance using data and predictive analytics

The convergence of physical and digital worlds significantly impacts the configuration of businesses in the organizations. Traditional organizations have never experienced the power of the cyber system, and they must deal with many unknowns and uncertainties. Their only option is transitioning to digital businesses to be relevant in the digital economy.

Digitalization and New Opportunities

The troika of the modern cyber system has created two significant impacts:

1. The emergence of digital age customers with different needs, wants and expectations, which, irrespective of their age, are unique.
2. The convergence of physical and digital worlds that leads to an entirely new breed of products, services and things which are intelligent and nearly autonomous.

Hence, organizations are expected to create new business models and operating models to create new revenue streams, new markets and achieve high operational efficiency.

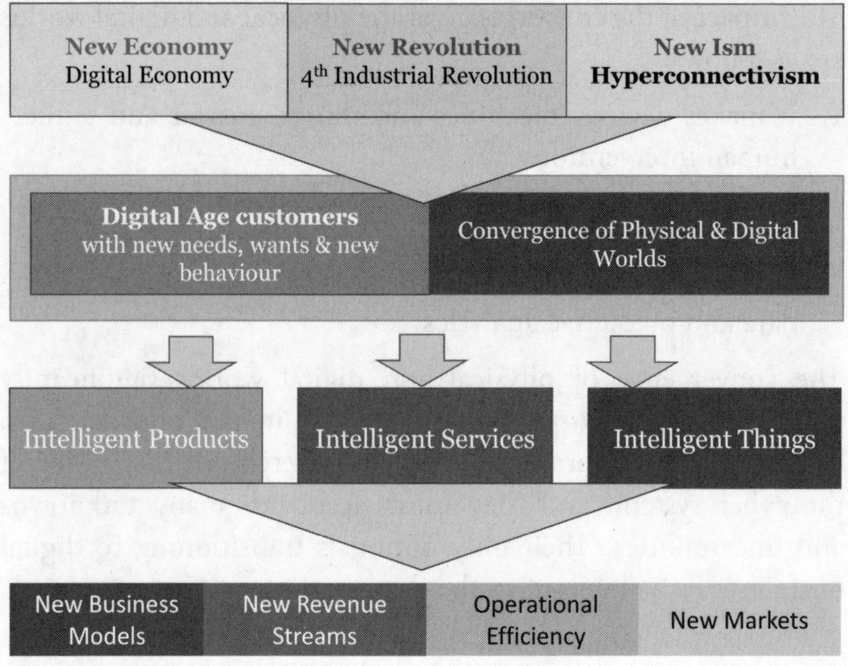

Source: Author

Figure 4.1: Why Digitalize?

Digitalization helps organizations become future-proof and future-ready.

Digital natives like Amazon, Google, Airbnb and Uber have the best leveraged cyber system. The organizations that have been there for many decades need to understand the potential of the cyber system and its capabilities to become digital. These organizations are called **digital immigrants**.

The cyber system is a game changer. Usually, the board of directors, chief executive officer (CEO) and business executives of digital immigrants have an attitude that technology is not their cup of tea and feel that the chief information officers (CIOs) and chief technology officers (CTOs) are responsible for the same.

Since CIOs and CTOs were not part of the board of directors, also due to skill gaps, they could not play strategic roles to transform their organizations. In a few organizations, CIOs report to CFOs and chief operating officers (COOs). Technology initiatives in the past were mainly implemented for process automation, lights on business and not for creating new business innovations, products, services, business models and revenue streams.

However, with the emergence of new capabilities from the cyber system, it is imperative for every employee—from top to bottom—to understand and acquire the basic knowledge of new technologies and the power of fusion of these technologies.

"I know everything, we do all things" attitude may cost the organizations growth and image. Hence, it is imperative for everyone in the organization to understand the basics of the cyber system and its potential to digitalize the organizations that become future-proof and future-ready!

The following example depicts the power of the cyber system (fusion of technologies) in the automotive industry. If the business executives of the automotive company do not take advantage of the cyber system promptly, their organization may sunset in a few years. A connected car is a classic example of how it has made cars more intelligent, customer-friendly and safe.

1. The car console is used as a screen to show information, alerts and warning messages to drivers.
2. Wireless hotspots, enabled by 4G LTE connectivity, connect various devices within the car. The car manufacturer ties up with a telecom company to get this feature.
3. Connected cars now have millions of lines of software code, thousands of sensors and microchips that help in alerting the driver about problems within cars. For example, when the fuel motor, battery or starter motor fails.
4. When other vehicles are nearer than the allowed limit, car breaks are applied automatically to avoid accidents.

5. With automatic breath analyzers, drunk drivers could be identified and prevented from driving under influence.
6. It is possible today to remotely start the engine, open and close car doors and control the AC.
7. Car manufacturers are now partnering with insurance companies. The latter fix the premium rates based on the data received from the car on the driving pattern of the driver. Rough driving would lead to a higher premium. This is called usage-based insurance.
8. When a driver crosses the speed limit set by government agencies for a particular location, auto alerts will be sent to the car console. Some cars may also be equipped with the feature to automatically reduce the speed of the car if the driver does not do so after repeated warnings.
9. Cars can communicate with other cars and flash a message about traffic congestion or weather conditions at a location so that other car drivers take a different route.
10. Car manufacturers are also partnering with various retail stores and hotels. If any retail store or hotel is nearby, the driver will be alerted about various discounts or sale offers through the car console.

Some organizations have understood the changing attitude and expectations of their customers and responded efficiently and effectively to meet the needs of digital age customers. Here are a few examples of organizations which have responded well to digital age customers.

1. Working moms are often worried about their sick babies at home. Smardii has developed an intelligent baby diaper which has sensors, a microprocessor and embedded software which connects the mother's mobile phone through 4G. The mother can monitor the body temperature, skin irritation, and other health details of the baby from the office on her mobile phone and take appropriate action. **Conventional baby diaper manufacturers will sunset very soon.**

2. Eight Sleep Pod is the smart mattress intelligent enough to offer the following features:
   a. Automate cooling and heating of the bed
   b. Custom temperature for each side of the bed
   c. Track of time slept and deep sleep
   d. Provide health reports that include body fatigue, respiration and heart rates
   e. Gentle vibration at chest level to wake you up
   f. A mobile app is used to control the pod's settings and access sleep reports and insights from anywhere
3. Tesla has manufactured software-defined electric cars, which are semi-autonomous. It maintain the cars by remotely updating the relevant software patches.
4. Garanti Bank, Turkey, has developed a mobile app to provide personalized services. Based on the previous spending pattern of the user, the user's favourite brands are identified by the app. The bank offers exclusive campaigns to the user for those brands. It has also integrated Foursquare and GPS.
5. Swiggy in India takes orders from customers using their software-defined app and delivers food from various restaurants.
6. Airbnb is a digital platform, which has been challenging conventional hotels and hotel chains and has gained a lot of momentum.
7. Uber generates revenue by bringing both cab drivers and travellers onto a software-defined platform. It creates a network effect—as more cab drivers enrol on the platform, travellers get better cab service with more options, and as more travellers enrol on the platform, there will be more income for the cab drivers and Uber. **The entire taxi industry has been disrupted by Uber.**
8. With the help of the cyber system, organizations can reach out to **new customer segments**. For example, corporate hospitals through telemedicine can offer services to a new segment of rural customers

9. Virtual reality (VR) and augmented reality (AR) provide foreign tourists with an immersive experience of the tourist spot remotely from their locations. They may book the tickets based on their satisfaction with the virtual experience. Tourism companies can attract more foreign customers this way.

Hence, digitalizing businesses is imperative due to the following reasons:

1. To meet the digital age customer's needs, wants and expectations.
2. To create new businesses aligned with the digital economy.
3. To solve business, social, environmental and economic problems relevant to the digital economy.
4. To release digital products and digital services aligned with the digital age customers to create new revenue streams.
5. To leapfrog the new competition that will emerge from an unknown corner.
6. To create digital capabilities that are relevant to the digital economy.

Therefore, in the digital era:

1. The **environment** you live in is different.
2. The **experience** you gain is different.
3. The **exposure** you get is different.

Digitalization Mantra

1. The fusion of technologies creates the modern cyber system.
2. The modern cyber system digitalizes the organization.
3. Digitalized organizations can uniquely create new business models, products, services and operating models.
4. Digitalization creates new customer segments, revenue streams, markets and improved profits.

ARE YOU READY?

PART 2

Acquire a Digital Mindset

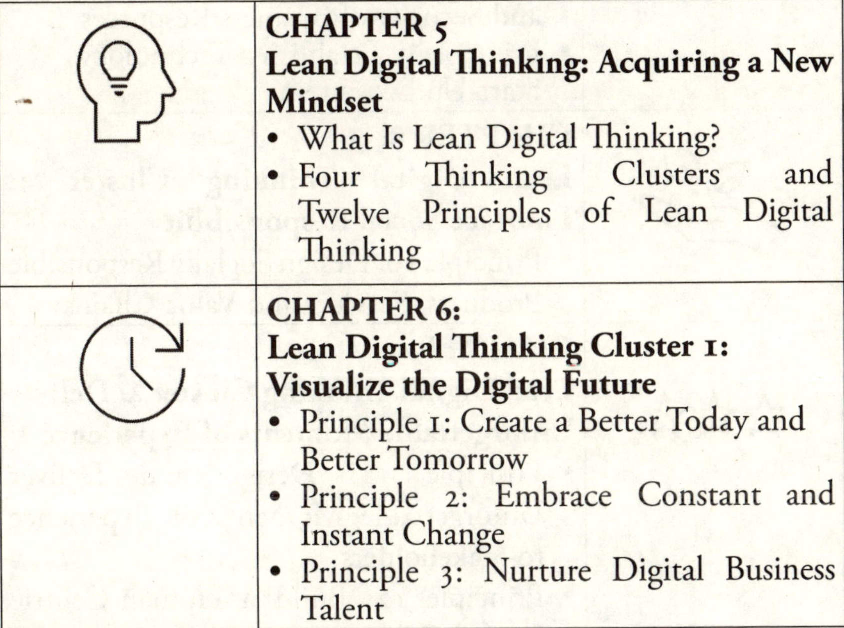

	CHAPTER 5 **Lean Digital Thinking: Acquiring a New Mindset** • What Is Lean Digital Thinking? • Four Thinking Clusters and Twelve Principles of Lean Digital Thinking
	CHAPTER 6: **Lean Digital Thinking Cluster 1: Visualize the Digital Future** • Principle 1: Create a Better Today and Better Tomorrow • Principle 2: Embrace Constant and Instant Change • Principle 3: Nurture Digital Business Talent

	**CHAPTER 7:** **Lean Digital Thinking Cluster 2: Achieve Digital Execution Excellence** • Principle 4: Identify and Eliminate Non-Value-Added Activities • Principle 5: Predict and Improve Business Performance • Principle 6: Optimize Human Interventions • Principle 7: Establish Real-Time and Contextual Communication and Collaboration • Principle 8: Safeguard Privacy, Safety and Security of Business Resources • Principle 9: Establish a Technology Start-Up Ecosystem
	**CHAPTER 8:** **Lean Digital Thinking Cluster 3: Embrace Social Responsibility** • Principle 10: Design Socially Responsible Products, Services and Value Chains
	**CHAPTER 9:** **Lean Digital Thinking Cluster 4: Deliver Unforgettable Moments of Experience** • Principle 11: Design and Deliver Unforgettable Moments of Experience to Stakeholders • Principle 12: Build a Human-Centric Digital Culture

CHAPTER 5
Lean Digital Thinking: Acquiring a New Mindset

What Is Lean Digital Thinking?

Lean digital thinking prepares the minds of employees, leaders, suppliers and ecosystem partners of an organization for new thinking and to manage their business effectively in the digital era.

Traditional thinking does not help in building the systems required for the digital economy and Fourth Industrial Revolution. Hence, organizations must acquire a new DIGITAL MINDSET.

With a **digital mindset**, you could create new businesses, industries, products, services, operating models and many other innovations. Also, this mindset enhances your ability to bring the capabilities mentioned in Figure 5.1 into your organization.

Predictability	Optimal Human Interventions	Immersive Experience	Omnipresence

Source: Author

Figure 5.1: Key Digital Capabilities

With a **lean mindset**, you could enhance the value of your organization by eliminating wasteful elements. This mindset was introduced in the era of the Second Industrial Revolution. In the era of the Fourth Industrial Revolution, the lean mindset must be contextualized. By embracing a lean mindset, you use methods, tools and techniques to eliminate wastes (non-value-added activities) and make the enterprises in the digital era more efficient and effective.

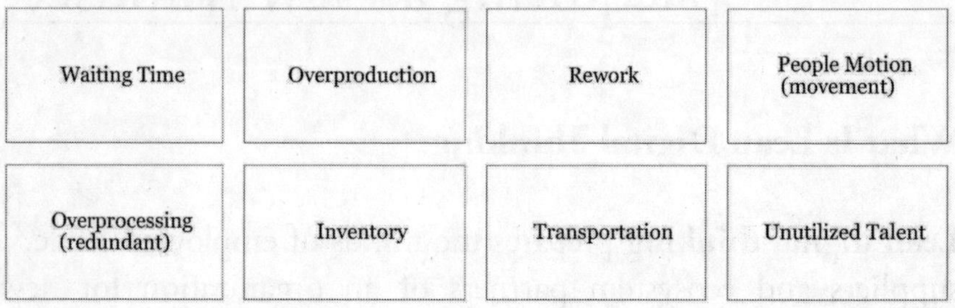

Source: Lean Methodology

Figure 5.2: Eight Business Wastes

Without a lean mindset, digital thinking creates fat organizations, which are costly to build and manage.

[Lean Thinking + Digital Thinking]

=

[Lean Digital Thinking]

Along with the classic digital thinking, you need to acquire a lean mindset in order to unlock the potential of your organization in the era of the Fourth Industrial Revolution.

Four Thinking Clusters and Twelve Principles of Lean Digital Thinking

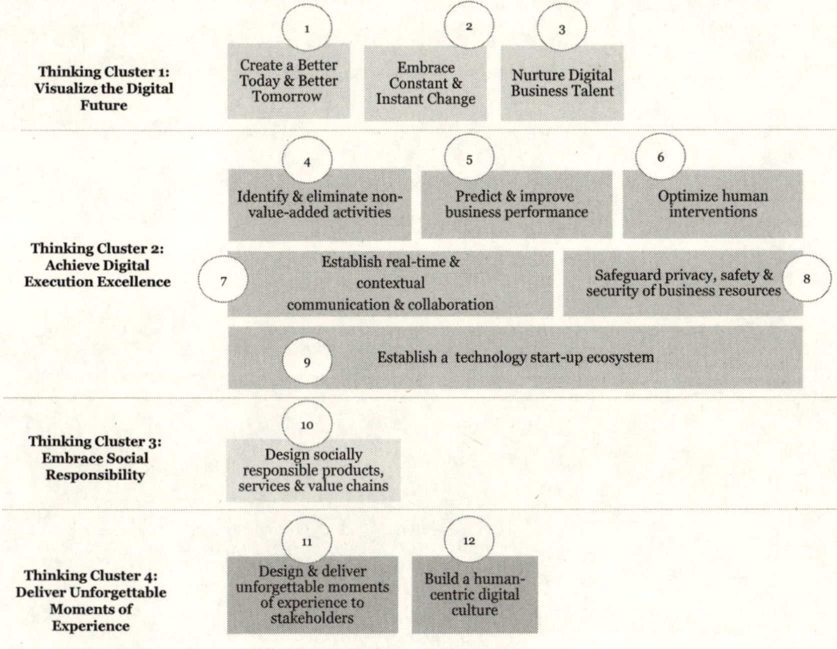

Source: Author

Figure 5.3: Four Thinking Clusters and Twelve Principles of Lean Digital Thinking

CHAPTER 6
Lean Digital Thinking Cluster 1: Visualize the Digital Future

Principle 1: Create a Better Today and Better Tomorrow

For many decades, organizations have leveraged the cyber system to bring operational efficiency and better customer services to their businesses. The cyber system is an enabler for these traditional businesses.

At present, the modern cyber system has been changing each and every business building block of every organization. The business battlefield is entirely different and software drives business. These digital businesses (software-defined businesses) are entirely different from traditional businesses. Table 6.1 compares the characteristics of traditional business and digital business.

Business Element	Traditional Business	Digital Business	Digital Business Examples
Economy	Service economy	Digital economy	In the digital economy, businesses are driven by the cyber system. A few examples are connected cars, digital healthcare, digital education, digital banking and digital retail.

Response to market	Marathon speed	Sprint speed	Due to changing trends and customer expectations, organizations are expected to release new products and services within a shorter time span. Hence, digital business must work at sprint speed. For example, today updated versions of mobile phones are released in short intervals at a fixed frequency.
Nature of Business	Stable, less risky and known competition	Unpredictable, evolving, uncertain, exploratory, highly risky and unknown competition	During the Fourth Industrial Revolution, the needs and wants of customers are changing very rapidly due to new technologies. Technology companies have been challenging traditional organizations. For example: 1. Google was never expected to be a competition for the automotive industry. 2. Airbnb has been challenging global hotel chains. 3. Samsung digital watches are competing with analogue watches, which have been there for many years. 4. Airlines consider telepresence a threat to their business.

Service access	Few channels	Omnipresent	Customer services are expected to be delivered anywhere and anytime using multiple devices like mobile devices, wearable devices, TV and others. For example, today grocery items can be ordered using a mobile phone, a web application or the car console screen of an internet-enabled car.
Products and assets	Static or semi-automated	Smart or digital	The cyber system powers products and assets by making them smart using software. For example: 1. Smart cars can park themselves or drive autonomously. 2. Predictive maintenance of equipment and assets through remote monitoring is possible with the cyber system.

Customer	Task needs; expects quality products and services	Task needs and emotional needs; expects unforgettable moments of experience with quality products and services	Digital business looks after the customer's task needs and emotional needs. For example, a customer visits an ATM to withdraw the money. 1. Task needs: The customer withdraws money from the ATM using biometric authentication. 2. Emotional needs: An intelligent ATM would know about the persona of the customer based on social media profile and past spending patterns. It starts a voice conversation with the customer. For example, based on social media conversations of the customer, an intelligent ATM will know that the customer does not have money to go to Canada to see her ailing daughter. Now ATM speaks with the customer and provides coupons or vouchers to purchase air tickets.

Employees	Baby boomers and Gen X—the office is the workplace	Baby boomers, Gen X, Gen Y, Gen Z—Digital workplace (anywhere and anytime)	In digital business, employees from different age groups and attitude must work together. Gen Y is mostly tech-savvy and mixes up professional and personal life. They may not like to work at the office from 9:00 AM to 6:00 PM and expect flexible working hours and workplaces. They do not hesitate to work from a coffee shop or ask questions and challenge the decisions made by their seniors. Gen X loves to work more formally. Most of them believe that working from the office is more professional and productive. They prefer a directive approach and expect subordinates to follow instructions without challenging them. There may be friction between Gen Y and Gen X employees if they are not effectively managed and prepared to collaborate towards a common purpose.

Source: Author

Table 6.1: Traditional Business vs Digital Business

Organizations cannot ignore traditional business to focus on digital business entirely. At the same time, organizations cannot ignore digital business.

"A Better Today and Better Tomorrow" principle highlights the importance of excelling in today's traditional business while preparing for digital business simultaneously.

Better today (future-proof): Traditional businesses must embrace current trends, resolve current problems and realize the current aspirations of customers to excel in today's business. **Traditional businesses (digital immigrants) are driven at marathon speed.**

Better tomorrow (future-ready): Software-defined digital businesses must embrace future trends, resolve unmet and unarticulated needs of customers, leverage the cyber system and work to realize future aspirations to survive and excel in the next few years. **Digital businesses must run at sprint speed.**

To create a better today and better tomorrow by using lean digital thinking, leaders must answer the following questions:

1. Are we ready to drive two-speed businesses at the same time within our organization?
2. Do we have any transition path to move from traditional businesses to digital businesses without cannibalizing the existing revenue streams abruptly?
3. Are we ready to build businesses that can address uncertain, exploratory requirements and unknown competition emerging from technology companies like Google, Amazon and Apple (banking, automotive and communication businesses)?
4. Do we have strategy, offerings, structure, talent and processes to drive traditional business at **marathon speed** and digital business at **sprint speed** at the same time?

Principle 2: Embrace Constant and Instant Change

A slanted S-curve is an apt representation of a typical business life cycle, as given in Figure 6.1.

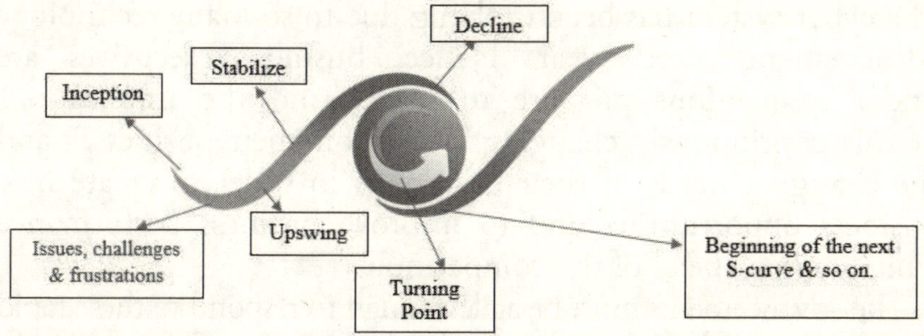

Source: Author

Figure 6.1: The S-Curve Representing a Business Life Cycle

In the past decades, a few organizations have successfully passed through many S-curves. For example:

1. IBM was founded in the year 1800 under the name, CTR.
2. In the early 1920s, CTR was renamed IBM.
3. In 1964, IBM brought out family computers which changed the world.
4. In the late 1980s and early 1990s, IBM faced a loss of US$8 billion as it failed to catch up with the personal computer revolution.
5. After many reorganizations, IBM is now one of the leading computer manufacturers in the world.

IBM has gone through many S-curves and created many turning points since 1920 and is still going strong. So many organizations like Kodak, Blockbuster, Circuit City and *Reader's Digest* are out of business today as they were not able to embrace the change at the right time by leveraging the cyber system. General Electric, which has successfully crossed many S-curves in the past, has been struggling to maintain the S-curve lately.

In the past, the time span of an S-curve was a few decades, and now it is much lesser about 1–3 years. This means change is constant and instant.

The cyber system has been evolving due to so many technology advancements every year. Hence, business executives are under tremendous pressure to understand the implications of this continuously changing system on their businesses and the changing needs of their customers in order to create new business opportunities and to improve business performance while staying ahead of the competition.

The advancements must be agile enough to respond to these rapid changes. Hence, leaders are expected to be receptive of change within a shorter time span and must be ready for a roller coaster ride.

To embrace constant and instant change by using lean digital thinking, leaders must answer the following questions:

1. Are we ready for the next S-curve of our business?
2. Have we identified turning points to take our business along the next S-curve?
3. Do we have the new shared vision and common purpose for all employees?
4. Do we have a business landscape prediction engine and actionable insights to design and implement change at the right time?
5. Do we have standardized tools, techniques, templates to absorb changes at a rapid pace?
6. Have we established rapid experimentation centres to prepare the proof of concept (for new business models, products, services, processes and organization structure) with the new capabilities of the cyber system?
7. Do we have agility in the organization structure to introduce new roles, culture and policies promptly to embrace change?
8. Have we defined continuous change management (CCM) processes and procedures?

Principle 3: Nurture Digital Business Talent

Leadership skills required to drive digital businesses powered by the cyber system are different from the skills required to drive traditional businesses. Organizations must introduce new roles, skills and behaviours to successfully drive digital businesses.

By implementing **lean digital thinking**, organizations will be able to design new roles, skills and behaviours in the areas given in Figure 6.2.

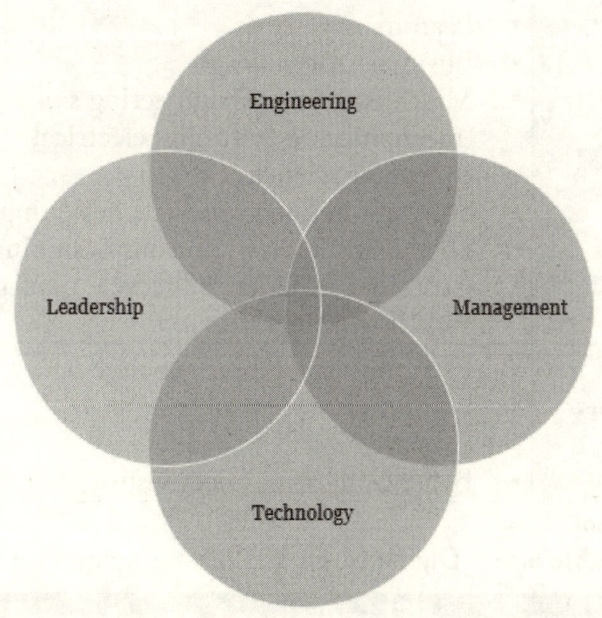

Source: Author

Figure 6.2: Digital Business Skill Mix

The following tables describe the roles, skills and behaviours for each of the areas illustrated in Figure 6.2. The roles given in the following tables are illustrative and may vary as per the organization's requirements.

Engineering	
Focus	• Leverage the cyber system to design intelligent and connected products, platforms and services
New Roles	• Chief digital product officer • Chief digital solutions officer
New Skills	• Complex problem-solving • Critical thinking • Creativity • Cognitive flexibility • Design thinking • Lean management • Business storytelling • Multidisciplinary engineering skills (mechanical, electronics, electrical, computer science and IT) • Software-defined products designing • Digital enterprise solutions designing • Customer experience lifecycle management and PLM integration
New Behaviours	• Flexibility to embrace change • Customer-first attitude and empathy • Belief in hyperconnectivism

Source: Author

Table 6.2: Digital Business Skills—Engineering

Technology	
Focus	• Establish and manage an integrated technology platform (cyber system) and centres of excellence using: o Information technology o Operational technology o User interface technology o Process automation technology o Communication and collaboration technology

New Roles	• Chief intelligence officer • Chief digital security officer • Chief business ecosystem officer • Chief human and things resources officer • Chief mobile officer • Chief blockchain officer • Chief IoT officer • Chief data officer • Business-savvy technology executives
New Skills	• Complex problem-solving • Critical thinking • Creativity • Emotional intelligence • Cognitive flexibility • Lean management • Communication, collaboration and coordination • Negotiation • Business strategy creation by leveraging the cyber system • New products/services identification by leveraging the cyber system • Cyber system management • Digital technology vision and strategy map • Digital technology architecture blueprint designing • Integrated application management (systems of record, systems of engagement and systems of insight) • Public, private and hybrid cloud life cycle management • Change management • Complex and multi-partner/technology start-ups integration and program management

New Behaviours	• Align with business executives • Collaborative troubleshooting/ideation • Integration of business units and sourcing partners • Rapid response (24x7-connected environment)

Source: Author

Table 6.3: Digital Business Skills—Technology

Management	
Focus	• Leverage the cyber system and establish new business models, policies, processes, procedures, tools and techniques for digital business performance
New Roles	• Chief digital operations officer • Chief customer experience officer • Chief analytics officer • Digital-savvy business executives
New Skills	• Critical thinking • Decision-making • Emotional intelligence • Communication, collaboration and coordination • Negotiation • Lean management • Business storytelling to engage customers • Customer experience management • Ability to interpret information and insight generated from data to take the right decisions at the right time • Ability to manage a crisis in social media • Ability to use technology to establish a flat and connected organization • Ability to use social media for customer service and brand management • Entrepreneurship • Technology-driven business management

New Behaviours	• Align with Gen Y and Gen Z culture (employees, customers, partners) • Collaborative troubleshooting/ideation • Work in a 24X7-connected environment • Every business is a technology business

Source: Author

Table 6.4: Digital Business Skills—Management

Leadership	
Focus	• Leverage the cyber system and create a digital strategy, new business opportunities, new revenue streams, acquire lifelong customers and leapfrog competition
New Roles	• Chief business technology officer • Chief digital strategy officer
New Skills	• Digital vision • Critical thinking • Creativity • Cognitive flexibility • Lean management • Digital business strategy map creation • Digital business models designing • Integrated business technology consulting • Ability to use crowdsourcing for products/services innovation
New Behaviours	• Customer empathy • Every business is a technology business • Collaboration with engineering, management and technology teams

Source: Author

Table 6.5: Digital Business Skills—Leadership

The other roles, which organizations must focus on to leverage the cyber system, are (indicative list):

1. Digital business process manager
2. Digital technology architect
3. Cloud infrastructure specialist
4. DevOps engineer
5. DevOps architect
6. Engineering manager
7. Site reliability engineer
8. Agile coach
9. Scrum master
10. Product manager
11. Edge computing infrastructure specialist
12. Unified data specialist
13. Digital security specialist
14. Hybrid networks specialist
15. Internet of things (IoT) specialist
16. Omnipresence specialist
17. User experience specialist
18. Digital apps developer
19. Prescriptive analytics specialist
20. Predictive analytics specialist
21. Descriptive analytics specialist
22. Mobile apps developer
23. Mobile device management specialist
24. Enterprise app store manager
25. Enterprise API manager
26. Social/collaborative apps developer
27. Hadoop, Map Reduce and NoSQL specialist
28. Blockchain specialist/engineer
29. Mixed reality developer

To nurture digital business talent by using lean digital thinking, leaders must answer the following questions:

1. Do we have the strategic, tactical and operational skills in product engineering to innovate digital products, digital platforms and digital services?
2. Do we have the strategic, tactical and operational skills in digital technology to develop digital software applications and digital technology infrastructure?
3. Do we have the strategic, tactical and operational skills in business management to manage digital business operations effectively?
4. Do our business leaders have the strategic, tactical and operational skills to prepare for the digital future and the Fourth Industrial Revolution?
5. Do we have the required tools, techniques methods and learning system to nurture digital skills at all levels?

CHAPTER 7
Lean Digital Thinking Cluster 2: Achieve Digital Execution Excellence

Principle 4: Identify and Eliminate Non-Value-Added Activities

Robert Johnson, the CEO of a manufacturing company, is annoyed with the cost escalations in the organization. Based on the data received from various sources, he and his team found the following problems during their review:

1. They had been waiting for a longer time than expected to get approvals for procuring materials.
2. They were taking a lot of time for machine's changeover (with different tools) to support the production of a variety of products.
3. They were not maintaining their material inventory effectively, and the non-availability of material was one of the main reasons for the delay in production.
4. Their machine downtime was too long, and time taken to restore operations was too much.
5. They had been producing products, which were not as per the actual customer demands, hence inventory costs were escalating.
6. Their production planning was ineffective, faulty and had been causing both overproduction and underproduction.

7. Some of their products had features which their customers were not looking for.
8. They observed that their rework cost had recently become high due to poor quality of production.
9. There were too many accidents at the worksite and their workers were frequently getting sick.
10. Their processes had not been automated yet, and they were putting in a lot of manual effort. A few of their policies were very complex and redundant causing delays and cost escalations.
11. They needed to review their store and warehouse locations as the transportation cost was in the upswing.
12. They had excellent specialists and competent leaders but observed that they were not optimally utilizing their talent for product innovations and operational excellence.

Robert Johnson expresses deep concerns that their organization was losing edge, and the competition was leapfrogging them in terms of pricing, quality and customer retention. They need to find ways to address all the problems which they had been facing; else they may not have a place in the business world in the coming years.

Rob and his leadership team discuss and agree to induct a specialist named Brian to improve the current processes, policies and systems. Brian spends three weeks gathering information from various leaders, managers and workers. Brian strongly believes that bringing **lean philosophy** was the only way to resolve the problems in the organization.

Many organizations have successfully implemented lean principles, tools and techniques and got tremendous benefits for the last five decades.

In the Fourth Industrial Revolution, each enterprise is digital, and each digital enterprise is lean.

Lean philosophy is not something new; it has been there since the Second Industrial Revolution. Lean is nothing but continuously identifying and eliminating non-value-added activities in businesses. However, today, we are in the Fourth Industrial Revolution and we need to contextualize the lean principles and tools to adapt to the Fourth Industrial Revolution.

Building digital businesses without integrating lean philosophy is very costly to manage and improve on an ongoing basis. Hence, digital businesses must become **lean digital businesses**.

What is Lean?

Lean means:

1. **Define the value** that must be delivered to the customer.
2. **Create a value stream** to identify step-by-step activities, information flow, material flow and other data points like cycle time to deliver value to customers.
3. **Optimize the value stream** by removing the non-value-added activities from the value stream.
4. **Improve the flow** with the right sequence of value-added activities in the value stream, remove interruptions and constraints, reduce time delays between the hand-offs in the value stream.
5. **Enable pull system** with just-in-time delivery of products and services to customers (as per the actual customer demand and not based on forecasting).
6. **Bring perfection** by continuously improving the processes, information flow and material flow in the value stream.

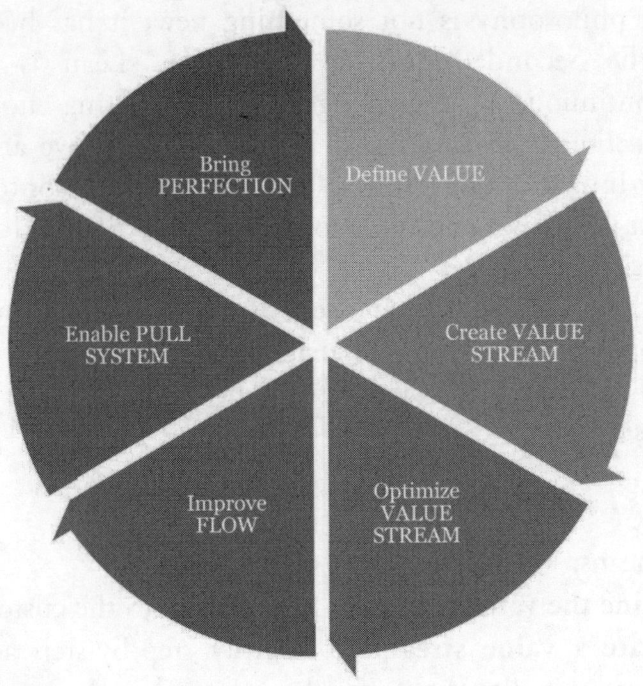

Source: Author

Figure 7.1: The Lean Circle

Define Value

Identify the **value** of a specific product or service in the following dimensions:

1. **Value dimension 1**: Deliver **benefits** or expected outcomes (price, quality, speedy response, reliable service delivery, etc.)
2. **Value dimension 2:** Resolve customer **pain points**
3. **Value dimension 3:** Meet customer **needs and wants**
4. **Value dimension 4:** Enable customers to do their work/tasks **effectively and efficiently**

The **value** could be any or mix of the options given in Table 7.1 for which the customer is ready to pay.

Value	Value Description	Value Dimension
Use value	Usefulness of product or service for the identified purpose	Resolve customer **pain points** Meet customer **needs and wants**
Esteem value	Enhancement of social status, image while using the product or service	Meet customer **needs and wants**
Time value	Speed at which the product or service is delivered to the customer	Deliver **benefits** or expected outcomes
Cost value	Affordable and optimal product or service cost	Deliver **benefits** or expected outcomes
Exchange value	Exchange of a product with something else (sale value)	Deliver **benefits** or expected outcomes

Source: Author

Table 7.1: Different Dimensions of Business Value

Create Value Stream

A **value stream** is nothing but the steps involved in delivering value to the customer. Designing a value stream includes the following:

1. Identify the sequence of activities of a process to be carried out for the identified product or service to deliver its **value**
2. Identify the flow of the following items from one step to the next step in the value stream:
    a. Flow of materials (manufacturing value stream)
    b. Flow of designs (design and development value stream)
    c. Flow of customer needs (services value stream)

In the following example (Figure 7.2), a customer wants to purchase an apartment and has been looking for the following benefits:

1. Best suitable apartment as per their preferences (use value)
2. Hassle-free and speedy bank loan processing (time value)
3. Shorter purchase cycle (time value)

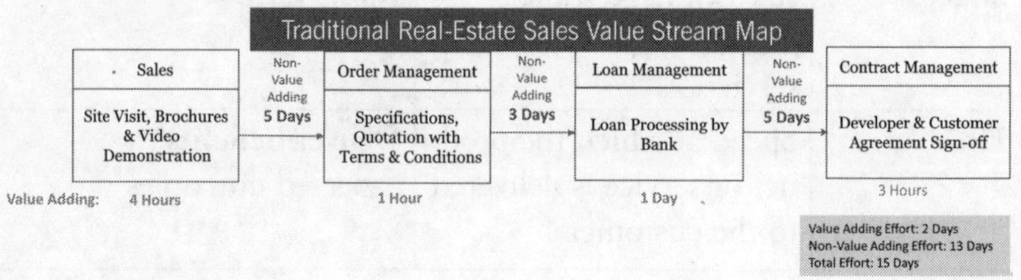

Source: Author

Figure 7.2: Traditional Value Stream Map for Real Estate Sales

In the value stream illustrated in Figure 7.2, the total purchase cycle is 15 days. That is because the loan processing by the bank is cumbersome and involves a significant manual validation process to approve the loan.

The non-value-added activities (increased waiting time due to process inefficiencies within the real estate development organization) is also considerably high. It makes up almost 87% of the total effort. In the 15-day cycle, 13 days are spent waiting to progress from one step to another! Here, the need is to find solutions to reduce these 13 days, which is non-value added!

Optimize Value Stream

Optimizing value stream means eliminating **non-value-added activities** in the value stream (design value stream, production value stream, sales value stream and maintenance value stream). The non-value-added activities in the organization are mentioned in Figure 5.2.

Waiting Time

Waiting time is one of the significant wastes or non-value-added activities in the organization which must be minimized.

Examples:

1. Waiting for instructions from supervisors
2. Waiting for repairs on faulty machines
3. Waiting for the completion of the previous operation
4. Waiting for testing software after development
5. Waiting for IT infrastructure to deploy the software

Overproduction

Creating or delivering products and services more than and before they are required by the customer. Inaccurate and incomplete customer demand information causes overproduction. Overproduction results in high inventory, transportation and quality inspection effort.

To minimize overproduction, organizations must produce or deliver services based on the actual demand, i.e., based on the orders or demand from the customers.

Examples:

1. Production of components before being ready for the next stage
2. Poor demand-planning and ineffective production-planning
3. Building too many features into a software product which customers are not looking for
4. Adding too many VFX features in a movie

Rework

Rework is nothing but putting a lot of effort into fixing the defects or problems to align with the requirements or plan because it was not done as per the specified requirements or plan in the first place. Products which are not built as per a customer's requirements lead to a lot of rework. It is very important to identify defects early in the life cycle of a product.

Rework is mainly caused because of lack of training, skill, standards, checklists, tools and procedures, low-quality materials received from suppliers and damages caused during transportation. Low quality work requires a lot of rework, high consumption of resources, material and too much documentation. This escalates costs and leads to loss of customers. Low quality work and rework may create serious legal tangles with customers.

Manual quality inspections are time-consuming and ineffective. Preventing defects and rework is very essential for better customer service and high-quality work.

Examples:

1. Recalling products due to incorrect design and poor reviews
2. Inspection and replacement of damaged materials
3. Correction of defects due to the poor design of products
4. Correction of defects due to poor workmanship
5. Non-compliance with standards and regulatory compliance practices

People Motion

Business waste from people motion is due to the unnecessary movement of people from one location to another location when doing work.

Examples:

1. Ill-planned workplace layout and long walk between work areas to perform the work or moving material/products from one location to another location
2. Improper design and location of cabinet or stores and searching for right files and materials that takes a lot of time
3. Ineffective and suboptimal design of processes, procedures and methods that causes extra movement of people from one location to another location
4. Strategically significant resources and specialist resources are highly distributed leading to travel to different far-off locations for any mission-critical engagements
5. Complex folder structures in computers and clicking through many folders to get the required documents

Overprocessing

Overprocessing means doing work that does not add value to the customer. It is considered to be a significant waste and reduces operational efficiency. Following are a few reasons behind overprocessing in the organizations:

1. Lack of standard operating procedures (SOPs)
2. Immature and incapable business processes
3. Low or no automation of processes
4. Lack of templates, checklists, guidelines, standards, tools and techniques
5. Reinventing the wheel and no knowledge-sharing platform within the organization

Examples:

1. Defining redundant and stringent policies and business rules, which cause many delays; for example, multiple approvals for even a small amount
2. Defining tight tolerance limits for the components which do not require that much precision
3. Painting the parts and surfaces which are not visible, and clean or polish the components, products beyond the specified level
4. Asking the customer to provide their details at multiple places for the same request (help desk, website and in person)
5. Conducting too many medical tests on patients than required
6. Collecting user requirements beyond the scope of the software application being developed
7. Unnecessary or too many inspections, meetings and management reviews

Inventory

Inventory is maintained for raw materials (stores), work-in-progress goods (shop floor) and finished goods (warehouse). It requires a lot of space, storage containers, administrative staff, electricity and cooling which escalate costs considerably. The inventory cost may also include the cost of damages to items and insurance. Hence, optimal maintenance of inventory is very essential to minimize waste/cost.

Examples:

1. The pile-up of raw materials and products caused due to interruption in the flow of production in the manufacturing industry
2. Large batches of products or components have to be stored temporarily at a location on the shop floor to carry to other machine or warehouse using robots, forklifts, etc.
3. Machine breakdowns, change over time of products from one machine to another machine, quality rejections cause accumulation of work-in-progress products in the production flow which occupy a lot of storage space
4. Transportation delay in supplying raw materials creates work-in-progress products which have to be stored in the production line and wait to move to the next operation
5. Creation of excess promotional material (videos, audios, flyers, brochures)
6. Overstocked medicines in hospitals or pharmacies take up a lot of inventory space
7. Production and storage of too much food that goes to waste due to low demand
8. Unused or rarely used machines or equipment
9. Procurement of more office stationery items than required
10. Pile-up of customer orders which have not been processed yet

Transportation

Without transportation, organizations cannot function at all. Movement of people, tools, equipment, products and materials from one location to another is required to produce and deliver products and services. There are two kinds of transportation within an organization:

1. Transportation within the boundaries of the organization among different departments (vehicles, forklifts, cranes, etc.)
2. Transportation outside the organization to/from suppliers and customers at different locations (air, water, pipe, rail and road)

Examples:

1. Cost of fuel, electricity and repairs to maintain transportation equipment and vehicles
2. Cost of transportation between stores, warehouses and shop floor when they are located far from each other
3. Cost of damage of items in transit during transportation
4. Movement of files, information and documents from one department to another

Unutilized Talent

Most of the times, organizations underutilize or fail to utilize the talent available to them. Due to this, organizations fail to deliver the products and services promptly. The quality of work is low as well and there is a lot of rework due to skill gap. Lack of innovation in the organization maintains the status quo and the organization lags behind the competition.

Examples:

1. There is a mismatch in role and job, due to which employees fail to perform their work effectively—we can't expect a fish to climb a tree
2. Employees work in silos, and there is no knowledge sharing among them, which makes the employees reinvent the wheel for similar problems or issues
3. There is no learning strategy and plan to nurture the skills of the employees as per the changing needs; hence, there is a lack of productive work
4. Employees and leaders are busy with day-to-day operational and administrative work and get less time for creative thinking
5. The work environment is not right and employee engagement is low

Eliminating Non-Value-Added Activities Using Lean Tools

Many lean tools and techniques have been designed by great people over more than six decades. Covering all those tools is beyond the scope of this book. Some of the tools have been mentioned below.

Lean Tools

1. Value stream mapping
2. Target value design
3. Level scheduling
4. Kanban
5. Single minute exchange of dies
6. Visual control management
7. Standardized work
8. Structured problem-solving
9. Mass customization and personalization
10. Built-in quality and error-proofing
11. First-run studies
12. Just-in-time execution
13. 5S (sort, set in order, shine, standardize, sustain)
14. 5 Whys
15. 5w1h (who, what, when, where, why, how)
16. Kaizen

Eliminating Non-Value-Added Activities by Leveraging Digital Technologies

Not limited to just lean tools and techniques, digital technologies play a vital role in optimizing the value stream by eliminating the non-value-added activities from the value stream. There are several ways to eliminate the business wastes discussed earlier in the chapter. Some of them are mentioned here.

Minimize Waiting Time by Using Digital Technologies

1. Sensors in IoT-enabled machines and equipment send data like temperature, pressure, noise and extent of wear and tear. Predictive analytics could help predict when a machine or equipment might fail and take proactive steps to avoid machine and equipment downtime, thereby avoid waiting time. This will reduce the waiting time for repairs
2. Most of the time, it is difficult to locate equipment, forklifts and other tools present at the worksite. Using sensors and IoT will make it easier to find their location and, therefore, waiting time could be minimized.
3. Waiting to get approvals causes considerable delays. Blockchain smart contracts automate the policies of an organization with the consensus mechanism and optimize human interventions through automatic approvals.
4. With the availability of the right information at the right time through mobile devices, wearable devices or web, faster decisions can be made, which would reduce waiting time.

The modern cyber system enables semi-autonomous or autonomous operations like blockchain smart contracts, IoT, artificial intelligence (AI), machine learning (ML) and deep learning (DL). Hence, human interventions can be optimized. For example, if the value of the purchase order is higher than US$15,000, the following could be done explicitly by an approving manager using a software application:

1. The supplier has an excellent performance track record.
2. The supplier is within the country (domestic).
3. There is a volume discount.

However, by using blockchain smart contracts, all the above-mentioned checks can be done automatically without any approving manager with the help of sophisticated cryptographic algorithms. Hence, there will not be any delays caused by approving managers.

Most of the time, production work gets stopped due to non-availability of materials in the stores. Installation of intelligent cameras in the bins/containers of the material/parts would help in this case. When the material/parts are below the threshold, the sensors would send alert messages or material shortage notice to the concerned manager in real time.

In the production line, changeover time between machines causes a lot of waiting time. Intelligent machines with sensors and software, smart enough to identify which product to produce, can automatically load the tools and programs without manual intervention. In this way, we could minimize the waiting time for machine changeover for different products.

Minimize Overproduction by Using Digital Technologies

Sophisticated algorithms based on the real-time information on resource capacity utilization, inventory and customer orders can be used to create an ideal production plan to ensure just-in-time production

Minimize Rework by Using Digital Technologies

For example, preventive and predictive inspections by leveraging analytics, AI, ML help in predicting possible defects/failures while producing a product or delivering a service. AI algorithms provide actionable insights (recommendations), and based on the recommendations provided by the system, workers can take proactive preventive steps to minimize defects and rework.

Robotic visual inspections using intelligent drones or cameras enable high-speed inspections with comprehensive coverage and help in containing the defects. Detecting and containing the defects before delivering to customers improves their happiness.

Minimize People Motion by Using Digital Technologies

People movement can be reduced with hyperautomation of work areas. By leveraging technologies, the safety and security of the employees can be improved, especially when they work or move in hazardous environments.

1. Sensors can be used to proactively send alerts on gas or chemical leakage to avoid people working in a dangerous work environment.
2. Jammers and warnings by sensors can help avoid vehicles hitting each other at the worksite.
3. VR and AR can be used to train employees before working on hazardous worksites by allowing them to practice the tasks in a simulated environment.
4. IoT-enabled workplaces (chairs, tables, water bottles, mobile devices, etc.) can send alerts on long working hours, stress, work-life balance and share appropriate recommendations with the employees.
5. Unmanned aerial vehicles like drones help in performing the work even in unsafe workplaces.

Minimize Overprocessing by Using Digital Technologies

Omnichannel solutions reduce processing overhead costs. For example, to get a loan, once the customers provide their details to the bank's call centre agent, there is no need to give the same information again at different interaction points like the bank branch, website or mobile app. This means there will be no data entry overhead for the customers while availing the loan from the bank.

Blockchain smart contracts have been enabling autonomous operations. For example, an organization's policies could be translated into smart contracts (intelligent code) so that without human interventions, policies could be implemented. In this way, organizations could minimize delays and multilevel approvals. These organizations are also called decentralized autonomous organizations (DAOs).

Optimize Inventory by Using Digital Technologies

Inventory management can be improved using digital

technologies, as mentioned below:

1. AI algorithms with IoT sensors help in managing inventory very effectively and efficiently. With the help of IoT, mobile devices and analytics, real-time information and alerts will be sent to concerned people to maintain the just required inventory.
2. IoT technologies help in locating the machines and equipment very easily without wasting time. For example, the hospital staff could track the medical equipment available (which floor, which room, etc.) at the time of emergencies.
3. Using AI algorithms, organizations can forecast nearly accurate customer demand and plan only for the required/just-in-time inventory.
4. Automatic changeover from one machine to another in smart factories is possible with a well-defined sequencing/routing algorithm, which will minimize inventory.

Minimize Transportation by Using Digital Technologies

AI algorithms can help design the best possible shop floor layout and process flows. This will minimize the piling up of materials or products within the factory between one machine to another.

When the products or raw materials are transported from one location to another, a mobile app could be used to

1. provide an optimal route plan to reduce the transportation distance,
2. provide details about the nearest gas stations and
3. provide vehicle health in real time for preventive maintenance, to reduce the maintenance cost of vehicles.

In the knowledge-intensive industry, digitalization of processes helps in accessing files and documents anywhere and anytime using any device without physical movement.

Minimize Unutilized Talent by Using Digital Technologies

1. Video- and audio-based personalized learning using VR and AR can help in filling the skill gap effectively and efficiently.
2. Leverage mobile, social and analytics technologies to trace the employee journey map within the organization and to design superior experience at each employee interaction point. This helps in improving employee engagement and happiness, thereby productivity.
3. Establish a knowledge management platform using mobile, social and analytics technologies, using which employees can exchange their ideas, suggestions and solutions in real time.

Example: Smart Real Estate

The traditional value stream map in Figure 7.2 was found to be inefficient. By leveraging digital technologies, the value stream map can be made smart. A smart value stream map delivers benefits and experience at a phenomenal speed. Please refer to Figure 7.3 for the modified value stream map.

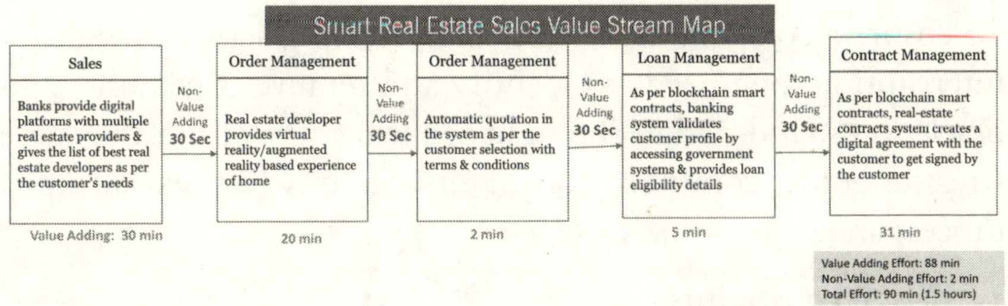

Source: Author

Figure 7.3: Digitalized Value Stream Map for Smart Real Estate Sales

1. **The best suitable apartment as per their preferences (use value)**: Banks are offering non-banking services; for example, banks enrol real estate developers at a particular location and obtain the details of all buildings (villa,

apartment, retail mall, etc.) in the area. Using this data and the best possible matchmaking algorithms, the customer can be offered suitable options from the list of various real estate developers with a comparison sheet (price, value and other benefits).
2. **Hassle-free and speedy bank loan processing (time value):** Using a blockchain-based system, customer validation and background check can be done in a few minutes to facilitate hassle-free loan processing.
3. **Shorter purchase cycle (time value):** With the smart value stream map, it takes around 1.5 hours to complete the whole purchase cycle against the 15 days in the traditional value stream map.

Hence, non-value-added activities must be permanently eliminated with continuous improvement of products, processes and people.

Improve the Flow

Improving the flow means ensuring work moves predictably and steadily to enhance the reliability of delivery of services. To ensure a continuous flow of value-creating activities without interruption, one needs to identify and remove constraints and bottlenecks, and design the right sequence of steps in the value stream. Some common constraints faced while executing a process are:

1. design constraints,
2. management constraints,
3. time constraints,
4. resource constraints,
5. legal constraints,
6. economic constraints,
7. environment constraints,
8. technical constraints and
9. social constraints.

For example, to manufacture a product, if the significant constraints are unavailability of specialists and delays in getting approval from environmental authorities, then they must be tracked until closure before the production starts. Organizations must maintain a constraint log and track the constraints until closure to bring **continuous flow** without interruptions. Look at Table 7.2 for a constraint log template.

S. No.	Constraint	Category	Owner	Target Date	Age	Status	Health
1	Unavaialability of specialists	Resource constraint	Karthik	15 Dec 2019	10 days	Closed	**GREEN**
2	Getting approval from environment authorities	Environment constraint	Venkatesh	30 Dec 2019	5 days	Open	**RED**

Source: Author

Table 7.2: Constraint Log Template

Kanban for Continuous Flow

Another essential element in maintaining a continuous flow of value stream is **kanban**. Kanban means depicting the value stream in multiple stages and the transition of items from one stage to another stage.

At each stage of the value stream, to perform the work, there must be sufficient capacity and resources. But in real-world, there will be many constraints and roadblocks. Hence at each stage, kanban puts a limit to the number of items, which helps in regulating the flow and avoid piling up of items in each stage.

Table 7.3 depicts an example of a kanban card for software development.

S. No.	Development (5)*	Testing (3)*	Sign Off (2)*	Roll-out (1)*
1	User Story 1	User Story 2	User Story 2	User Story 3
2	User Story 2	User Story 3	User Story 3	—
3	User Story 3	User Story 1	—	—
4	User Story 4	—	—	—
5	User Story 5	—	—	—

Source: Author

Note: *Number within brackets shows the maximum capacity at that state (e.g. there is a limit of 5 user stories in the development).

Table 7.3: Kanban for Software Development

A user story is a small and self-contained feature of a software that is designed to accomplish a specific goal of the overall software product or application. User stories are always written from the user's perspective in the following format: As a [user], I want [something] so that [I can get this benefit].

For example, as a chief executive officer, I want to access the business performance dashboard through a chatbot, so that I can get real-time visibility of business issues and risks.

Digital technologies can be used to create a kanban card automatically and send proactive alerts to improve the flow of work. Users can access the kanban cards through a mobile phone or web application. Based on the capacity and resources available, the work limit is defined at development, testing, sign off and roll-out stages. This enables a continuous flow without the piling up of the work at a particular stage

Enable Pull System

The pull system is a potent tool to avoid developing products or services which are not required by customers at that moment. To achieve **pull**, it is essential to understand customer wants and needs continuously.

Nearly accurate production-planning can be done by applying AI algorithms based on the current needs and wants of the customers. With just-in-time production, all the customer needs and wants are met to the dot (nothing more and nothing less).

Bring Perfection

A continuous review of customer requirements and needs to improve products, services, business models, value stream, operating model and structure helps organizations achieve perfection to meet customer needs and wants.

Achieving perfection is not a one-time activity. Every employee of the organization must strive towards perfection.

To continuously identify and eliminate non-value-added activities by using lean digital thinking, leaders must answer the following questions:

1. Do we have a lean strategy to eliminate all the non-value-added activities in the business value chain?
2. Do we have the system to digitalize value streams?
3. Do we identify opportunitics to use digital technologies to eliminate non-value-added activities?
4. Do we have the required tools, techniques and templates to implement lean philosophy within our organization?
5. Do we have a plan to make our digital products, digital services and organization structure lean?

Principle 5: Predict and Improve Business Performance

Redefine Business Performance

Digital immigrants have been driving their businesses for many decades with identified key performance indicators (KPIs). When traditional businesses transition to digital businesses, the current KPIs may not correctly represent the performance of digital businesses. Hence, new business metrics need to be defined. A few examples are:

1. Number of new revenue streams created due to digital products and services
2. Percentage of human effort reduced within an organization for acquiring US$1 revenue
3. Cost reduction per customer acquisition due to digital channels
4. Percentage reduction in privacy compromise incidents reported by customers
5. Number of internal digital start-ups/incubators established
6. Percentage of leaders ready to take up the roles of digital business

Data: A Strategic Resource

For many decades, organizations have been producing a lot of data as mentioned below:

1. Data from transactional systems like enterprise resource planning (ERP), customer relationship management (CRM), supply chain management (SCM), product lifecycle management (PLM) and others
2. Content from PowerPoint presentations, Word documents and Excel sheets
3. Content from websites, emails, social media sites and apps
4. Call records data from the help desk
5. Data from machines, equipment, vehicles and other things

In the past, technology was not mature enough and was able to process only the structured data generated from transactional systems which is not more than 15% of the overall data generated in the organization. However today, technology advancements are mature enough to process 100% of the data generated from various sources.

Data is the new oil; data is the new gold mine! Data is an asset and a unique resource. Those organizations which realize this fact and best utilize data to improve the business performance will excel in the era of the Fourth Industrial Revolution. Making decisions only based on experience and gut feeling without depending on real data will make organizations inefficient and ineffective.

Data warehouse, big data, AI, ML, DL, cloud computing and edge computing technologies have been enabling organizations to mine data. This means classifying, clustering, correlating the structured, semi-structured and unstructured data to provide hindsight, insight and foresight concerning business performance in real time.

Algorithms for Predictable Performance

There are so many algorithms available today which can be applied to the data generated from various sources to predict future scenarios of business. A few algorithms are mentioned below:

1. Naive Bayes classifier algorithm
2. k-means clustering algorithm
3. Support vector machine algorithm
4. Apriori algorithm
5. Logistic regression algorithm
6. Artificial neural networks
7. Random forests
8. Convolutional neural networks
9. Recurrent neural networks
10. Deep Boltzmann machine
11. Long short-term memory networks

A few areas in which organizations could focus on improving business performance are described in Table 7.4. This is possible by acquiring the hindsight, insight and foresight against a performance indicator using data technologies by applying various algorithms mentioned earlier. This data-based performance management would help organizations to understand the past, present and future allowing them to take the right decisions at the right time. Without obtaining real-time business intelligence, organizations will fail to perform in the digital era.

<table>
<tr><th>S. No.</th><th>Performance Theme</th><th>Hind-sight</th><th>Insight</th><th>Foresight</th><th>Autono-mous</th></tr>
<tr><td>1</td><td>Financial performance</td><td rowspan="14">What hap-pened?</td><td rowspan="14">Why did it happen?</td><td rowspan="14">What will happen and what action will be taken?</td><td rowspan="14">Auto-matically imple-ment the recom-menda-tions and actions without human interven-tions</td></tr>
<tr><td>2</td><td>Performance of customer channels</td></tr>
<tr><td>3</td><td>Customer relation-ship performance</td></tr>
<tr><td>4</td><td>Effectiveness of pricing models</td></tr>
<tr><td>5</td><td>Effectiveness of revenue streams</td></tr>
<tr><td>6</td><td>Product performance</td></tr>
<tr><td>7</td><td>Service performance</td></tr>
<tr><td>8</td><td>Supplier performance</td></tr>
<tr><td>9</td><td>Partner performance</td></tr>
<tr><td>10</td><td>Employee performance</td></tr>
<tr><td>11</td><td>Policies performance</td></tr>
<tr><td>12</td><td>Process performance</td></tr>
<tr><td>13</td><td>System performance</td></tr>
<tr><td>14</td><td>Others</td></tr>
</table>

Source: Author

Table 7.4: Business Performance Themes

A few scenarios are explained below:

1. **Simulation-based performance improvement:** Organizations can implement simulation models to see their future business performance and take the right decisions. Digital twins are the virtual representations of physical objects that gather real-time data using drones, cameras and sensors, and predict the performance of physical objects. For example, a digital twin of their jet engine helps General Electric to know the real-time problems and performance so that preventive maintenance and improvements can be done.
2. **Product enhancements:** Technology-embedded smart products and services generate a lot of data that can be mined. Intelligent algorithms find patterns in this data and automatically recommend new features required for a product or service to improve customer satisfaction or operational performance. For example, collated information on customer sensitivity about a product using the data acquired from social media, call records and CRM system can be used to obtain recommendations for product quality improvement.
3. **Customer purchase decisions:** Organizations could give recommendations and insights to customers to help them take the right decisions while purchasing products or getting services. Here are some examples:
    a. Amazon makes product recommendations using the browsing history, ratings and reviews provided by various users on products.
    b. Netflix uses the browsing history of viewers to recommend shows and movies based on their taste.
    c. YOOCHOOSE sends personalized notifications and product recommendations through newsletters based on the buying behaviour and shopping pattern of customers.

4. **Field staff service recommendations:** Providing guidance and insight to field staff improves staff performance. For example, the field team can directly go to the location using GPS and, based on the trouble ticket, get a suitable solution/fix recommendation on their mobile app or chatbot.
5. **Employee performance recommendations:** Using gamification techniques added within software applications, employee performance can be automatically assessed objectively and performance improvement actions can be recommended. For example, the sales manager has revenue targets and the system automatically tracks the sales performance. Points and rewards are given according to the following criteria:
    a. Not met the sales target: 2 points; sends the quarterly performance report with improvement recommendations
    b. Met sales target: 3 points; sends the certificate of recognition
    c. Exceeded sales target by 20%: 4 points; sends a gift voucher
    d. Exceeded sales target by 30%: 5 points; sends an invitation to meet the MD of the company for a cup of coffee and a cash award of US$5,000
6. **Process performance:** Digitalizing the process flow in such a way that real-time insight/foresight are provided as the task is carried out and will help users take appropriate action. For example, while issuing a purchase order in the procurement process, the buyer will get details on the past performance based on historical data and also predict the future performance of the suppliers based on predictive algorithms. Now the buyer can place the purchase order based on the past performance and expected future performance of the supplier.

Example: Predict and Improve Financial Performance

Performance Dimension	Actions
Hindsight	Get the information on past performance, revenue, profit and ROI
Insight	Conduct root-cause analysis and identify the reasons for the high or low performance
Foresight	Visualize the future performance of revenue, profit and ROI based on the predictive modelling techniques applied to the data collected; recommendations and actions are based on the future performance scenarios of revenue, profit and ROI
Autonomous	Automatically execute the recommendations or actions for the identified context and future performance scenario of revenue, profit and ROI

Source: Author

Table 7.5: Financial Performance Prediction and Improvement

For example, a hotel may utilize data and improve its revenue according to the performance dimensions given in Table 7.5 as:

1. **Hindsight:** It is observed that there is a decline of 30% revenue in hotel business due to the Airbnb digital platform.
2. **Insight:** There are so many accommodation options available to guests. Based on the sensitivity analysis done on social media, many guests have been expressing issues about the quality of service.
3. **Foresight:** Hotel expects to see a further 10% revenue decline in the next 12 months. The system provides a few recommendations:

a. Introduce new discounts and pricing models to leapfrog the competition.
b. Introduce guest service app, which can be accessed through mobile phone or TV in the hotel room, to improve the quality of service.
c. Leverage MR technologies to provide immersive experience within the hotel.

4. **Autonomous:** System automatically introduces discounts and optimal pricing options after automatic analysis of Airbnb and other hotel pricing packages and displays them on the hotel web page and mobile app. The system automatically sends personalized advertisements to potential guests using different channels.

To predict, recommend and improve overall business performance by using lean digital thinking, leaders must answer the following questions:

1. Have we defined new KPIs to measure the performance of the digital business?
2. Do we have the system to consolidate the structured, semi-structured and unstructured data generated from different data sources?
3. Do we have the system to get real-time hindsight, insight and foresight from various data sources on different KPIs to take the right decisions at the right time (data-based decisions)?
4. Do we have simulation models to visualize the performance of various KPIs and what-if scenarios?

Principle 6: Optimize Human Interventions

In traditional businesses, even after considerable automation of business value chain, there is still significant manual work involved, which increases the cost, reduces productivity and causes delays in the delivery of services and products.

Thanks to digital technologies, "things" like machines, equipment, devices and assets are becoming intelligent. People and smart things have been working together to perform business operations effectively and efficiently. People can perform their tasks anywhere and anytime using any device due to the internet and powerful communication and collaboration tools. Following are a few digital technology interventions that optimize human interventions.

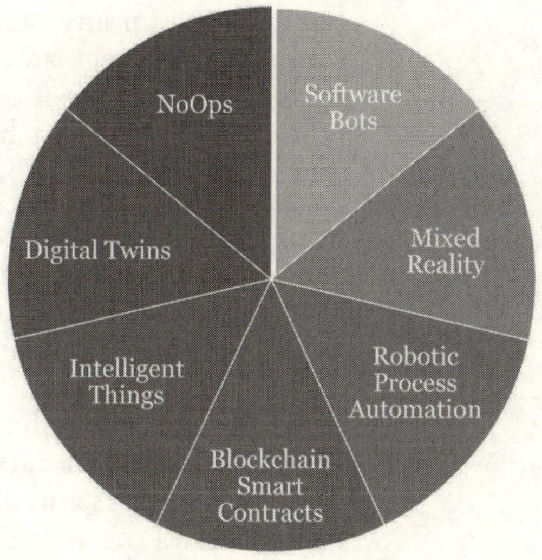

Source: Author

Figure 7.4: Optimizing Human Interventions—Key Areas

Digital Technologies	How to Optimize Human Interventions
Software bots and virtual assistants	Software bots and virtual assistants provide information, clarifications, perform routine tasks, book appointments, schedule meetings and take meeting notes, all without any human interventions.

Virtual reality (VR), augmented reality (AR) and mixed reality (MR)	VR, AR and MR help in providing an immersive and augmented experience to perform tasks without human interventions.
Robotic process automation (RPA)	RPA runs software applications as per pre-configured business rules, which helps in executing the processes automatically without any human interventions.
Blockchain smart contracts	Blockchain smart contracts avoid middlemen for any validation, approvals and regulatory compliance checks; they automatically enforce the same without any human interventions.
Intelligent things	Intelligent things like connected cars and digital equipment enable predictive monitoring and control without human interventions.
Digital twins	Digital twins are replicas of physical objects/entities like cars, rotors, blades, machines and cities. Physical entities send the actual data collected by sensors and images to its digital twin. By applying predictive models on this data, the digital twin helps in proactively taking necessary actions for the operational efficiency of the physical entity.
NoOps	NoOps means automated deployment of software applications and management of IT infrastructure without human interventions.

Source: Author

Table 7.6: Digital Technologies for Optimizing Human Interventions

Chatbots and Virtual Assistants

Chatbots and virtual assistants use AI to communicate with users and perform day-to-day business tasks which minimize human interventions. While both chatbots and virtual assistants leverage AI, there is a subtle difference. Chatbots are relatively less intelligent and are used to clarify questions, extract information and make appointments. The user interface of chatbots uses audio and text.

Virtual assistants are more intelligent and help in performing business tasks more effectively. They can book appointments, manage to-do lists, send reminders, take calls, read messages and emails, check-in for flights, call a taxi, among other things. For example, Marks & Spencer's virtual assistant helps customers to troubleshoot around 70% of their issues without human interventions.

Augmented Reality, Virtual Reality and Mixed Reality

Augmented Reality

Augmented reality (AR) augments the real-world environment and objects with computer-generated information (visual, audio, haptic, etc.). Organizations have been using augmented reality for education, training, branding, sharing information and other purposes. For example:

1. Today marketing flyers are designed using AR. The flyer content includes text and images relating to a product or service. When an AR mobile app is used to view the product image on the flyer, a video or audio clip starts playing automatically. This provides information about the product, which helps customers make the right purchase decision. In this way, AR has been used to promote products and services that enhance the brand.
2. Using an AR app, guests at a restaurant in Japan can read the Japanese menu items or signboards in their native language. A

Telugu tourist can choose to read the Telugu translation superimposed on the Japanese menu through the AR app.

Virtual Reality

On the other hand, virtual reality (VR) means simulating a non-existent environment or object and providing an immersive experience to people as if they were in the real environment. VR provides two senses, like sight and sound, out of five human senses (sight, sound, touch, taste, smell). For example, by leveraging VR, tourists can immerse into a tourist location remotely using a VR headset before booking the ticket.

Mixed Reality

Mixed reality (MR) means the seamless merging of digital and real worlds to produce an entirely new environment in which physical and digital objects coexist. MR will change the way we view and interact with the world. For example:

1. Quality inspections can be done very effectively with the help of MR. By wearing an MR headset (like HoloLens), one can see the 3D drawing of a room superimposed on the physical room. Deviations can be easily spotted using the headset, and the quality inspector gets information and recommendations to fix the deviations. Here, the digital 3D drawing coexists with the physical room when viewed through a headset.
2. MR also helps in creating a virtual human body with additional information and provides a great learning experience for students to learn human anatomy.

For MR and VR, users use the headsets and for AR, users usually use mobile phones or tablets.

Robotic Process Automation

There are many business processes within organizations. A few business processes handle high volume, repetitive and business rules-based structured tasks. Humans have been performing these tasks for many years. With the power of AI and ML, software bots perform the functions with no/fewer human interventions. This is called robotic process automation (RPA). RPA tools capture data, retrieve information, communicate with other systems and process transactions without human interventions.

Software bots can automatically log into the software applications and mimic the human worker to enter data, perform calculations and complete tasks. After completion of the work, the software bot logs out automatically. Software bots are trained to understand the tasks of business processes within software applications and adjust to changing circumstances.

Process automation requires natural language processing, speech recognition, computer vision technology and ML to comprehend the vast amount of data. Software bots learn on the go and intelligently automate processes.

For example, let us see how an invoice is processed using RPA:

1. Configure the invoice management process workflow and business rules in any of the RPA tools like UiPath or Blue Prism.
2. Once the invoice management application is opened, RPA enables automatic login to the system.
3. The mail and the attachment sent by the supplier opens automatically.
4. Manually scan the QR code on the invoice.
5. Based on the details scanned, the corresponding purchase order is retrieved automatically.
6. The invoice data is compared and reconciled to the purchase order data to highlight any discrepancies.
7. The taxes are calculated automatically.
8. Instructions are sent to the accounting department via email with detailed instructions and attachments.

In the earlier example, except for scanning the QR code, all the other tasks are performed automatically without human intervention as per the workflow and business rules configured into the RPA tool.

Blockchain Smart Contracts

In organizations, there are many business rules and contractual clauses in place to make sure that operations are performed effectively and efficiently. Unfortunately, many middlemen are part of each business operation and consume a lot of time and money. Some examples are:

1. Reporting manager to approve leave request
2. Procurement manager to approve purchase orders
3. Bank manager to verify the financial transaction before transferring money to others
4. Registration officer to verify the authenticity of land records before registering land
5. Charted accountant to audit and certify counts

All these middlemen follow their organization's policies, business rules and contractual obligations while performing their assigned tasks. Translating these policies, business rules and contract clauses into a clever software code removes the middlemen. This software code is called a smart contract. Smart contracts are essential elements in the journey of making an organization nearly autonomous.

1. A smart contract is a pre-written computer algorithm based on policies and rules (computer code).
2. The computer code is stored and replicated on distributed computer nodes (e.g., a blockchain).
3. The computer code is executed/run by a network of distributed computers that must deliver the same outcome (little programs that are executed, run and verified by many computers to ensure trustworthiness).

4. It completes a transaction by updating the ledgers at each computer node (cryptocurrency payments, asset transfer, approvals etc.).

Blockchain smart contracts help in building autonomous businesses with which organizations can minimize human interventions significantly. The following example is a simplified version of such contracts.

1. **Develop** a software code as a business rule or policy. For instance, if there is a delay in the delivery of items against a purchase order, send a penalty notice to the supplier every 15 days, with a US$100 per day penalty.
2. **Store** the smart contract code, as mentioned above, at each computer node in the distributed ledger network. Let us assume that there are three nodes identified by this organization as per their standards and guidelines.
3. **Execute** the smart contract code at each computer node and expect that the outcome is the same at each node.

If the delivery of items by the supplier is delayed by 30 days, after the execution of the smart contract, the following outcomes are expected at each node. If the outcome is different at any of the nodes, the transaction will be rejected.

1. Node 1: Generate a penalty note with US$3,000 fine and send an SMS, WhatsApp message and an email to the supplier.
2. Node 2: Generate a penalty note with US$3,000 fine and send an SMS, WhatsApp message and an email to the supplier.
3. Node 3: Generate a penalty note with US$3,000 fine and send an SMS, WhatsApp message and an email to the supplier.

If the outcome is the same at all three nodes, then only will the action be taken to send the penalty note. In case anybody hacks the smart contract at any of these three nodes and change the code, say at Node 3, then the outcome will be different.
Node 3: Do not fine any penalty for delays in delivering the items by the supplier.

Since the outcome is different from Node 1 and 2, the total transaction will be rejected and an alert will be sent to the concerned authorities.

There are only three nodes in this example; it could be any number of nodes as per the organization's requirement. Multiple nodes are required for consensus among them as we don't have any human being to approve or send messages.

Intelligent Things

In a hyperconnected world, the IoT makes things intelligent, which include sensors, electronic microprocessors with embedded code, short-range and long-range networks and analytics powered by AI. When things are intelligent, they perform most of the activities themselves without any human intervention. At present, intelligent things have been effectively carrying out repetitive tasks which do not need high cognitive capabilities possessed by human beings.

There are intelligent cars today which automatically lift the accelerator to stop the driver from going over the speed limit when he does not stop even after repeated reminders sent through voice and text messages. Integrated breath analyzer stops a drunk driver from starting the car.

Most importantly, intelligent things bring preventive and predictive maintenance that reduces operations cost tremendously. By applying predictive models, an ATM with IoT can predict when it will break down based on the sensor data on the wear and tear of the belt, temperature and other details.

Digital Twins

A digital twin is the replica of a non-living or living physical object. Digital twin and physical objects coexist. Digital twins are used to design, build and test the physical entities like cars, wind turbines, shafts, rotors, blades, cities, human bodies, MRI machines and washing machines.

Physical entities gather a lot of data using the sensors and cameras about real-time status, position, working condition, temperature, noise, vibration and others. This data is sent to its digital twin and using predictive models and simulations, the digital twin provides an insight into the physical condition of its counterpart and recommends various actions for predictive and preventive maintenance.

For example, Singapore has created a digital twin for the whole island-city (including buildings, roads, malls, streetlights, etc.) to manage the city services effectively. Similarly, General Electric uses digital twins for its wind turbine configuration and shares the configuration information with stakeholders to reduce rework while constructing the same. Digital twins are also used for human bodies and the actual health details of a person are sent to his digital twin to predict any major health issues in the future.

Using digital twins, General Electric has achieved the following benefits:

1. Increase in reliability in less than two years: 93–99.49%
2. Reduction in reactive maintenance in less than one year: 40%
3. Reduction in time to achieve outcomes: 75%
4. Savings in production by detecting and preventing three failures: US$11 million

Hence, digital twins help in minimizing rework during designing, building and maintenance of various physical entities.

NoOps

NoOps, or no operations, means automating deployment, monitoring and management of software applications and the infrastructure on which these applications run.

In the past, availing computing power was very expensive. With cloud computing, elastic supercomputing infrastructure, storage space and internet broadband are available on-demand today. Elastic in nature means the computing, storage and internet broadband capacity are upgraded or downgraded as per the load and demand and hence, is very affordable.

Also, there are many automated tools available to install software applications with no or less human interventions. In this way, software applications and IT infrastructure management are automated so that IT operations are optimized.

Example: Optimizing Human Interventions while Purchasing a House

For example, in the real estate industry, a customer who wants to purchase a home from a builder must follow the steps given in Figure 7.5.

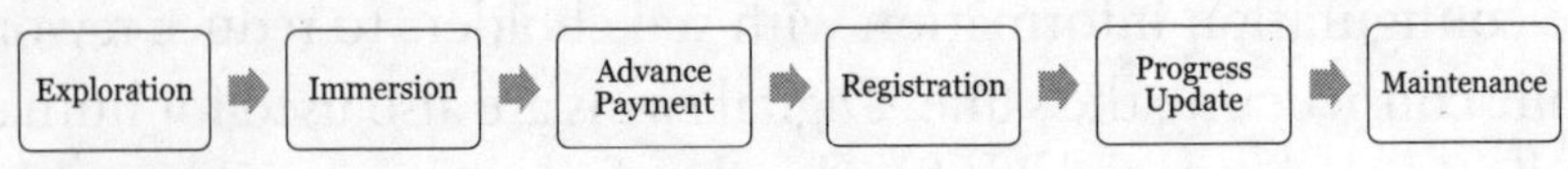

Source: Author

Figure 7.5: Steps to Purchase a House

Table 7.7 mentions the different technologies used to minimize human interventions at each step (as shown in Figure 7.5). This is a hypothetical example and is given only for a better understanding.

S. No.	Step	Optimizing Human Interventions
1	Exploration	Using software chatbots, the customer gets detailed information about the real estate venture and house details.
2	Immersion	Using a VR app with a headset, the customer immerses into the virtual house from a remote location. He/she can inspect various aspects of the house like tiles, walls, colour, garden, kitchen, bedroom, etc.

3	Advance payment	With a blockchain decentralized app, the customer uses cryptocurrency to make the payment. There is no bank in the middle. A few countries have some restrictions on cryptocurrencies and hence, legal policies need to be checked before making any payment.
4	Registration	Using the blockchain decentralized app, the land asset can be verified and transferred to the buyer without involving any government sub-registrar or registration official.
5	Progress update	Digital bands (sensors) on each worker and a geofence ensures that the builder gets updates about: • the number of workers hired to work at a location and • the number of workers present within the geofence. Intelligent cameras (IoT solution) and drones (where required) give the customer construction progress update based on 3D rendering and image analytics.
6	Maintenance	During the maintenance/operations phase, the digital twin of the house and amenities proactively identify the maintenance problems using sensors data, analytics and fix them before any crisis arises.

Source: Author

Table 7.7: Digital Technologies Used While Purchasing a House

To optimize human interventions by using lean digital thinking, leaders must answer the following questions:

1. Do we have a strategic plan to digitalize organization products, services, processes and workplaces to optimize human interventions?
2. Do we invest in research and development (R&D) and innovation initiatives to re-engineer our organization in such a way that it becomes nearly autonomous with less human interventions?
3. Do we improve the digital skills of the leadership team and employees to work in an autonomous business environment?
4. Do we have an automated system to design, build, deploy and manage the software applications and IT infrastructure to support autonomous business operations?

Principle 7: Establish Real-Time and Contextual Communication and Collaboration

The real or physical world has many limitations. The digital world reduces the distance between people, locations, and removes communication barriers. Through communication, organizations and people impart or exchange information, news or data in real-time.

Non-availability of the right information at the right time in the hands of stakeholders creates a lot of challenges. Establishing the right communication channels is very essential. Timely communication minimizes issues, escalations and crisis.

The survival of organizations depends on **collective intelligence**, and hence, collaboration among various stakeholders is crucial. Collective intelligence improves innovation, execution excellence, learning, productivity, etc. Collaboration is nothing but the participatory and collaborative work among distributed teams using collaborative tools to create something useful. If communication and collaboration are not done in real time and are not contextual, it would be very difficult to unlock the true potential of the organization.

Real time means access to information, insight and execution of tasks over the internet **anywhere** and **anytime** using **any device**.

Contextual means concerning the location, task, time, circumstance, family history, lifestyle, business environment, economic, social, cultural, legal and technological situations. Context also means weather, health, standing position, in the car, in the train, in-flight and others. Businesses need to factor in the context while communicating and collaborating with others.

When I was writing this chapter, I was on a train and listening to my favourite Telugu songs using my mobile device connected to the internet. **This is real-time entertainment!**

Many communication and collaboration tools are available today to improve business performance. Some of them are mentioned in Figure 7.6.

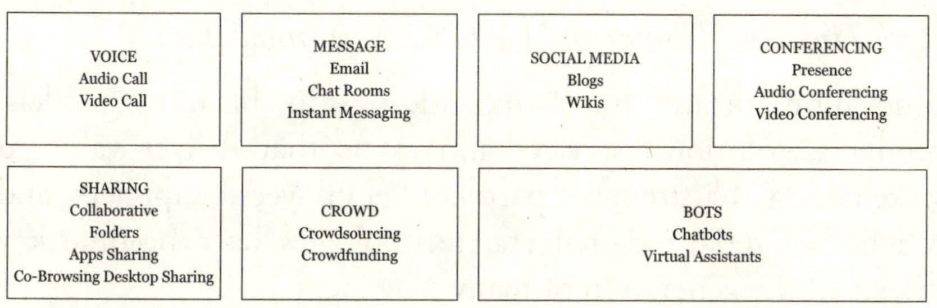

Source: Author

Figure 7.6: Real-Time Virtual Communication and Collaboration Tools

Real-Time Innovation Excellence, Execution Excellence and Learning Excellence

Innovation, execution and learning without real-time and contextual collaboration and communication do not yield optimal results.

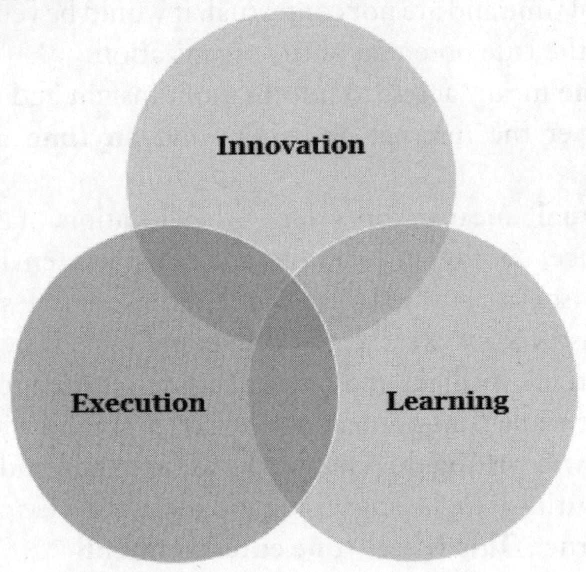

Source: Author

Figure 7.7: Real-Time Innovation, Learning and Execution

Real-Time and Contextual Collaborative Innovation

Innovation means translating ideas into business models, products, solutions, services and assets that deliver value to stakeholders (customers, partners, employees, suppliers and others). In today's digital era, technologies have become key drivers in the generation of many new ideas.

Due to unimaginable changes in technology and trends occurring over short periods, organizations have been struggling to build new skills that help them introduce innovations into their business. Closed innovation with the help of a few specialists within the organization with many skill gaps does not support the organizations to prepare for the digital future. Hence, open innovation has become prominent in which innovation happens in collaboration with specialists from across the world who need not be a part of the organization.

For example, Co-Innovator is a real-time and contextual collaboration and communication platform for innovation. Co-Innovator is a crowd innovation platform developed by BT&BT and OSI that helps organizations implement **open innovation** by bringing **solution seekers** (business or supporting units in an organization) **and** globally distributed **solution providers** together.

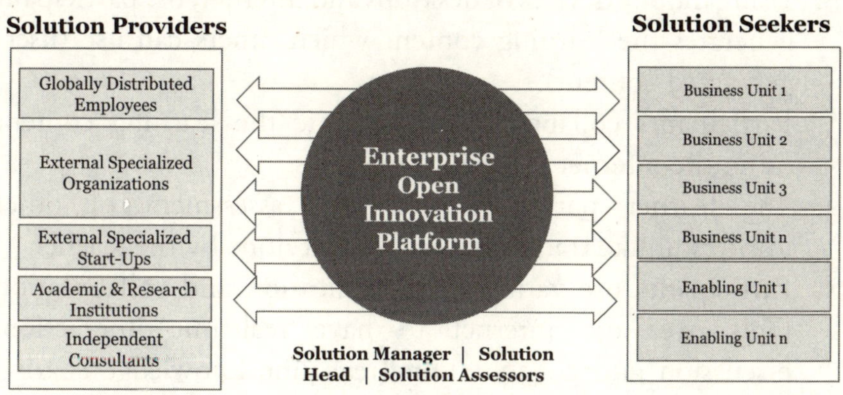

Source: Author

Figure 7.8: Open Innovation Digital Platform

In today's digital world, customers have different needs, expectations and aspirations. Hence, organizations must **innovate** newer products, services, methods, processes, tools, technologies, governance models and strategies for growth and operational excellence.

Organizations have been facing serious skill gaps which hinder their preparations for the digital era. They have been failing to bring innovations to meet the changing needs of the digital economy. **Innovation has now become boundaryless**, enabling organizations to bring external talented brains onto a common platform to enable crowd-based innovation. Hence, this co-innovation approach provides many alternative solutions for an identified problem/challenge faced by organizations.

Real-Time and Contextual Collaborative Learning

Learning changes the behaviour of people. Learning is done through various methods to improve knowledge and skills. Real-time and contextual collaboration and communication improve learning efficiency and effectiveness, as given here:

1. Participants play the dual role of teacher and learner as per their knowledge and skill level.
2. Using mobile devices or desktops and internet, the participant generates the learning content which others can use (user-generated content).
3. Participants collaboratively co-create the learning content using the internet and wiki-type tools.
4. The learner participates in peer assessment of other learners unlike traditional assessment done by the teacher.
5. Participants use various communication and collaboration tools over the internet to have real-time interaction, discussion and collaboration (real-time knowledge sharing and learning).

Organizations must focus on building their own learning platforms like Coursera, edX, XuetangX and Udacity for their organizations. Employees, external specialists and partners could enrol in these platforms for continuous learning in real time, in a highly collaborative environment.

For example, edX is one of the best real-time and contextual collaborative learning platforms available today. Following are a few features:

1. edX has built a platform to transform employees into lifelong learners.
2. edX educates employees, customers and partners effectively.

3. edX maintains business-relevant content.
4. edX improves learner performance using the learner's behaviour analytics and provides contextual recommendations.
5. edX implements social learning tools for real-time collaboration among learners and establishes internal cohorts to enhance collaborative learning.

Real-Time and Contextual Collaborative Execution

Execution means performing the work within the allotted time, budget and at the expected quality. There are two areas of **execution** in any organization as mentioned in Figure 7.9.

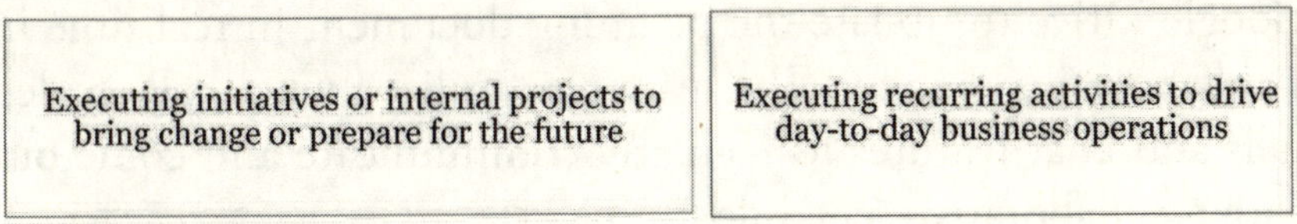

Executing initiatives or internal projects to bring change or prepare for the future	Executing recurring activities to drive day-to-day business operations

Source: Author

Figure 7.9: Two Areas of Business Execution

Executing initiatives or projects and executing operations (recurring tasks) both are entirely different and need different types of leadership and mindset. In both cases, collaboration and communication play a very vital role. A few examples are given below:

Examples: Real-time and contextual collaborative execution

Planning a Project

Project planning is a critical phase of every project which must be done in collaboration with many teams and experts. Sunco is a company that has embarked on a journey to expand their business to different locations in Europe. Their experience in business expansion in India had been terrible; nothing had happened as per their plan.

The plan was incomplete and ineffective; hence, there were many delays and cost overruns. Therefore, the Sunco leadership team wanted to make sure that the planning is done effectively. They wanted to involve all the critical stakeholders distributed across different locations while preparing a detailed plan.

Sunco rented a crowdsourcing platform called Co-Innovator which offered a collaborative workplace that could be accessed by the Sunco stakeholders anywhere and anytime.

The Sunco team used the **virtual post-it notes feature** of the Co-Innovator platform. They brainstormed by placing the virtual post-its in different sequences and finally derived the most feasible and optimal project schedule. The team utilized Google Drive to update the planning document in real time in a collaborative manner. The team also utilized voice calls, video calls and chat features to regularly communicate and come out with a collaborative schedule.

In the above example, the following communication and collaboration tools were used:

1. voice call,
2. video call,
3. chat,
4. virtual post-its and
5. Google Drive.

Bus Repair

A bus repair company, Zen Road Transportation, had been facing problems as their buses had been breaking down due to maintenance issues. It took them 15–20 days to make a bus operational. This was because the field staff could not get the required spare parts promptly. A few times, experts were also not available to sort out the issues or incidents. Due to the repeated delays, the company had been losing much revenue.

So, the company developed a collaborative digital platform to bring together all the field staff, experts and related department staff from different locations to a single platform. With this solution, the company was able to get buses up and running in 2–3 days as against 15–20 days.

The **context** here is the location, that is, by a national highway. The vehicle is in non-working condition and field staff has to go to that location to repair the vehicle. Field engineers were now able to collaborate and communicate in real time using the digital platform.

When field staff fails to get the required spare parts from their nearest stores, they publish the spare part requirements on the digital platform. The incharge of other stores confirm the availability of spare parts in real time in their store. This significantly reduces the lead time taken to repair the bus.

To troubleshoot complex problems or incidents, field staff publishes them on the digital platform, and experts distributed in a different location offer the best possible solutions. The field staff communicates with them in real time using chat, integrated video and audio conference features of the digital platform with improved coordination.

In the above example, the following communication and collaboration tools were used

1. chat,
2. audio conferencing,
3. video conferencing and
4. private social media.

Patient Health Diagnosis

A patient named John uses IoT-enabled intelligent medical wearable device to record his health details like blood pressure, glucose levels, temperature and others at home. These details can be seen on the Happy Health app. This app transmits the details to hospital cloud servers hosted at Amazon Web Services (AWS) using the internet.

Dr Rita at Happy Care Hospital gets a notification about the data received from her patient John. The doctor opens the Happy Health app and checks John's health data, which is alarming. She thinks that it is better to consult another doctor for a second opinion on John's health. She uses the same app to call Dr Robert and discusses John's health.

She also uses the chat feature in the app to ask a few more questions to get clarifications. For more clarity, she video calls John and gets all the clarifications and informs John about his health.

Dr Rita thinks John's health is not good and he needs immediate attention. She puts a request to six other doctors in the hospital for a conference call at 04:00 AM on the same day. She gets confirmation from three doctors and discusses John's health condition with them over a video call.

Dr Rita also uses the Happy Health chatbot to get the details of the past patients who had suffered from a similar illness, the medicines they used and detailed information on side effects. Finally, Dr Rita sends the prescription through the Happy Health app, which sends it to both John and the medical store via email and WhatsApp.

Integration of communication and collaboration tools within the patient management process workflow in the Health App enables useful conversation between doctor and patient, which helps in improving the diagnosis and patients' confidence.

In the above example, the following communication and collaboration tools were used:

1. notification,
2. voice call,
3. chat,
4. video call,
5. video conferencing,
6. chatbot,
7. WhatsApp and
8. email.

To establish real-time and contextual communication and collaboration by using lean digital thinking, leaders must answer the following questions:

1. Do we have the required digital technologies and tools to establish a real-time and contextual communication and collaboration platform in our organization?
2. Do we leverage digital technologies and tools for real-time crowd innovation by leveraging the collective intelligence of the distributed employees?
3. Do we leverage digital technologies and tools for real-time personalized learning collaboratively with other employees and specialists?
4. Do we leverage digital technologies and tools for providing real-time alerts, information, insight and bringing the team together to complete a task/work more effectively and efficiently?

Principle 8: Safeguard Privacy, Safety and Security of Business Resources

In the digital era, **people to people**, **people to things** and **things to things** connections are established in real time using digital technologies. This means many cybersecurity, safety and privacy issues will be encountered by various organizations. Hence, preventive and corrective cybersecurity practices must be implemented.

In today's hyperconnected world, people are very active on the internet. Every minute amazing things happen on the internet.

Source: https://twitter.com/lorilewis/status/1103687442727616523?lang=en

Figure 7.10: Internet in 60 Seconds

IPv6, sensors and communication technologies have been connecting intelligent things/devices to the internet. Following are a few global statistics[1]:

1. 20.4 billion IoT devices will get connected to the internet by 2020.
2. Over 64 billion IoT devices will get connected to the internet by 2025 (as per *Business Insider*).
3. Every second, 127 IoT devices get connected to the internet (as per David Evans).
4. Companies will invest up to US$15 trillion in IoT by 2025.
5. US$53.45 billion will be spent on home IoT market by 2022.

1 Kaylie Gyarmathy. 2020. 'Comprehensive Guide to IoT Statistics You Need to Know in 2020'. Available at: https://www.vxchnge.com/blog/iot-statistics

6. IoT healthcare market is expected to reach US$140 billion by 2024.
7. Customized stores with IoT devices are gearing up to provide a personalized experience to customers.

With the emergence of cloud computing, documents, data, information and other sensitive information are stored in data centres, which can be accessed over the internet. The whole world is now connected and hence, organizations will encounter severe privacy, safety and security challenges that could jeopardize businesses, governments and individual lives.

Privacy, Safety and Security

Privacy means you have the sole control over the collection, sharing and usage of your information, data, feelings, work and beliefs. If anybody intrudes into your personal life, then it is a serious privacy violation. For example, somebody hacks your health records and uses them for the wrong reasons then, it is a serious privacy issue.

Security means being free from the danger of malicious intentional internal or external attempts to harm the organization's resources. For example, if a cybercriminal intentionally hacks into your organization's system and shares your intellectual property with the public leading to financial loss, it is a serious security breach.

Safety means freedom from unintentional and accidental incidents that may happen due to negligence, poor maintenance and non-compliance with safety standards. For example, if vehicle maintenance is not done regularly, the brakes may fail and lead to an accident.

Following are the business resources that encounter serious privacy and cybersecurity issues in a hyperconnected world:

1. People
2. IoT devices
3. Software applications
4. IT infrastructure (servers)
5. Information
6. Intellectual property
7. Physical property

Security Threats in a Hyperconnected World

S.No.	Security Threats	Details	Examples
1	Malware	Malware is a malicious software code, which is intentionally written to damage, disrupt and gain unauthorized access to a computer system.	WannaCry malware has impacted 200,000 computers across 150 countries. With its malicious code, this virus has encrypted documents, photos, videos and databases on various computers and created a show-stopper to prevent recovering these files and data. Emotet, Kovter and ZeuS are other few prominent malware.

2	Phishing	Phishing is sending mails with wrong links and messages from disguised parties to trick people into providing their personal information, credit card details, passwords, etc.	A phishing mail looks just like any legitimate mail just like the one given below: Subject: Pay 10 Bitcoins Immediately—You Are in Danger Mr Ken, We know that you have been visiting a few adult websites for the past few months and we have captured all your viewing history with proofs. You seem to have a very good taste. Pay us 10 bitcoins immediately to avoid exposing your private information. Sincerely, Robert Johnson
3	Password attacks	Password attacks include cracking the passwords and getting unauthorized access to the systems and tampering with the data or asking for ransomware.	A few years ago, 164 million LinkedIn accounts and 360 million Myspace accounts were hacked by a hacker known as Peace.

4	Denial-of-Service (DoS) attack	A DoS attack is carried out by overflowing the network with a high volume of traffic using existing network resources and preventing legitimate users to use the network resources. It also includes the overflow of connection requests and consuming the operating system resources making the system nonresponsive.	GitHub was hit with thousands of autonomous systems resulting in massive traffic of 1.35 terabits per second which disrupted its operations. In 2012, banking operations of the Bank of America, JP Morgan Chase, Citi Group, PNC Bank and US Bancorp were disrupted when hundreds of hijacked servers created more than 60 gigabits of traffic per second.

5	Malvertising	Malvertising means malicious advertising that spreads malware. When somebody clicks on the online advertisement link, malware enters into the computer system and disrupt the work. Online legitimate advertising networks are injected with malicious code by perpetrators. This malicious code redirects users to malicious websites.	In case of a fake offer by Amazon, like an Amazon Membership Rewards scam, users may receive following mail. From: ------------ Congratulations! Amazon.com user You have been selected as a lucky user for a chance to get the $1000 Amazon Gift Card, Apple iPhone X 256G or Samsung Galaxy S8! Please click OK for a chance to GET your reward. Amazon Membership Rewards virus is an online scam that tricks people into filling a fake survey in which the user reveals personal information. This may lead to serious privacy-related issues later. This scam is caused by an adware.

6	SQL injection	SQL injection attack means executing malicious SQL statements to control the database server behind web applications by bypassing the application security measures. SQL injection attacks allow hackers to access customer details, credit card information, employee information and other confidential data of an organization.	In 2009, the sixth-largest payment processor in the US announced back that its systems were breached in 2008. But within a few days, it was classified as one of the biggest ever cybercrime that exposed the credit card data of 100 million users and compromised more than 650 financial services companies at the cost of hundreds of millions of dollars.
7	Robocall scams	Robocall is a phone call that uses computerized auto-dialers to deliver the messages. Scammers use caller ID spoofing due to which the receiver gets the calls from a fake local code, where they or their friends and relatives live.	These include fraudulent robocalls, claiming to be from charity organizations asking for donations or coming from government agencies threatening license cancellation. In the US, around 48 billion robocalls were made in the year 2018, and in 2019 the number was 58.5 billion. The third busiest cancer hospital in the US, Moffitt received 6,600 robocalls in 90 days. They were claimed to be from the Department of Justice and doctors were warned about the problems with their medical license number.

8	IoT devices hacking	In a hyperconnected world, IoT devices connected to the internet, other systems and devices become highly vulnerable to hacking. Hackers could take control of IoT devices and perform operations, which are not intended to be completed by the IoT device. When vulnerable IoT devices are connected to software applications, security is compromised. IoT device providers are not systematic and do not release regular patches to strengthen the security of IoT devices.	A botnet is several internet-connected devices and each device runs one or more bots. Botnets control IoT devices for illegal purposes. Botnet usually performs distributed denial-of-service (DDoS) attacks. Connected cars are prone to such security threats. Hackers can hack the car from remote locations and take control of car door locks, brakes, music system, engine, etc. IoT devices being used for payments are under serious security risk. Hacking an IoT-enabled vulnerable irrigation system of a city may cause a complete disruption of water distribution in the city.

9	Blockchain security breach	Blockchain security attacks are a serious concern today. Blockchain code developed by many open-source enthusiasts may not follow the required security standards.	Hacking the private keys of a blockchain wallet to carry out illegal transactions. Malicious code is injected into smart contracts enabling illegal transactions. Hacking into blockchain cryptocurrency exchanges and stealing them. Acquiring too much of mining power (e.g., 51% of mining power) in the blockchain network may allow hackers to tamper with the distributed ledger data saved at various nodes. Some examples of such attacks are: • Wallet-based attacks • Smart contract-based attacks • Consensus and ledger-based attacks • Peer-to-peer network-based attacks

10	Algorithmic bias	Unrepresentative, incomplete training data and dependency on flawed information generates wrong algorithmic models and has a serious impact on people and businesses and may lead to serious legal tangles too.	Amazon has used an experimental machine-learning hiring tool to rate the job-seeking candidates. But the company has realized that the new system was not rating the software developers and other technical job seekers in a gender-neutral way. This was because the ML algorithm was trained with the resumes, which were mostly received from men in the last 10 years. The algorithmic model has created a gender bias by downgrading the women resumes, and hence, there was a very low intake of women employees. Use of automated risk assessments by judges to determine sentencing limits and bails have led to incorrect and biased conclusions about people of colour. This implies wrong and non-representative data sets are used to train the machine, which produced wrong results.

11	Privacy breach	AI algorithms require a lot of data sets. Some companies use their customer's viewing history on the web and their social media data without their consent. This compromise the privacy of the customers.	One of the companies had analyzed the personal data of millions of people's Facebook profiles without their consent and used for political advertisement in one of the developed countries. In India, a hacker hacked the smart TV camera of a couple in the bedroom, recorded them in compromised positions and uploaded the video on adult websites. A hacker hacked the laptop camera and recorded a girl in the hostel room while she was changing her clothes and used the video to blackmail her. With **deep fake technology**, many more dangerous things have been carried out. For example: • A fake video of a high-profile political leader was created by using this technology to make her appear drunk. • Celebrity's photos are fused with videos and pictures containing adult content and uploaded on adult websites.

12	Unmanned air vehicles (UAVs)	UAVs or drones have been becoming a serious threat to people and organizations. Aviation regulators and law enforcement agencies have expressed serious concerns about security threats. Remotely controlled drones may destroy buildings or injure human beings.	In September 2019, drones attacked Saudi Aramco's Abqaiq refinery and sparked massive fires. This attack disrupted its oil production and export capacity. Arms dropped by UAVs in Punjab from India-Pakistan border have caused serious security concerns.
13	Spambots	Spam is an unwanted and inappropriate message sent to a large number of users. Spambots are computer programs that execute repetitive tasks and usually operate over the internet. Spambots automatically send a lot of messages to a large number of users. Spambots are also fake followers for social media accounts	Bots have been posting spam links and promoting propaganda. Some political leaders have been using bots to spread propaganda too. Facebook has removed 3 billion fake accounts between October 2018 and March 2019, and most of them were spambots.

14	Breaking data encryption	Quantum computers can solve some problems, which take conventional computers years, almost instantaneously. Due to the high computational power quantum computing has the potential to break the current encryption models.	Quantum computers may be made available to the general public around 2025. Google, IBM, Amazon Braket and the Alibaba Group have been experimenting and using the first generation quantum computers.

Source: Author

Table 7.8: Security Threats in a Hyperconnected World

Cybersecurity Management

Ecosystem Security Management: In the digital world, an organization is nothing but an ecosystem that includes so many suppliers and partners. These partners and suppliers again build their own ecosystem. Hence, an organization has become an "ecosystem of ecosystems" and is highly connected using IoT, internet and other technologies. Cybersecurity vulnerabilities are high due to tightly interconnected systems. The management has to focus on the organization's security beyond its boundaries.

Governance: Implement cybersecurity policies, initiatives and security insurance for products. Cybersecurity skills of the employees and leadership team must be enhanced as well. Take necessary steps to give confidence to consumers, employees, shareholders and regulators that they are secure. Cybersecurity officers must be appointed at different levels of the organization. Adhere to the General Data Protection Regulation (GDPR) as applicable to uphold data privacy.

Design: Design products, services, software applications and technology infrastructure to be not or less vulnerable to easy cyberattacks.

Vulnerability Inventory and Threat Management: Maintain an inventory of cybersecurity vulnerabilities covering the whole value chain of the organization and manage them proactively. Implement a system for ongoing cybersecurity threat modelling and threat intelligence to take preventive actions.

DevSecOps: With DevOps, there is high collaboration and communication between the software applications development and the technology operations team throughout the software life cycle. This has improved applications and technology infrastructure performance, and enabled the timely release of the software. DevSecOps integrate development, operations and security teams throughout the software life cycle.

To safeguard privacy, safety and security of business resources by using lean digital thinking, leaders must answer the following questions:

1. Do we have the strategic, tactical and operational structure and plan to safeguard the security of connected products, services, software applications and technology infrastructure?
2. Do we have the required cybersecurity policies, tools, techniques and methods to take preventive steps and to respond to cyberattacks within a shorter period?
3. Do we have the strategic, tactical and operational skills and team to respond to cyberattacks effectively?
4. Do we have a strategy and plan to strengthen the cybersecurity policies and system among extended partners and business ecosystem?

Principle 9: Establish a Technology Start-Up Ecosystem

In traditional businesses, stability and predictability are the driving forces (Model 1). However, digital businesses require agility, speed and explorative work environment (Model 2). Technology is the driving force for digital business; hence, companies must be driven like technology start-ups with an entrepreneurial mindset. Companies must manage uncertainties and risks with an attitude to **fail small, fail fast and scale up soon.**

Technology has been changing rapidly and impacting businesses significantly. For example, between 2010 and 2015, digital technologies like social media, mobile, analytics and cloud have transformed businesses radically. Since the year 2016, AI, ML, DL, AR, VR, IoT and blockchain technologies have been changing the business rules.

Business executives are flooded with so many technologies and are under tremendous pressure to embrace these technologies to create new business opportunities, to leapfrog the competition and to improve business performance.

As mentioned earlier, today, products and services are software-defined like digital cars, digital washing machines and social media-based customer services. This means every business is driven by technology and is a technology business.

In the coming years, it will be no surprise if organizations keep the technology budget of around 25%–30% of their overall business budget, which is significant. That means business executives must learn to drive their core businesses as technology businesses.

Every Business Today Is a Technology Business

Technology has changed the rules of the business and organizations have been facing the following changes:

1. Rapid changes in customer behaviour and requirements—expectations of continuous newness in products and services
2. Uncertain future (rapid emergence of new technologies and impact on businesses)
3. Unknown competition (Google is the competition to the automotive industry, Tesla has changed the automotive industry, Apple and Samsung are now becoming the wristwatch providers)

Without integrating technology into their products, services and processes, organizations will sunset in the digital era. Following are a few examples:

1. Connected cars powered by software
2. Software-enabled baby diapers
3. Software-enabled bedsheets
4. Social media and chatbot based customer services

5. Robot-driven help desks
6. Realtime telehealthcare (digital healthcare)
7. RPA for appointments, patient data entry and billing
8. RPA for updating orders, shipping and tracking products
9. Smart contract (blockchain) based autonomous leave approval processes

Irrespective of the industry, segment, products, services and processes must be embedded with technology/software. Also, due to the changing needs of digital age customers, unexpected competition and continuous technology innovations, organizations are under pressure to introduce new products and services within a shorter span of time.

Rigid structures, processes, policies and business models do not work, and continuous improvements are required. A real start-up environment must be exploratory and evolve with high agility. Hence, every business today must be a technology business and must be driven like a technology start-up to prepare for the digital future.

Establishing a Technology Start-up Ecosystem

There are so many units within an organization like sales, marketing, logistics, HR, procurement and others. All these units must embrace digital technologies and establish technology start-ups in their respective units. New initiatives in these units must be nurtured until they reach a mature state. A few examples are given in this section.

Social Media Command Centre

A consumer goods company maintains Twitter and Facebook accounts in which millions of people share their suggestions, complaints, grievances and emotions in real time. A social media command centre is established in a separate room where big screens display the status of customer responses: positive, negative or neutral based on the conversations happening on social media. Social media analytics and mining tools carry out sensitivity analysis.

If the sensitivity of the people is negative for a longer time, it implies that there is something wrong happening. This is visible on the screens displayed in the social media command centre. The company immediately looks into the details and performs a root-cause analysis to take preventive and corrective actions. One such measure could be immediately releasing a video by the CEO of the company on social media channels to minimize the impact of the crisis.

The social media command centre is driven like a technology start-up to manage the initial uncertainty and unknown requirements with less structural and process overheads. Later, once the command centre reaches a stable state, it would be integrated with the business operations of the marketing department.

Farm-to-Fork Solution Development

When suppliers deliver fruits or vegetables from farm to the restaurants, many a time, these fruits and vegetables get damaged in transit, as the suppliers fail to maintain the required temperature, ventilation and storage environment while moving from the farm to restaurants. The supply chain department of the organization establishes a technology start-up to address this problem.

In this new start-up, they design a **farm-to-fork** solution using IoT, blockchain and analytics technologies. When packaged, each packet of fruit or vegetables is embedded with sensors, microchip and embedded software to track the temperature, ventilation and any damage that might happen to the supplies. The data from the sensors is stored in a distributed ledger using blockchain technology so that it becomes difficult to tamper with the actual data generated by the sensors.

An early warning system, designed as part of this solution using analytics, sends real-time alerts and recommendations to the concerned stakeholders on any deviations in the packet parameters.

Any deviation from the threshold value of various parameters like temperature or ventilation means there is something wrong. In this way, prompt actions could be taken that reduce the cost of wastage. This solution also ensures that high-quality fruits/vegetables reach the restaurant.

Once this farm-to-fork solution reaches its mature state, then the technology start-up seamlessly integrates with the regular supply chain operations of the company. In a similar way:

1. The HR department could set up a technology start-up to deliver unforgettable moments of experience to its employees using social media, mobile, IoT and AI technologies.
2. The customer service management department could set up a technology start-up that builds a solution with which intelligent chatbots use AI, ML, DL and analytics to perform help desk operations with no or optimal human interventions.

In this manner, all departments must realize the fact that without technology interventions, they fail to prepare for supporting digital businesses. Hence, it is imperative for organizations to establish a start-up ecosystem within each department and sophisticated central governance to monitor and transition the outcomes of the start-ups into business operations.

Technology units driven by chief information officers (CIOs) have been failing to support the business units on technology initiatives. Hence, each business unit has been focusing on its own technology initiatives, which consume the lion's share of the business budget. Technology units must become system integrators, facilitators and compliance managers and provide full support to various business units to successfully drive their technology start-ups.

To establish a technology start-up ecosystem by using lean digital thinking, leaders must answer the following questions:

1. Do we have trained leaders at strategic, tactical and operational levels to drive our businesses as technology start-ups?
2. Do we have the required structure, policies, processes, tools, techniques and governance to drive a few of our businesses as technology start-ups to align with the Fourth Industrial Revolution?
3. Do we have the required governance to seamlessly drive our technology initiatives as start-ups in collaboration with the technology units under the CIO?
4. Do we have active technology units under the CIO that would perform the activities of system integrator, facilitator, compliance manager and technology innovators to help us drive our digital business strategy?

CHAPTER 8
Lean Digital Thinking Cluster 3: Embrace Social Responsibility

Principle 10: Design Socially Responsible Products, Services and Value Chains

While we have been progressing well in many dimensions in the world, there are some alarming situations that need the attention of industries, enterprises, governments and individuals. Organizations must be more responsible while introducing their products, services and performing day-to-day operations.

In the next 20–30 years, we may create a world where people suffer from many health ailments and cannot live comfortably due to ecological imbalance and serious environmental issues. There is a greater need for each organization to create socially responsible products, services and value chains.

Some alarming statistics collected from various sources are mentioned in Table 8.1 that emphasize the importance of each organization to be socially responsible.

S. No.	Impact on the Environment and Society	Source
1	Carbon dioxide (CO_2) levels are at their highest in 800,000 years. Average monthly level of CO_2 in the atmosphere exceeded 410 parts per million (PPM). Humans have not experienced this much pollution in the past, which is hazardous to health.	World Economic Forum*, 9 May 2018
2	In 2018, CO_2 emitted from fossil fuel combustion (burning) was equal to about 75% of the total US anthropogenic GHG emissions.	US Environment Protection Agency, Inventory of US Greenhouse Gas Emissions and Sinks (1998–2018)**, April 2020
3	Fossil CO_2 from energy use and industry reached around 37.5 gigatons per year.	Emissions Gap Report, 2019 by the UN Environment Programme
4	Around 1 million animal and plant species are threatened with extinction due to serious negative impacts on the environment.	United Nations Report, May 2019

5	If we do not reduce the greenhouse gas emissions to 55% lower than the current levels by 2030, global temperatures will increase beyond 1.5 degrees Celsius. This will be over the threshold set at the Paris Agreement (2015).	Intergovernmental Panel on Climate Change, 2019
6	420 million hectares of forest have been lost through conversion to other land uses since 1990.	UN Environment Programme, May 2020
7	Around 705,145,388 people live in extreme poverty which is 11% of the world's population.	World Poverty Clock, September 2020 (11 September 2020, 9:23 AM IST)
8	Air pollution causes 4.2 million deaths every year while household exposure to smoke from dirty cook-stoves causes 38 million deaths every year. About 91% of the world's population lives in places where air quality is inferior and not within the prescribed limits by WHO.	WHO, May 2018

9	785 million people live without basic water service and 144 million people live on surface water. Around 2 billion people use a drinking water source contaminated with faeces.	WHO, June 2019
10	In 2016, around 2.01 billion tons of solid waste was generated by cities around the world and which will shoot up to 3.4 billion by 2050.	World Bank, September 2018

Source: Author

Notes: *Available at: https://bit.ly/31wf7rs

**Available at: https://bit.ly/37v4xVI

Table 8.1: Impact of Industrialization on the Environment and Society

In a capitalist society, most organizations give importance to delivering high value to customers, shareholders and ignore the impact of their products, services and operations on society at large.

Through corporate social responsibility (CSR) initiatives, organizations are expected to self-regulate their business so that they become socially accountable. Unfortunately, in most organizations, CSR initiatives have not been integrated with the regular stream of business.

Organizations establish their non-profit arm in the form of a society or foundation and sponsor CSR initiatives that help in creating social and economic impact. However, organizations are not able to direct CSR initiatives towards contributing to their growth and profitability. CSR initiatives have also failed to create a significant social impact.

Hence, finding business opportunities by solving social problems leveraging digital technologies would make the organizations socially responsible. That means organizations develop products, services and value chains that resolve social issues, which also help them grow their business and profits.

Digitally powered products, services and value chains effectively address the following social problems (illustrative list) and create new revenue streams:

1. Irresponsible usage of natural resources
2. Solid waste
3. Poor air quality
4. Poor water quality
5. Careless usage of available land
6. Deforestation
7. High carbon emission
8. Low biodiversity

Examples: Socially Responsible Business Solutions

Digital Car Parking Service to Reduce Carbon Emissions

Due to serious parking issues in cities, every year

1. thousands of gallons of gas is consumed by vehicles while driving on the road to find a parking place,
2. tons of carbon dioxide is emitted and
3. the search for a parking place increases traffic on the roads.

Internet of things (IoT) and analytics solutions with digital parking meters at private and public parking places with sensors and a mobile app that provides the parking occupancy details can solve this problem.

Drivers can reserve the parking place from anywhere using the mobile app and could directly go and park their car at the pre-booked location. In this way, fuel-saving and low carbon emissions are possible.

Digitalized Water Pipes to Reduce Water Contamination

A water pipe manufacturer with IoT can solve this social problem while getting revenue. Please see the following example.

Water pipelines with embedded sensors (IoT) can quickly identify whether any chemicals or contaminants are present in the drinking water. The precise location where drainage water is mixing with drinking water can also be identified.

Mining of the sensor data using artificial intelligence (AI), machine learning (ML) and analytics help in preventing many health issues with the help of proactive alerts and recommendations.

Digitalized Socially Responsible Business Value Chain

A value chain is a mix of core and enabling (supporting) processes. Orchestration of core and enabling business processes within the organization delivers value to customers and improves operational excellence. For example, core processes and supporting processes for a manufacturing organization are mentioned in Table 8.2.

Core Processes	Enabling Processes
1. Inbound logistics 2. Production 3. Outbound logistics 4. Sales and marketing 5. After-sales service	1. Procurement management 2. Human resources management 3. Financial management 4. Compliance management 5. Information technology management 6. Alliance and partnership management

Source: Author

Table 8.2: Core Processes and Enabling Processes

For example, the chemical production process is critical in the manufacturing business. While producing chemicals, hazardous waste is generated, which is usually pumped into rivers and oceans, causing the death of underwater species and also many health problems to humans. Hence, the production process needs to be modified to reduce the impact on the environment and health of human beings.

To make the production process socially responsible, the following additional activities could be implemented to reduce the impact on the environment and human health:

1. Using IoT, identify the hazardous parameters and using analytics, generate recommendations to effectively dispose of the chemical waste.
2. While pumping the chemical waste into rivers or dumping yards, process them to reduce adverse impacts on the environment.
3. Purify the water in the rivers or other dumping yards every month to remove any chemical waste.

Hence, it is essential to modify business processes (value chain) with additional activities/tasks to make sure that the business operations of an organization are environment-friendly.

To create socially responsible products, services and value chains by using lean digital thinking, leaders must answer the following questions:

1. Do we have a system to continuously assess the impact of our policies, processes, products and services on the environment?
2. Do we have the strategic, tactical and operational skills and team to reduce the adverse impact on the environment and society?
3. Do we have the focus to bring the practices of compassionate capitalism and shared value within our organization?
4. Do we continuously monitor the impact of our organization on the environment and society and take corrective and preventive actions in a timely manner?

☆☆☆

CHAPTER 9
Lean Digital Thinking Cluster 4: Deliver Unforgettable Moments of Experience

Principle 11: Design and Deliver Unforgettable Moments of Experience to Stakeholders

People in the digital economy have different needs, expectations and aspirations. Selling **unforgettable moments of experience**, along with products and services, is essential. The stakeholders here are customers, employees, suppliers and partners. In the digital era, emotionally connecting with people while performing their work has become imperative.

Acquiring lifelong customers is the aspiration for every organization today for which just meeting the transactional or task needs is not enough. Understanding the **emotional needs** and meeting them while delivering products or services is the need of the hour.

Competition is fierce and, most of the time, unknown. For example, a few of the customer interactions for banking transactions have been taken over by Google Pay, PayPal, Paytm, Amazon and others. Banks are not able to provide the superior front-end experience to customers and hence most of their customers are using other channels to get an excellent experience.

Similarly, the new workforce joining the organizations are digital natives and they have different attitude, behaviours and expectations. Significant changes must be brought in the HR policies, work environment and learning strategies to emotionally engage this new-age workforce. The current generation expects unforgettable moments of experience at their workplace. They hate monotonous jobs.

The following sections cover how unforgettable moments of experience is designed and delivered to customers at each interaction point throughout their journey. Similarly, a better experience is designed and provided to employees and other stakeholders at each interaction point as well.

Offering Better Experience at Each Interaction Point

For many decades, organizations have been offering services and products by meeting the **task needs** or **transactional needs** of the customers. Now, due to changes in customer attitudes and behaviours, customers expect emotional engagement and faster response, while getting products or services.

It is important to understand the customer's journey from **acquisition to utilization** of the products and services. The journey of a customer with organizations spans from the **selling of products/services** to **offering support** for the sold products/services.

In today's hyperconnected world, customers interact with people and intelligent things of an organization to get their work done. There are so many channels like websites, mobile devices, social media, wearable devices and chatbots for the customers to obtain services, products and information.

Customer experience management means understanding customer attitude, behaviours and emotions at each interaction point and implementing the appropriate response to delight them at the identified interaction points.

Customer experience management is the sum total of positive **perceptions** of a product/service at all interaction points. It is an ongoing process because of which customers become loyal, repeat business and become strong advocates of the products/ services.

Customers now discuss their buying and support experience very openly on social media. The positive and negative experiences spread like wildfire, which impacts the buying behaviours of other prospective customers.

It is very important to identify significant interaction points/ touchpoints in the customer journey that can **make or break** an organization. Based on the buying and service experience at each interaction point of customers, they are highly engaged or wholly disengaged. These interaction points are called **Moments of Truths**. At each Moment of Truth, the experience must be designed effectively.

Nine Steps to Deliver Unforgettable Moments of Experience

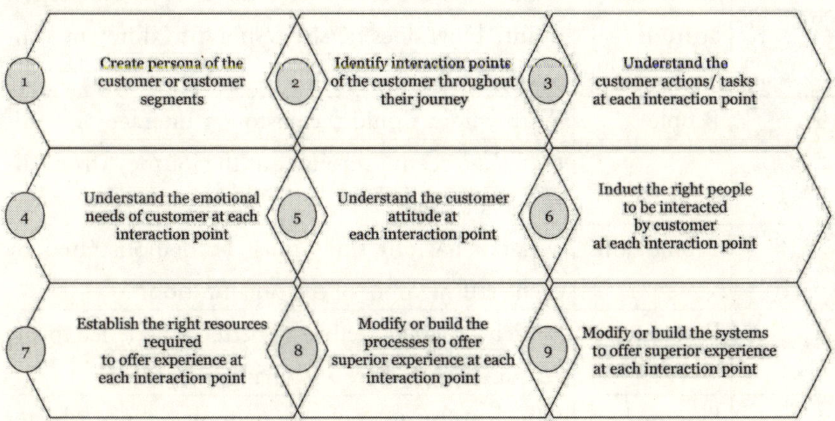

Source: Author

Figure 9.1: Nine Steps for Delivering Unforgettable Moments of Experience to Stakeholders

S. No.	Elements of Experience	
1	Persona of the customer(s)	Customer name Buying preference (channels) Age Gender Lifestyle What motivates the customer to buy? What is the value he/she expects in general (time, cost, comfort, convenience, joy, support, etc.)?
2	Interaction points	What are the touchpoints at which the customer interacts with the organization throughout their journey like website, branch, store, email, social media, etc.?
3	Customer actions	What are the tasks/actions carried out by the customer at each interaction point?
4	Emotional needs	What are the feelings of the customer (fear, joy, sadness, surprise, disgust, anger or inspiration) at each interaction point?
5	Customer attitude	What does the customer think at each interaction point? How does he/she respond to different situations?
6	People	With whom would the customer interact with (directly/indirectly) throughout the journey (front office, back office)?
7	People skills	What behaviour skills must be demonstrated by each staff member of the organization?
8	Processes	Which processes deliver a better experience to the customer? Are there any changes required?
9	Systems/tools	Which software applications/tools deliver a better experience? Are there any changes required?

Source: Author

Table 9.1: Nine Steps for Delivering Unforgettable Moments of Experience to Stakeholders

Example: Sample Customer Journey Map

Ashok decides to apply for a loan at XYZ bank. Figure 9.2 depicts a sample customer journey map for Ashok. The journey map shows the multiple stages and the channels he uses while getting the loan from the bank. The circles mark the interaction points of the customer with the bank and the colour of the circle signifies whether the experience was good or bad for Ashok.

Source: Author

Figure 9.2: Customer Journey Map for Availing a Loan

	Needs are met, attitude and emotions of customers are managed very effectively and customers are engaged and feel extremely happy.
	Needs are met, but attitude and emotions of customers are not effectively managed and customers are partially engaged.
	Needs are not met, attitudes and emotions of customers are not effectively managed and customers are not engaged and are very unhappy.

Out of the seven interaction points in this map, the bank failed to provide a better experience to its customer at four of them.

Current Customer Experience at Each Interaction Point

Elements of Experience	Current Experience Description
Persona of the customer(s)	• Ashok • Prefers mobile phone for purchasing • Age: 30 • Male • Gen Y • Active on social media • Likes music • Less paperwork motivates Ashok to buy • Expects speedy process
Interaction points	Website, face-to-face, email and help desk
Customer actions	• Searches websites for various loan options from different banks • Has conversations on social media with friends and opinion leaders about the best loan options and banks • Selects a bank • Meets the bank representative and submits documents • Responds to the bank to complete the verification of documents and credentials • Follows up on the delay in the loan approval • Receives loan approval information in an email/direct mail • Follows up for correction of his last name using email/phone • Checks bank account for loan disbursement • Shares experience on social media

Current emotions	• Gets frustrated due to long waiting time and not being attended by the bank staff • Feels cheated because the bank staff intentionally hid the loan terms sheet • Is hurt by the rude behaviour of the staff • Feels unhappy and disrespected because the loan amount is low
Customer attitude	• This is one of the worst banks • The bank did not meet my needs • It was horrible to follow up many times • I can't wait for this much time • I raised my voice and complained to the bank manager
Staff involved	Branch manager, sales executive, loan verification agent, etc.
Skills gap	• Rude behaviour • Lack of customer centricity • No listening skills • Non-receptive to feedback
Process gaps	The current loan management processes have not factored any emotional needs of the customers while processing the loan
Systems/tools gaps	Siebel CRM and Oracle Financials focus only on transactional needs and ignore customer engagement and emotional needs—these features are completely missing

Source: Author

Table 9.2: Loan Service—Current Customer Experience

New Experience design at Each Interaction Point

In the previous example, it is quite clear that the bank failed to provide a better experience to Ashok. After a detailed analysis of his journey, task needs, emotional needs and attitude, the bank improves his experience as given in Figure 9.3.

	CUSTOMER JOURNEY							
INTERACTION POINT	DECISION TO TAKE LOAN	EXPLORE	SELECT BANK	SUBMIT DOCUMENTS	SUPPORT DURING VERIFICATION	RECEIVE LOAN APPROVAL	RECEIVE DISBURSMENT MESSAGE	SHARE FEEDBACK
CHANNELS: WEBSITE								
HELP DESK								
EMAIL								
BRANCH								
MOBILE DEVICE / APP								
SOCIAL MEDIA								
CHAT BOT								

Source: Author

Figure 9.3: Revised Customer Journey Map for Availing a Loan

Elements of Experience	New Experience (Improvements/Changes)
Persona of the customer(s)	• Ashok • Prefers mobile phone for purchasing • Age: 30 • Male • Gen Y • Active on social media • Likes music • Less paperwork motivates Ashok to buy • Expects speedy process
Interaction points	Website, face-to-face, email, help desk, mobile phone and chatbot

Customer actions	Establish a system and introduce the following changes: • Develop an early warning system within the bank loan management process to proactively inform the customers about potential delays • Apply AI algorithms to verify the name from multiple sources (driving license, credit card, etc.) • Allow customers to perform the tasks using multiple channels like mobile phone, wearable devices, chatbots and web application • Allow customers to perform the tasks using different interfaces like gestures and touch
Current emotions	Establish a system in such a way that customer get following emotions while dealing with the bank • This bank respects my time • The staff is very transparent and explains all the loan terms • The staff is very polite and treats me with respect • The staff is very responsive • Integrate gamification techniques within the loan management process to motivate the customers • Tailor the relationship mode (face-to-face, automated) based on the category of customers (Gen Z, Gen Y, Gen X and baby boomers)
Customer attitude	• Wow! This is a great bank • I think XYZ bank exceeds my expectations • First time right, no follow-ups required • Very less waiting time • Wow, at last, I got my loan
Staff involved	Branch manager, sales executive, loan verification agent, etc.

Skills gap	Improve bank staff skills in the following areas: • Business storytelling • Reflective conversations • Active listening • Receiving feedback
Process gaps	Identify interaction points in the loan management process and integrate features to meet the emotional needs of the customer
Systems/ tools gaps	Integrate social media analytics, call data analytics and other personalized features to meet the emotional needs of the customer

Source: Author

Table 9.3: Loan Service—Improving Current Customer Experience Using Nine Steps

To design and deliver unforgettable moments of experience to stakeholders by using lean digital thinking, leaders must answer the following questions:

1. Do we have the vision to deliver unforgettable moments of experience to customers to make them lifelong customers?
2. Do we record the journey map for each customer segment and all the interaction points in the journey?
3. Do we pay attention to take care of the emotional needs of the customer at each interaction point in addition to the task needs?
4. Do we record current experience being offered to customers at each interaction point in their journey and identify the improvement areas?
5. Do we design new experience at each interaction point of the customer journey to enhance their happiness?
6. Do we continuously update people skills, processes and systems to deliver unforgettable moments at each interaction point?

Principle 12: Build a Human-Centric Digital Culture

Without the digital culture, it is nearly impossible to design and manage businesses in the era of the Fourth Industrial Revolution and the digital economy. Digital culture is different from past cultures due to digital technologies.

Baby boomers, Gen X, Gen Y and Gen Z cultures are significantly different from each other, and hence, there are cultural conflicts among these generations. Now, everyone, irrespective of their generation, must embrace the digital culture to be relevant in the digital era. Contextualizing culture in the digital age is essential. As shown in Figure 9.4, digital culture has seven essential elements.

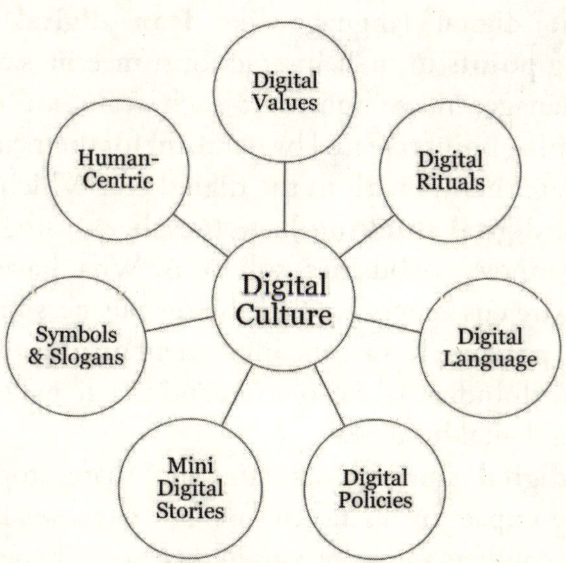

Source: Author

Figure 9.4: Elements of a Human-Centric Digital Culture

The elements of a human-centric digital culture must be incorporated into organizations in the following manner:

1. **Digital values:** Define digital values on the top of Core Values of the organization. Apply digital values in different

situations that would impact the attitude and behaviour of the organization and individuals in the era of the Fourth Industrial Revolution.

2. **Digital rituals:** Establish habits and practice them regularly (daily, weekly, monthly, yearly) to promote digital culture. For example, start every day with the anthem of the organization. Set aside three hours as thinking time every month to explore various digital business innovations.
3. **Digital language:** Design a digital language. Use this inspirational and motivational digital language consistently in meetings, reviews and discussions to continuously remind and ignite the spark in the minds of stakeholders about the Fourth Industrial Revolution and the digital economy. For example, an organization can consistently use the digital language like **lean digital thinking**, **turning points** in their interactions, meetings and events. The manager may inquire at each team meeting about the turning points created by her team for their customers to accelerate their growth in the digital era. Which principles of **lean digital thinking** have they been using mostly to deliver innovative business solutions? What have they done to prepare customers for their better business future?
4. **Digital policies:** Define and implement policies aligning with the Fourth Industrial Revolution and the digital economy to support digital businesses.
5. **Mini digital stories:** Maintain and share stories on the positive impact of digital culture and scenarios in handling various conflicts and situations in the era of a hyperconnected world and the Fourth Industrial Revolution. Mini stories include sharing best practices, lessons learned and knowledge with employees, suppliers and partners.
6. **Symbols and slogans:** Define visual representation of digital culture in the form of symbols, posters and slogans. Display them at various places in the organization. These symbols and slogans consistently remind the employees, suppliers and

partners about the culture to be followed in the era of the Fourth Industrial Revolution.

7. **Human-centric:** There is a misconception that the digital world emotionally disconnects people and lacks the human touch. That is not true. With virtual reality (VR), AR, AI, ML, chatbots, virtual assistants and analytics, the cyber system has been creating situations and scenarios with which human beings get an immersive and emotionally engaging experience. Hence, an essential element of the digital culture is to sustain the human touch and emotional connect.

Designing and institutionalizing a lean digital culture requires committed effort:

1. **Design** and co-create rituals, values, common language, policies, stories, slogans and symbols with a human touch in collaboration with all stakeholders.
2. **Educate** employees, suppliers, partners and other stakeholders about lean digital culture.
3. **Pilot** lean digital culture within identified groups and based on the feedback, take corrective actions.
4. **Operationalize** and institutionalize lean digital culture throughout the organization, once the piloting is complete.
5. **Monitor** adherence to lean digital culture, **reward** the champions and share the success stories across the organization.

Example: Establishing a Digital Culture

Digital processes, organization structure, digital business models and digital strategy are not enough. Establishing a **digital culture** is imperative for organizations to excel in the era of the Fourth Industrial Revolution. Each organization must create a digital culture for itself as per their needs. See Table 9.4 to understand how the elements of a human-centric digital culture can be used to do so.

Digital Culture	Description	Remarks
Digital values	• Customer experience • Data-driven • Anywhere, anytime work • Real-time collaboration • Real-time response • Eliminate non-value-added work	Apply these digital values, while taking decisions, responding to situations and crisis, and addressing conflicts in the organization.
Digital rituals	• Daily telepresence meeting • Monthly 60 min ideathons using open innovation digital platform • Monthly virtual conference (telepresence/video conference)	Digital rituals help employees, partners and suppliers to acquire the habits by practising the rituals at regular intervals. Geographic and social aspects also to be factored in while designing the rituals. A religion may not accept the rituals of other religions.
Digital language	• Augmented business intelligence • Unforgettable moments of experience • Human intervention optimization • Lean	Use this language consistently in meetings, reviews and discussions to promote a positive digital culture within the organization.
Digital policies	• Enterprise reputation management policy in social media • Cybersecurity policy • Digital content policy • Blockchain smart contracts policy • Connected devices policy	To align with the digital vision and digital business strategy of the organization, new policies must be implemented.

Mini digital stories	• Mini Digital stories are nothing but sharing best practices, lessons learned, core values adherence, case studies within five minutes using audios or videos. • Following is an example of a mini digital story as explained by Catherine to her team, which can be viewed at the organization's internal social media channel. The mini-story goes like this; on 15 November 2016, just one day before, one of the senior digital architects expressed his inability to conduct the workshop with customer due to some personal exigencies at his end. That disturbs the overall relationship with the customer in New York. Considering one of the organization's core values "Customer Experience", the workshop was not cancelled and a telepresence session with three experts from London, India and Germany was arranged instead. This mini-story highlights how adherence with organizations' core value satisfies the customers.	Collate mini stories like this, which reflects the organization's values and inspires many others to follow.

Symbols and slogans	• Symbols and slogans for digital rituals, digital values and digital policies • For example, the slogan "Digitally Yours" could be used at appropriate places • A symbol for the Fourth Industrial Revolution and digital economy • A symbol for real-time collaboration	Slogans, symbols in the form of posters, wallpapers and flyers help in building the digital culture.
Human touch	• Employees working from geographically distributed locations can use telepresence to have effective meetings as if they were in the same room • For example, an intelligent HR system with a virtual assistant for each employee on their mobile phone advises them about their current stress level and recommends taking preventive healthcare precautions • Before joining an organization, prospective employees can get an immersive experience (MR) from remote locations and emotionally get connected to the organization even before joining	Leverage the cyber system to promote human touch within the organization and introduce various features as part of the digital culture.

Source: Author

Table 9.4: Example—Establishing a Human-Centric Digital Culture

To build a human-centric digital culture using lean digital thinking, leaders must answer the following questions:

1. Do we understand the cultural changes required to be successful in the Fourth Industrial Revolution?
2. Do we factor human emotions while designing a culture that is highly influenced by digital technologies?
3. Do we implement the required structure, policies, processes and procedures to institutionalize lean digital culture within our organization?
4. Do we induct the right leadership and team to design and institutionalize lean digital culture within our organization?
5. Do we establish culture champions who continuously educate the employees and monitor the progress of culture institutionalization in the organization?

PART 3

What to Digitalize?

	CHAPTER 10 **Digitalization Cluster 1: Digitalize Business Blueprint** • Four Clusters and Twelve Building Blocks for Business Digitalization • Building Block 1: Business Vision • Building Block 2: Business Model • Building Block 3: Business Strategy • Building Block 4: Business Outcomes
	CHAPTER 11 **Digitalization Cluster 2: Digitalize Technology Foundation** • Building Block 5: Technology Strategy • Building Block 6: Technology Governance • Building Block 7: Technology Architecture
	CHAPTER 12 **Digitalization Cluster 3: Digitalize Offerings** • Building Block 8: Products • Building Block 9: Services

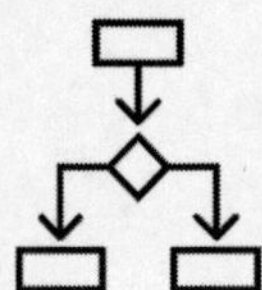

CHAPTER 13

Digitalization Cluster 4: Digitalize Business Operations

- Building Block 10: Workplace
- Building Block 11: Business Processes
- Building Block 12: Business Operating Model

CHAPTER 10
Digitalization Cluster 1: Digitalize Business Blueprint

Four Clusters and Twelve Building Blocks for Business Digitalization

Apple, Google, Microsoft, Amazon, Netflix, etc., are digital natives. However, more than 95% of organizations in different segments like manufacturing, automotive, banking, retail, hospitality, telecom, utilities and insurance today are not **digital natives** and are digital immigrants. Digital immigrants must start a major digital transformation journey to become future-proof and future-ready. Digital immigrants have many unanswered questions like:

1. Why digital?
2. What is digital?
3. Where to start the digital journey?
4. When to start the digital journey?
5. How to start the digital journey?

The organization's leadership and employees must understand the fact that significant changes are required in the 12 building blocks of business to prepare for the digital journey. Transitioning to digital is not just limited to the development of a few mobile apps, artificial intelligence (AI) or machine language (ML) algorithms or erecting internet of things (IoT) devices. Applying **lean digital thinking** and its 12 principles on the 12 business building blocks will help digital immigrants become digital systematically and optimally. The 12 business building blocks are grouped into 4 clusters as mentioned in Figure 10.1.

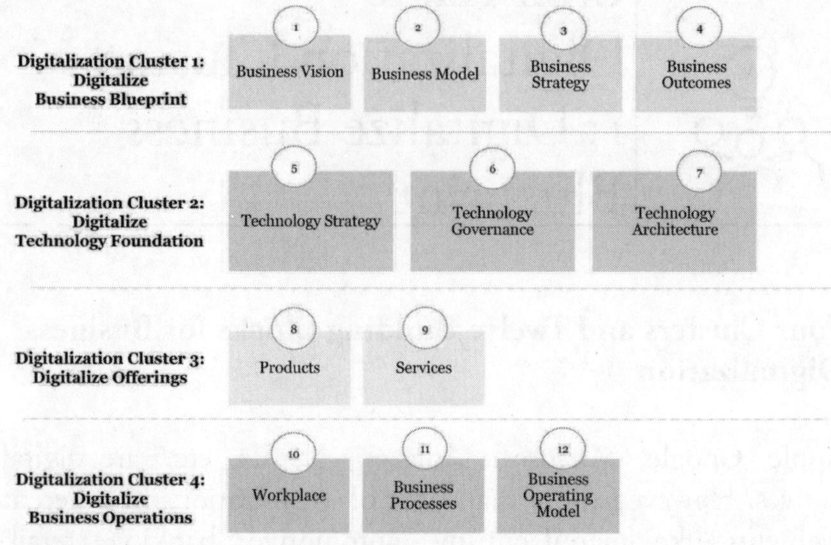

Source: Author

Figure 10.1: Four Digitalization Clusters and Twelve Business Building Blocks

If an organization wants to unlock the true potential of digitalization, the 12 business building blocks must be digitalized within approximately 3 years.

Digital Transformation = [Digital Business Transformation + Digital Technology Transformation]

Digital business transformation means digitalizing business blueprint, offerings and business operations.

Digital technology transformation means digitalizing technology strategy, technology governance and technology architecture.

Building Block 1: Business Vision

One cold December morning, while taking hot coffee, Sue had been profoundly thinking about the future of Maxima Bank. Sue had recently joined as the CEO of Maxima Bank, which has been doing reasonably good business as a midsize bank for the last 30 years.

She observed that companies like PayPal and Google Pay had been impacting the customer-facing functions like payments. These tech companies had been providing superior customer experience, which banks had been failing to do. Sue was worried that in a few years, most of the customer-facing functions might be taken over by the active tech companies and banks will be forced to focus on back-end operations.

Most of the systems in Maxima Bank were legacy systems and hence, caused delays in offering various services, which was unacceptable to their current customers.

Sue had a broad idea that both customer-facing functions and core banking functions must be modernized with digital technologies. This will ensure that Maxima Bank is geared up to meet the needs and wants of digital age customers.

The first big challenge she wanted to take up was to co-create a digital vision for the bank with her senior colleagues. She appointed Tom, senior vice president, Strategy to drive this task as a mini project so that every senior colleague would actively participate in designing the digital vision of Maxima Bank.

Tom planned to conduct a workshop with all his senior colleagues to explain to them Sue's expectations for their digital vision. During the workshop, Steve, the head of payments, raised a concern that since they were all core banking experts, they did not have a clear idea about how digital technologies would impact their business. In this scenario, creating a digital vision may not be effective. Other senior leaders also expressed similar concerns.

Tom noted all the points raised by the senior colleagues and updated Sue about the status. Sue had been in touch with Rao, who was a lean digital specialist. Sue was a co-panel member in a few of the digital banking panel discussions along with Rao. She spoke with Rao in detail and explained the situation. Rao happily agreed to take up this challenge to mentor and guide the senior leaders of Maxima Bank to define their digital vision.

Need for a New Business Vision in the Digital Era

On 23 May 2014, Rao conducted a meeting with the senior leaders of Maxima Bank, which Sue also joined. Rao explained to the participants that digital technologies have been changing the market dynamics, economy, customer expectations, competition landscape and business dynamics. Hence, driving businesses with an outdated vision will prepare organizations for a successful failure.

Rao also emphasized that, in the past, once the vision was defined, it used to be there for many decades. Now, in the digital era, organizations are expected to revisit their vision every 3–5 years.

Digital business vision means a dream or envisioned future and the desired state of an organization in the coming years after embracing digital technologies. Without a digital business vision, digital transformation will be directionless, and organizations will not achieve the desired business results. **A digital business vision is like a navigator**.

Rao shared a few examples of digital vision, which was set by a few other organizations as reference.

Few Examples of Digital Business Vision

Simplify and digitize our business and provide Do It Yourselves (DIY) digital options for customers through our direct channels.

—National Australia Bank

Digital has to be the way we can tell our story, build our brand and have a relationship with our customer.

—Adam Brotman
Chief Digital Officer, Starbucks

The digital world multiplies the ways our brands can create an emotion-filled relationship with their customers.

—L'Oréal

Senior leaders of Maxima Bank and Rao agreed to organize a detailed workshop to co-create the digital vision for Maxima Bank. They requested Rao to use a templatized approach to define the digital vision. Tom agreed to coordinate with everyone to arrange the workshop soon.

Applying Lean Digital Thinking to Design Digital Business Vision

Tom arranged a workshop on 31 May 2014 outside the office. Sue and all senior leaders joined this workshop. The chief information officer (CIO) of Maxima Bank, Richards, knew that they have already started implementing a few digital technologies like mobile, social, analytics and cloud technologies. Hence, he felt that Rao's presence was only an overhead and did not add much value to Maxima Bank.

Richards' team had been observing the competition and had developed many mobile apps for payments. But unfortunately, these apps had not become very popular. They were sort of in a race with their competition and put their energies to beat the competition with newer apps instead of providing high value to customers.

Rao started the session with a provocative statement that building a digital Maxima Bank was not just about implementing a few digital technologies. It means

1. delivering unforgettable moments of experience to the customers,
2. introducing mobile internet-based new services that enable customers to fulfil their requirements without visiting the bank,
3. introducing new business models to add new customer segments and revenue streams, and
4. reducing the cycle time in delivering services like opening a bank account to a few minutes.

Rao emphasized the point that new thinking was required among senior leaders of Maxima Bank, which could be acquired with **lean digital thinking**.

Rao mentioned that in the digital era, the vision must be written like a narrative for which **lean digital thinking** comes in handy. **Lean digital thinking** prepares the organization's leadership to create a 360-degree vision which would navigate the whole organization through the digital future.

The irrelevant, ambiguous and incomplete vision put the organization in danger. Hence, the co-creation of a digital business vision was essential.

William, the head of customer relationships, asked Rao how would **lean digital thinking** help in devising a new vision for Maxima Bank. He said that they were confused because of jargon, and there was no common standard or language to understand the meaning of digital. Rao assured William and other leaders that after the orientation on **lean digital thinking**, they will start thinking differently, which would help them define an accurate digital vision for Maxima Bank.

Rao explained in detail each principle of **lean digital thinking** in the context of a bank as mentioned in Table 10.1.

S. No.	Lean Digital Thinking	Focus Areas
1	Create a better today and better tomorrow	To coexist with the current traditional banking business and slowly move to the digital banking business
2	Embrace constant and instant change	To build an agile bank which embraces changes regularly
3	Nurture digital business talent	To identify new skills, new roles and new behaviours to drive the digital business vision of the bank
4	Identify and eliminate non-value-added activities	To build a **lean** bank in the era of the Fourth Industrial Revolution
5	Predict, recommend and improve overall business performance	To redefine the bank's business performance with new metrics in line with the digital era
6	Optimize human interventions	To build nearly autonomous bank
7	Establish real-time and contextual communication and collaboration	To establish a technology-driven collaborative banking business ecosystem with employees, partners, suppliers and customers
8	Safeguard privacy, safety and security of business resources	To build cyberattack proof secured bank
9	Establish a technology start-up ecosystem	To establish an internal technology start-up ecosystem within the bank that accelerates the digital journey

10	Create socially responsible products, services and value chains	To build socially responsible bank services and banking processes
11	Design and deliver unforgettable moments of experience to stakeholders	To deliver unforgettable moments of experience to bank stakeholders for a lifelong association
12	Build a human-centric digital culture	To build a human-centric digital culture within the bank

Source: Author

Table 10.1: Applying Lean Digital Thinking Principles to Digitalize Business Vision

Digital Business Vision Template

Rao believed that a digital business vision template would help the Maxima Bank senior leaders to brainstorm and devise their bank's vision. For organizations to be successful in the digital era, they must create a vision in multiple dimensions. Following is the template distributed by Rao to create the digital business vision of Maxima Bank.

Vision for Lean Business	Vision for Autonomous Business	Vision for Two Speed Business	Vision for Digital Talent Gardening	Vision for Superior Experience to Stakeholders	Vision for Human Centric Digital Culture

Vision for Real Time Collaborative System	Vision for Cyber Attack Proof Organization	Vision for Social Responsibility	Vision for Digital Business Performance	Vision for Driving Business as Technology Start-up	Vision for Agile Organization to embrace change

Source: Author

Figure 10.2: Digital Business Vision Template

Digital Business Vision 2026 for Maxima Bank

After a detailed discussion, the senior leaders and the CEO of Maxima Bank came up with 12 digital vision statements as mentioned in Figure 10.3, which will become the navigator for their digital journey. Rao requested Richards to establish a digital technology ecosystem to realize this digital vision. Richards was convinced and started thinking about his next big digital game plan.

Vision for Lean Business	Vision for Autonomous Business	Vision for Two Speed Business	Vision for Digital Talent Gardening	Vision for Superior Experience to Stakeholders	Vision for Human Centric Digital Culture
By 2023, we will eliminate all non-value adding activities (eight wastes) in our digital businesses.	By 2025, we will make our organization ecosystem near-autonomous. *(That means our products, assets, services, workplaces, processes and operations become highly algorithmic driven with less human interventions.)*	By 2025, we will completely evolutionarily transition into digital business, without cannibalizing our existing business.	By 2023, we will reskill all our employees and leaders with digital talent and prepare them for taking new roles while driving digital businesses.	By 2025, we will establish a lifelong association with customers, employees, suppliers and partners by providing unforgettable moments of experience at each interaction point. *(That means we recognize the needs and wants of different segments of people like Baby boomers, Gen X, Gen Y, Gen Z.)*	By 2023, we will completely re-engineer our business culture, rituals, principles, which are essential to become future proof and future-ready in the era of the 4th Industrial Revolution.
Vision for Real-Time Collaborative System	**Vision for Cyber Attack Proof Organization**	**Vision for Social Responsibility**	**Vision for Digital Business Performance**	**Vision for Driving Business as Technology Start-up**	**Vision for Agile Organization to embrace change**
By 2023, we will minimize face-to-face meetings and travels by establishing a real-time collaborative business ecosystem with employees, partners, suppliers and customers.	By 2021, we will establish a robust cybersecurity and privacy system to secure our business resources, *(Prevent cyber attacks, address the security and privacy vulnerabilities and respond to the crisis promptly.)*	By 2026, all our products, services, assets and processes will be environmentally friendly and will contribute to society and the economy.	By 2022, we will establish a new performance management system with data sources to effectively manage digital businesses. *(Continuously re-imagine and redefine metrics to measure business performance as per the changing needs and technological innovations.)*	By 2024, we will prepare our organization in such a way that each business unit can set-up and drive their new initiatives as technology start-ups.	By 2024, we will re-engineer our whole organization by leveraging digital technologies, which becomes agile and flexible. *(Absorb the changes promptly, which arise from the market needs, customer needs and technology innovations.)*

Source: Author

Figure 10.3: Digital Business Vision 2026 for Maxima Bank

Sue and the team thanked Rao for conducting a very detailed and useful workshop on **lean digital thinking** and digital vision. Tom thanked Rao and the Maxima Bank senior leadership team and concluded the workshop. Rao advised Sue to visually represent the digital vision statements and educate all her employees to feel like a part of this digital journey.

Building Block 2: Business Model

Michael, the chief strategy officer (CSO) of a global bank, Horizon, was given the responsibility to develop unique business models by leveraging digital technologies.

Banks are not known for performing non-banking activities. Michael had been thinking about how to generate more banking revenue by performing non-banking activities. Michael and the team were well-versed in creating business models. Still, he was concerned about the significant changes to be brought in the current business models by leveraging digital technologies.

Michael and the CEO of Maxima Bank were very close friends and had worked together in the past. Michael and Sue met at one of the CXO Roundtables and exchanged their views on digital banking. In their conversation, Michael told Sue that he had been looking for a digital specialist, who could help them create unique digital banking business models. Sue gave him the contact details of Rao, who is a lean digital specialist.

Michael has shared the challenge of creating a unique digital business model and requested Rao's help. Rao explained a new business excellence framework called **lean digital thinking**, with which organizations could acquire the new digital mindset that would help them create new business models very effectively.

Rao requested Michael to form a core team representing the different functions of the Horizon Bank and agreed to conduct a workshop for the core team on the **lean digital thinking** and its 12 principles to create a digital business model for the bank.

Need for Business Model Innovation in the Digital Era

On 15 June 2015, Michael and his team were excited to attend the workshop being conducted by Rao on **lean digital thinking** to create unique banking business models. The core team from different functions of Horizon Bank had also joined the workshop.

Considering the mix of experts from Horizon Bank, Rao wanted to contextualize business models within the digital era, with various examples not limited to the banking industry. He also wanted to make sure that the core team members understood the impact of digital technologies on various industries and how they were helping in creating new revenue streams using innovative business models. Rao shared a lot of information during the workshop. He went on to say:

"Business model is nothing but a framework for value and wealth creation for customers and organizations, respectively. In the past, the lifespan of business models was 12–15 years, and now new business models must be redesigned on an average every 3–5 years. This cycle time may go down further in the digital era due to changing market dynamics and customers' needs and wants. This means continuous business model innovation is required to survive and excel in the age of the Fourth Industrial Revolution. A business model is the visual representation of business architecture that includes value design, value delivery elements, channels to reach customers, revenue streams and cost structure. The visual representation helps all the stakeholders arrive at a common understanding of the business model."

Rao continued to explain that the business model design must be cohesive and multidimensional. He introduced a tool called the business model quadrant that helps organizations to gauge the maturity of their business model.

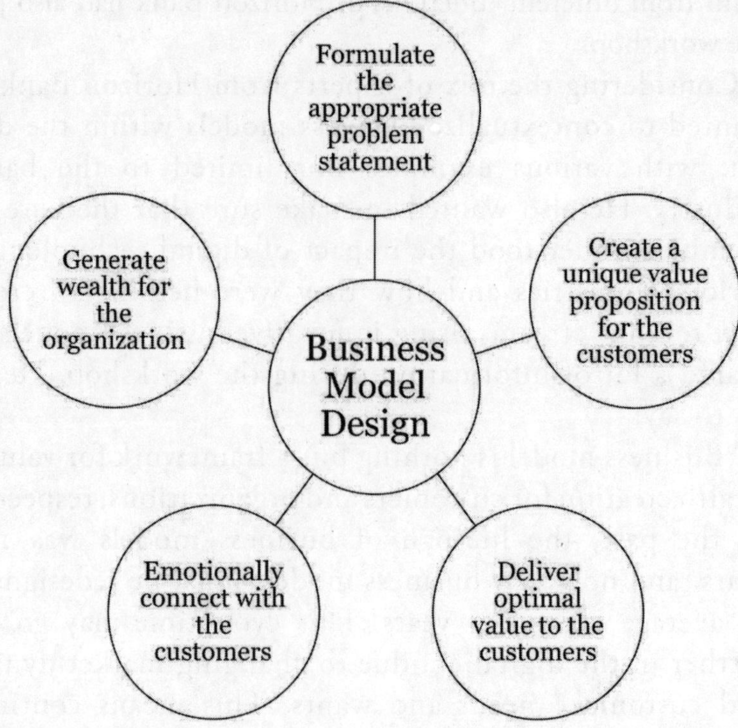

Source: Author

Figure 10.4: Business Model Design

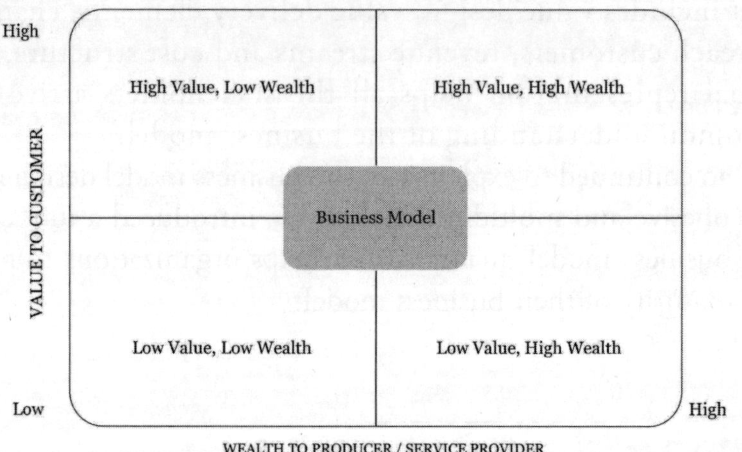

Source: Author

Figure 10.5: Wealth and Value Quadrant

The Wealth and Value Quadrant is a potent tool that helps in getting clarity on the capability of the business model at that moment to take corrective and preventive actions. A new business may initially offer a low-value proposition to the customer and low wealth potential to the organization. Continuous business model innovation is required to reach the **high value and high wealth** quadrant.

SVP (Retail Banking) Christopher interrupted Rao and said that it was too much theoretical knowledge and most of the elements covered by Rao were known to the leaders attending the workshop. He emphasized the point that they wanted to know more about digital business models and how they were different from conventional business models. Rao carefully listened to Christopher and responded to him:

"Christopher, I agree with you, what I covered till now is theoretical, but it was done intentionally. In most cases, senior leaders of an organization do not have a common understanding of the business model. A few people think that it is the revenue model, and a few others believe it consists of various revenue streams and customer channels. Some leaders interchangeably use the business model with the operating model, which is wrong. Hence, all leaders must understand the essential elements of a business model to participate in designing the digital business model for Horizon Bank actively."

Rao managed to convince the participants about the logic of his teaching, and they agreed to meet after lunch.

Examples: Digital Business Model Innovations in the Digital Era

Post lunch sessions are always tough to manage. So Rao decided to share a few innovative digital business models implemented by other industries so that participants become aware of the various innovations happening in the business model.

The cyber system (fusion of varied technologies) has been significantly impacting business models, and organizations have gotten an opportunity to introduce new business models, which never existed in the past. A few examples shared by Rao are given in Figure 10.6 and Table 10.2.

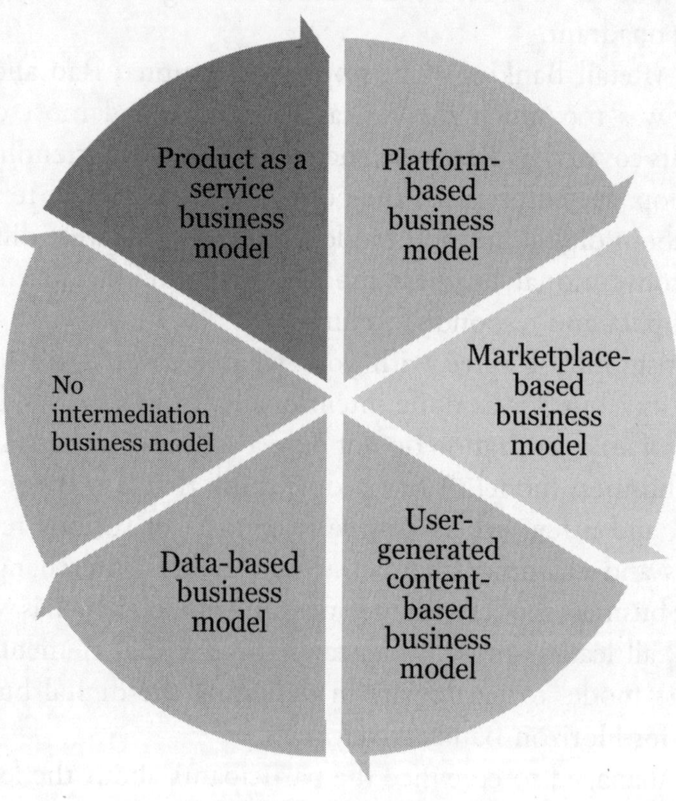

Source: Author

Figure 10.6: Digital Business Model Innovations

Business Model Type	Details

Platform-based business model	The platform-based business model is based on the following characteristics:
	Network effect: Enrol both producers and consumers into the digital software platform. The network effect means higher the number of producers, higher is the value received by consumers and vice versa. Producers and consumers communicate directly and exchange information using the digital software platform. Following are a few examples:
	• Airbnb (housing accommodation platform):
	  ◦ Producers = houseowners
	  ◦ Consumers = guests/travellers
	If there are many homes available for accommodation in the Airbnb digital software platform, guests get more options and better quality of service. At the same time, homeowners and Airbnb get more revenue.
	• Uber (cab service platform):
	  ◦ Producers = cab drivers
	  ◦ Consumers = passengers
	If there are many cabs available, passengers get more options and better quality of service using the Uber digital software platform. At the same time, cab drivers and Uber get more income.
	Note: Airbnb does not own any homes, Uber does not own any cabs.
	Long-tail business: Most of the time, organizations invest in a minimal number of customer segments and product suite, which generates maximum revenue. This means organizations do not focus on niche areas that experience a problem of diminishing returns (proportionately fewer profits with high investments). In the platform model, with fewer cost, organizations can reach very niche segments (long tail), which they had never tried in the past due to non-viability. A long-tail approach with niche products/services to minimize diminishing returns is the key here.
	• Today, even the owner of a small house in a remote location can become an accommodation service provider using the Airbnb platform. Airbnb reaches into remote areas which hotels usually do not want to pursue.
	• In the past, people used to travel using very limited taxis. Uber has provided an opportunity for each car owner to offer cab service and helped passengers get a cab of their choice within a shorter time.

<table>
<tr><td></td><td>Asymmetric competition: By leveraging digital software platform, even a small organization can easily compete and leapfrog large-scale organizations.
• Airbnb was a start-up in 2008 with a platform-based business model. There was no client base or brand and now it is an unbeatable competition to big hotel chains.
• Uber was a start-up in 2009 with a platform-based business model and now it is an unbeatable competition to the taxi industry.
Asymmetric growth: Multiply the business growth with complementary businesses instead of restricting the organization to core business.
• Uber, with its cab platform, enrolled millions of passengers for its service. Leveraging this consumer base, it has also started a complimentary business called Uber Eats for service delivery, which has increased Uber's revenue. This is called asymmetric growth.</td></tr>
<tr><td>Marketplace-based business model</td><td>In the marketplace-based business model, producers and consumers do not communicate directly, and the platform owner takes the responsibility of taking the orders and delivering the goods (supply chain). In a platform-based business model (like Airbnb or Uber), producers and consumers have a direct connection, and the supply chain is owned by the producer and not the platform provider like Airbnb or Uber.

For example, Amazon is built on a marketplace-based business model in which consumers can purchase the product of their choice, but they will never communicate with the producer/supplier. Amazon takes the whole responsibility.</td></tr>
<tr><td>User-generated content-based business model</td><td>After the emergence of social media, millions of users started having real-time conversations and communication, generating a massive amount of content. Companies like Facebook have more than a billion followers who produce a humongous amount of content. When billions of users access Facebook to view or contribute to the content, Facebook publishes many advertisements and generates revenue streams. The Guardian promotes citizen journalism and gets text and video-based news items from citizens directly and then publishes the content after verification and editing for consumption. In this way, The Guardian gets more unique news items that was otherwise not possible with a limited number of journalists. This attracts more readers and hence the revenue goes up.</td></tr>
</table>

Data-based business model	One of the top eye research institutes and hospitals in Hyderabad, India (not for profit organization), collects very important data about the eye from the millions of patients who visit them. Now, this data is useful for building sophisticated machine learning algorithms to bring predictive and preventive eye care. This institute has been working on a business model in partnership with software majors to develop eye care machine learning algorithms. At present, they are working on data privacy and patient consensus procedures. Once they establish a secure process, they will use it for their eye care research.
No intermediation business model	Manufacturing organizations usually have thousands of dealers who sell their products to consumers. Now, a new business model has emerged without these intermediaries. For example, Tesla has been selling electric cars directly online without any dealers in the middle.
Product as a Service (PaaS) business model	In the past, manufacturing organizations used to sell their products to enterprises at a very high price. Due to which, enterprises incurred a lot of capital expenditure. Now, due technological innovations, the manufacturing firms have been offering their products as a service for which enterprises pay as they use. There will not be any capital expenditure, only operational expenditure. In this way, manufacturing firms can access a bigger market and reach a wider customer base too. For example, Medtronic has introduced a business model with which they offer **medical devices as a service**. Hospitals can use Medtronic's medical devices and pay as per usage without any capital expenditure. Similarly, General Electric has been offering **jet engines as a service** to airlines and collect payments as per the usage of the jet engines.

Source: Author

Table 10.2: Digital Business Model Innovations

Senior leaders of Horizon Bank, who attended Rao's session, found it to be an eye-opener for them, and they understood what new business models are emerging due to the cyber system. But still, they were not sure about how they could design a unique business model for Horizon Bank. Now the time had come for Rao to introduce **lean digital thinking** to prepare the minds of the senior leaders of Horizon Bank to think **digital**.

Applying Lean Digital Thinking to Digitalize Business Models

In the era of the Fourth Industrial Revolution, it is possible to establish brand new industries and new businesses which was not possible in the past. These new industries and new businesses require entirely new business models, which cannot be designed if the leaders do not understand the implications of digital technologies.

Hence, it is imperative to apply the principles of **lean digital thinking** to devise digital business models that create value for the digital age customers and new revenue streams for the organizations.

Rao decided to conduct an interactive session in order to explain how each principle of **lean digital thinking** changes the anatomy of business models. Table 10.3 lists how each principle of **lean digital thinking** can be applied to digitalize business models.

S. No.	Lean Digital Thinking Principles	Changing the Anatomy of Business Model
1	Create a better today and better tomorrow	Create digital business models without cannibalizing the existing traditional businesses. Bring equilibrium and evolve incrementally.
2	Embrace constant and instant change	Continuously analyze the data received on the trends and shifts in the market, and performance of the digital business model. Conduct root-cause analysis on the deficiencies and improve in an evolutionary manner.
3	Nurture digital business talent	Nurture digital business leadership skills on a regular basis to refine digital business models as per changing needs.
4	Identify and eliminate non-value-added activities	Identify **activities** which do not add value to the business and eliminate them by leveraging digital technologies. This will improve profitability, productivity and quality.

5	Predict, recommend and improve overall business performance	Predict the digital business model performance by using various simulation tools and implement actionable recommendations to get the best out of them.
6	Optimize human interventions	Identify **critical activities** of the business which can be digitalized so that human interventions can be optimized significantly.
7	Establish real-time and contextual communication and collaboration	Establish real-time communication and collaboration system to virtually involve employees, partners, suppliers and other stakeholders anywhere and anytime to innovate digital business models.
8	Safeguard privacy, safety and security of business resources	Identify **digital solutions/product features** that safeguard physical resources, digital resources, IP, human resources and other resources from cyberattacks.
9	Establish a technology start-up ecosystem	Devise lean business models to drive business initiatives as technology start-ups.

10	Create socially responsible products, services and value chains	Identify **digital solutions/product features** that will enhance the environment and society at large.
11	Design and deliver unforgettable moments of experience to stakeholders	Identify **key interaction points** in the customer journey to deliver a superior experience at those interaction points by leveraging digital technologies. This improves customer loyalty and generates better revenue.
12	Build a human-centric digital culture	Establish a digital culture and set of values that promote continuous digital business model innovation.

Source: Author

Table 10.3: Applying Lean Digital Thinking Principles to Digitalize Business Models

Michael and the team, who are responsible for creating business models for Horizon Bank, requested Rao to share any templates available to design a digital business model for Horizon Bank. They were confident that they could now devise new business models with fresh thinking—**lean digital thinking**.

Digital Business Model Template

Rao distributed a template for a digital business model designed based on **lean digital thinking,** which had five essential elements, as mentioned in Table 10.4. Rao explained each element in detail and provided a detailed orientation on the template.

Senior leaders who attended the workshop asked many questions to get clarification. Rao gave all the required inputs to prepare the team to brainstorm on designing a unique digital business model for Horizon Bank.

S. No.	Digital Business Model Element	Description
1	Problem definition	Define the problem based on unmet and unarticulated needs of the digital age customer segments.
2	Digital value design	Innovate digital features to offer a unique value proposition to the identified customer segments.
3	Digital value creation	Establish new partnerships and new resources to acquire digital capabilities, digitalize core activities and introduce new cost structure to deliver digital value.
4	Real-time customer connect	Emotionally connect with the customers and maintain a personalized relationship. Introduce different digital channels to reach the customers as per their preferences.
5	Wealth creation	Identify new revenue streams from digital value delivery and provide flexible and omnipresent payment options to customers.

Source: Author

Table 10.4: Elements of a Digital Business Model

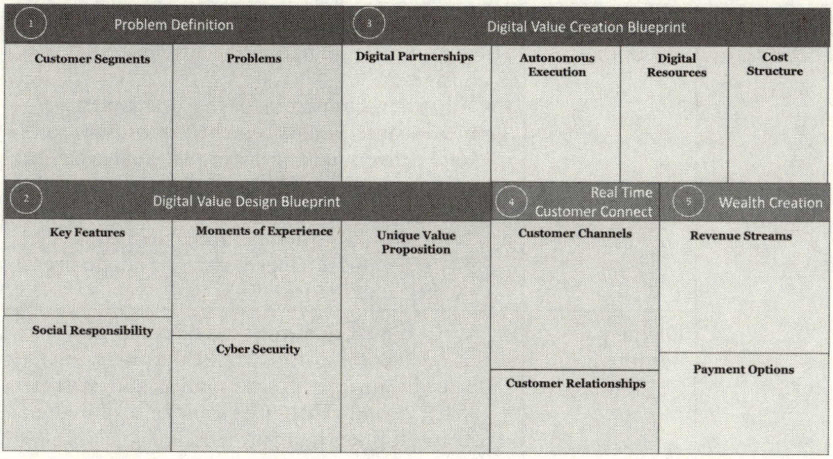

1 Problem Definition
Customer Segments
Problems
3 Digital Value Creation Blueprint
Digital Partnerships
Autonomous Execution
Digital Resources
Cost Structure
2 Digital Value Design Blueprint
Key Features
Social Responsibility
Moments of Experience
Cyber Security
Unique Value Proposition
4 Real Time Customer Connect
Customer Channels
Customer Relationships
5 Wealth Creation
Revenue Streams
Payment Options

Source: Author

Figure 10.7: Digital Business Model Template

Element	Sub-Element	Description
Problem definition	Customer segments	• Which customer segments do you want to deliver the value proposition to? • Which customer segments in the long tail do you want to address by leveraging digital capabilities? • Which intelligent things in your business could become customers? Things, like cars, refrigerators and machines, made intelligent with digital technologies could perform a few activities performed by humans. For example, when eggs are over, a smart refrigerator can place an order automatically to get the eggs. This means today intelligent things are digital customers. • What are the characteristics of the digital age customers you want to deal with?
	Problems	• What are the burning problems customers have been facing in the digital age? • What is the list of prioritized unmet and unarticulated needs of the customer segments to be addressed?

Digital value delivery	Autonomous execution	• What critical activities must be digitalized to minimize human interventions in the business? • What new core activities (e.g., data management, content management, continuous software delivery management, API management) must be added to integrate digital businesses with the business ecosystem? • What should be done to integrate people, processes and intelligent things to digitalize the key activities?
	Digital partnerships	• What digital resources (elastic computing power, elastic storage, elastic networks, etc.) must be acquired from suppliers and partners? • What digital capabilities must be acquired from suppliers and partners? • What key interfaces must be built with partners and suppliers to build the digital business ecosystem? • What type of digital start-ups must be acquired to enhance digital capabilities?
	Key resources	• What digital resources (electricity, internet, computing power, storage, networks, etc.) are required to deliver value to customers? • Which physical assets (buildings, machines, vehicles, equipment, etc.) must be digitalized to deliver value • What digital workforce (bots, robots, smart contracts, etc.) must be inducted to optimize the human workforce? • What new roles (e.g. chief digital officer, chief smart contracts officer) must be introduced to drive digital business? • What new digital financial instruments (crowdfunding, peer-to-peer lending, etc.) are available to drive the business? • What key machine algorithms and clever code (smart contacts) must be built to optimize human interventions?
	Cost structure	• What are the most important costs inherent in your digital business model? • What is the cost of acquiring key resources to drive your digital business? • What is the cost of digitalizing and managing core activities? • What is the cost of establishing new partnerships or acquiring digital start-ups to support digital business?

Digital value design	Key features	• What product and service features will you introduce leveraging digital technologies to provide multiplied value to customers?
	Social responsibility	• What product and service features you will introduce by leveraging digital technologies to reduce the impact on environment, health and society?
	Moments of experience	• Which customer persona you want to address to provide a better experience? There are customers with different persona that include habits, personality type and interests. Personalizing services by aligning with the persona of customers would satisfy the customer more. • Which customer interaction points are very critical for unforgettable moments of experience?
	Cybersecurity	• What product and service features will you introduce to safeguard from cybersecurity attacks and privacy issues?
	Unique value proposition	• What benefits do customers expect from your digital products and digital services? • Which needs and wants of customers your digital products and digital services will address? • What negative experiences, emotions and risks do you want to address to relieve the pains of your customers?
Real-time customer connect	Customer channels	• What channels—like social media, phones, web, wearable devices, chatbots, virtual expos and others—do digital age customers prefer to be reached by? • How are various channels like social media, phones, web, wearable devices, chatbots and virtual expos integrated to provide a seamless experience to customers? • Which channels are very cost-effective?
	Customer relationships	• What are the expectations of digital age customers? • What type of personalized relationship do you maintain with each customer segment (Baby boomers, GenX, Gen Y and Gen Z)? • While acquiring customers, how will you improve their comfort and convenience using digital technologies and intelligent devices? • How do you address customer requests and complaints in real time using multiple channels?

Wealth creation	Revenue streams	• What new revenue streams do you want to create with the digital offerings, packages and assets (e.g., data)? • What new pricing models, made possible by digital technologies, do you want to introduce?
	Payment options	• How do your customers prefer to pay (payment channels/options)?

Source: Author

Table 10.5: Detailed Description of Digital Business Model Elements

It was too much information overload for the senior leaders who participated in the workshop conducted by Rao. Michael requested Rao to give them two weeks to digest, understand and analyze **lean digital thinking** and digital business model. He also promised Rao that with the help of the core team who attended the workshop, they will come up with innovative business models before they meet again.

Digital Banking Business Model for Horizon Bank

On 23 June 2015, a week after the workshop, Michael arranged a brainstorming session with all the participants to start working on their digital business model. After some detailed groundwork with the team, Michael discussed a few sample questions to devise the first digital business model for Horizon Bank:

1. How to generate new revenue streams from existing customers?
2. Are there any new channels which can be used to reach new customers, who have never been approached before?
3. Are there any opportunities to perform non-banking activities that enhance banking revenue?
4. Can we establish a partnership with car dealers, real-estate developers and integrate their products and services with our

loan management system? In this way, our bank customers get access to purchase vehicles or houses of their choice. Once our customer chose a product, our bank could provide the loan. This means by performing a non-banking activity, we want to improve our loan business.

The strategy office of Horizon Bank formed three different groups with a mix of the following skills:

1. Loan business expert
2. Customer relations expert
3. Digital technology specialist
4. Business strategy specialist

Michael's team distributed the digital business model template provided by Rao and conducted a recap session for all three groups. He advised the core group members to conceptualize a digital platform-based business model from their brainstorming session and provided a few more inputs, as mentioned below:

1. Banks have been performing non-banking transactions to improve their loan business. Banks have millions of customers. With the digital software platform, Horizon Bank will be able to enrol real estate developers with their project details.
2. The existing customers can easily interact with real estate developers directly. By using matchmaking algorithms, they can find their choice of home.
3. In this case, banks act as an aggregator. Once the transaction between real estate developers and the customer is completed, banks provide the loan, which is the goal of this new business model.
4. This meant banks were going beyond the regular transactions and facilitating non-banking transactions (home purchase).

The three groups presented their version of business models from the brainstorming sessions. Michael's team took all their inputs and after a two-week-long detailed analysis, they released their first digital business model as given in Figure 10.8.

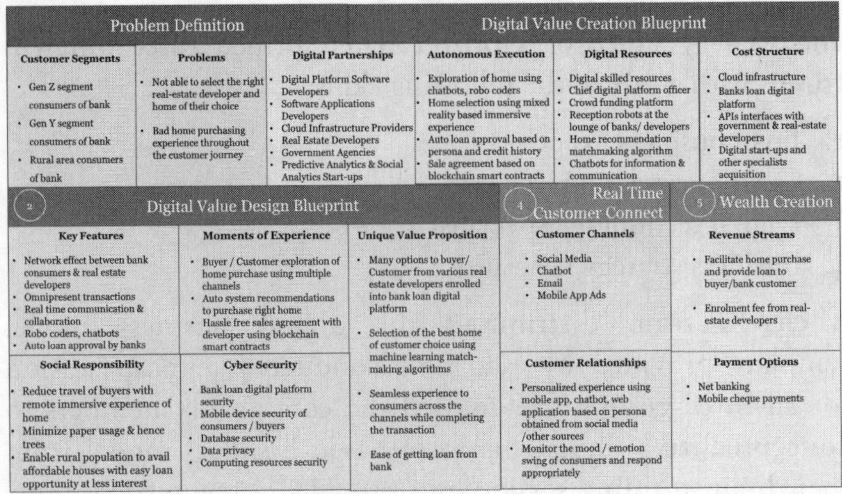

Source: Author

Figure 10.8: Digital Business Model for Horizon Bank

Michael was confident that they could now design a few more digital business models with the experience gained in the last two weeks from Rao's workshop. He was convinced that **lean digital thinking** will become a valuable game changer for Horizon Bank, making it relevant in the digital era.

Creating digital business models is not a one-time activity but a continuous task and must be done iteratively. It is also evident that business executives must have the technical acumen to conceptualize and understand the digital age business models. If business executives do not acquire technical knowledge, they will fail to introduce digital business innovations. At the same time, the CIO has to play an advisory and consultant role to guide the business executives appropriately.

Building Block 3: Business Strategy

Smart Inn hotel chain has been in business for the last 40 years and is one of the top hotel chains in the world. On 3 November 2014, Bryan, the CEO of Smart Inn, came to office and preoccupied with the discussion he had had with the board members and the frustration expressed by the board.

"Bryan, what's happening? We don't see any digital strategy for our organization, and we are losing our ground to new age competition. If you are not able to create a digital business strategy, we may be soon out of business. We have not yet gained confidence in our abilities to prepare for a digital future. We have already delayed creating a digital business strategy. We want to see a solid digital business strategy soon without further delay."

Bryan had an exceptional leadership team but the problem was that they were proficient only in driving typical hotel businesses. They did not have any experience and knowledge of leveraging digital technologies to meet the needs of digital age customers. The occupancy rate was going down and the number of guests' complaints was going up while the new competition was offering better services. Bryan's stress levels were also going up! The good news was Bryan was a vivid reader and knew how technologies had been impacting the hotel business.

Bryan arranged a meeting with his senior leaders on 5 November 2014. He shared his vision on how he wanted to see Smart Inn business in the next 3–4 years.

1. Smart Inn would no longer be a hotel chain but a technology company that offers hotel services.
2. We will not lose business to platform providers like Airbnb.
3. We will make at least 40% of our business operations autonomous (with low human interventions) which would include material supply chain, hotel operations, restaurants, room service and others.

4. We will allow the guests to get an immersive experience of the hotel and rooms using mixed reality technology.
5. We will not have any reception desk; guests will directly go to the room and open it with NFC-enabled phones. There will be a kiosk for help.
6. We will introduce a Smart Inn mobile app that would enable the guests to set the room temperature, music and lighting from anywhere so that when they lie down on the bed, the room is set as per their preferences.
7. A mobile tablet within the hotel room will make room service very simple by enabling guests to order their food using a chatbot.

Bryan was very keen to change the mindset of his team from traditional thinking to digital thinking so that they could drive the business like digital natives.

David, the CSO, had been working for the last year to collate information to devise a digital business strategy. He had realized the fact that without technology specialists, he will not be able to devise a digital business strategy. David cautioned Bryan that his vision of future Smart Inn was fantastic, but there were many resource constraints and capability gaps to be addressed.

Bryan took a deep breath and told David that they had lost enough time and instructed him to immediately acquire a tech start-up, which had been working in the hospitality industry. He emphasized that since they already had excellent subject matter experience and domain specialists, with the tech start-up, they could bring innovations that would help them provide unforgettable moments of experience to their guests.

Bryan expressed his unhappiness with CIO Catherine's lack of passion and enthusiasm in preparing for a digital business battle. He asked Catherine to be ready for more significant challenges in the future and cautioned her about her focus of limiting work to the data centre and software applications. She needed to develop her technology department as a digital innovator and strategic advisor. He emphasized that the current performance indicators like infrastructure availability, uptime and lead time to develop software applications must be changed as they do not contribute to the digitalization of the business.

He also instructed the head of Learning Academy, John, to immediately conduct workshops for all senior leadership on fundamentals of the digital economy, Fourth Industrial Revolution, digital technologies and digital hospitality.

After the meeting, Bryan had a new spark in his eyes; he now had a broader picture of what he was going to do for Smart Inn in the next 18 months.

CSO David, along with his core team, identified three start-ups and selected one start-up called Experience Design and completed the M&A (mergers and acquisitions) formalities.

On 3 April 2015, Smart Inn kicked off its digital hospitality programme. The first step of this programme was to form the core team with a mix of domain specialists, technology specialists, anthropologists, psychologists, customer executives and operations heads to create a digital business strategy. CSO David, along with the CEO of Experience Design, Peter, was set to drive this programme.

Bringing a Common Understanding of Business Strategy

David and Peter conducted a workshop on 2 May 2015. Everybody agreed that Smart Inn was a digital immigrant and competing with digital natives was a very complex task. This workshop helped the core team arrive at a common understanding of digital business strategy as mentioned below.

1. Baseline the **current state** of a Smart Inn in multiple dimensions.
2. Understand Smart Inn's digital vision, digital business model, digital strengths, digital weaknesses, opportunities and threats in correlation to the Fourth Industrial Revolution.

3. Design the **future state** of Smart Inn to be successful in the Fourth Industrial Revolution (**digital state**).
4. Quantify the future digital state and measure the pace and progress of various initiatives.

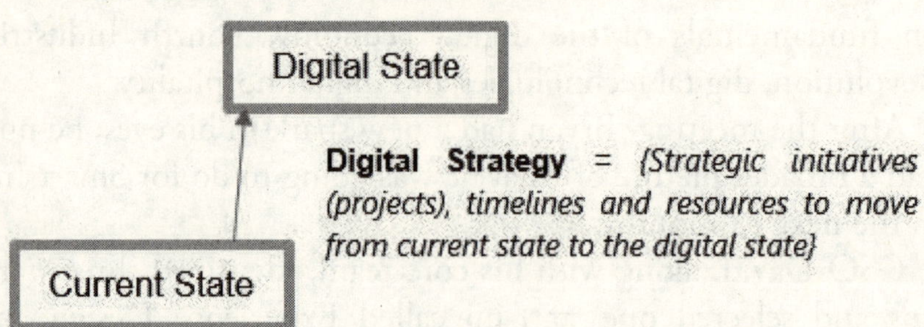

Source: Author

Figure 10.9: Current Business State to Digital Business State

David and Peter conducted various workshops and were able to bring about a shared understanding of their digital business strategy. Still, they did not have any clue about how to find the current state of the organization. Peter advised David to involve Rao, a lean digital specialist who had developed the lean digital quotient (LDQ) finder. LDQ finder is a digital business excellence assessment framework that helps in identifying digital strengths and weaknesses of the business building blocks of an organization.

On 4 May 2015, David and Peter met with Rao in a coffee shop and explained their digital vision and objectives. Rao assured them of giving a full digital-strengths-and-weaknesses report in 12 weeks, covering different business pillars of Smart Inn.

David invited Rao to arrange an orientation session with the Smart Inn core team on **lean digital thinking** and how it helps in designing a digital business strategy. On 3 June 2015, Rao clearly explained to them each principle of **lean digital thinking** in the context of the digital business strategy. One of the core members asked Rao, why were they using **lean digital thinking** now as they had been creating business strategies for many decades without it?

Rao responded with a smile that what they had been building was conventional business strategy, now they must create a digital business strategy, which requires different thinking for which **lean digital thinking** provides an excellent thinking framework. Rao explained to them how to apply **lean digital thinking** to business strategy.

Applying Lean Digital Thinking to Digitalize Business Strategy

S. No.	Lean Digital Thinking Principles	Changing the Anatomy of Business Strategy to Make it Digital
1	Create a better today and better tomorrow	Create a digital business strategy without cannibalizing existing traditional businesses. Bring equilibrium and evolve incrementally.
2	Embrace constant and instant change	Continuously analyze the market trends, shifts and the data received about the performance of the digital business strategy. Conduct root-cause analysis on the deficiencies and improve evolutionarily.
3	Nurture digital business talent	Nurture digital business leadership skills regularly to refine the digital business strategy as per the changing needs.
4	Identify and eliminate non-value-added activities	Identify improvement actions that help in eliminating non-value-added digital initiatives in the business.

5	Predict, recommend and improve overall business performance	Predict the digital business strategy performance by using various simulation tools and implement actionable recommendations to get the best out of the digital business strategies.
6	Optimize human interventions	Identify improvement actions that make the assets, machines, equipment and services, processes intelligent.
7	Establish real-time and contextual communication and collaboration	Establish a real-time communication and collaboration system to virtually involve employees, partners, suppliers and other stakeholders, anywhere and anytime, to innovate digital business strategies.
8	Safeguard privacy, safety and security of business resources	Identify improvement actions that safeguard intelligent assets, equipment, websites, mobile apps, data, servers and others from cyberattacks.
9	Establish a technology start-up ecosystem	Build an entrepreneurial mindset to devise unique business strategies.
10	Create socially responsible products, services and value chains	Identify improvement actions that make sure that the guest services are offered, and business processes are built with a low impact on the environment and society.

11	Design and deliver unforgettable moments of experience to stakeholders	Identify improvement actions that provide unforgettable moments of experience to guests throughout their journey using augmented reality, virtual reality, omnichannel, mobile and social media channels.
12	Build a human-centric digital culture	Establish a digital culture and set of values that promote continuous digital business strategy innovation.

Source: Author

Table 10.6: Applying Lean Digital Thinking Principles to Digitalize Business Strategy

Understanding the Current State of Business

Using the LDQ finder, Rao and the team carried out an extensive assessment of the 12 business building blocks of Smart Inn (e.g., business vision, business model, business strategy, services, processes, workplace and others).

On 10 August 2015, Rao and the team delivered a digital strength and weakness report of Smart Inn using their LDQ finder tool, which reflected the current state of Smart Inn.

Rao emphasized the fact that the tool considers both internal and external environmental factors and is comprehensive enough to understand the current state of Smart Inn. The tool gives both the digital strengths and weaknesses of Smart Inn. Here are a few illustrative examples of digital weaknesses that Rao's tool managed to find.

Digital Weaknesses of Smart Inn's Business Strategy

- In Smart Inn, little or no emphasis is laid on autonomous businesses using digitally intelligent things (products, equipment, devices, machines, etc.).
- Smart Inn does not have the vision to provide unforgettable moments of experience to customers so that they become lifelong customers.
- Smart Inn does not review current business models frequently to bring necessary changes/improvements to be successful in the digital era.
- Smart Inn does not have a long-term plan to design socially responsible services leveraging digital technologies.
- Smart Inn does not collect information about the intentions of the visitors/competitors visiting their websites/channels by using cybersecurity tools and techniques.
- Smart Inn does not optimize the usage of physical workplaces leveraging mobile, analytics, social media, IoT and other technologies.
- In Smart Inn, business processes are not digitalized to make it nearly autonomous.
- Smart Inn does not have any exploratory and experimental technology environment to translate ideas into proof of concept.

Deriving Digital Business Strategic Themes

On 15 September 2015, David and Peter called for a meeting with the Smart Inn core team. The attendance was too low. Peter was disappointed with the response. David smiled and shared with Peter a quote by Mahatma Gandhi—"First, they ignore you, then they laugh at you, then they fight you, then you win."

David and Peter spent considerable time meeting with the core team members one-on-one and created much interest in their minds. They also met with the senior leadership team of the respective units to motivate and inspire their executives in the core team to participate in the digital business strategy initiative actively.

On 29 September 2015, David and Peter got the approval from the steering committee, headed by CEO Bryan, on the digital weaknesses report shared by Rao. Collaboratively, the core team brainstormed and identified the areas that needed to be improved to overcome the digital weaknesses. These improvement areas are called strategic themes, which are derived from both digital strengths and weaknesses, and Table 10.7 lists some themes based on the digital weaknesses mentioned earlier. The strategic themes defined here are only representative, not comprehensive.

S. No.	Prioritized Digital Weaknesses	Strategic Themes
1	In Smart Inn, little or no emphasis is laid on autonomous businesses using digitally intelligent things (products, equipment, devices, machines, etc.).	Build autonomous business
2	Smart Inn does not have the vision to provide unforgettable moments of experience to customers so that they become lifelong customers.	Provide unforgettable moments of experience

3	Smart Inn does not have a long-term plan to design socially responsible services leveraging digital technologies.	Deliver socially responsible services
4	Smart Inn does not collect information about the intentions of the visitors/competitors visiting their websites/channels by using cybersecurity tools and techniques.	Secure business resources
5	Smart Inn does not optimize the usage of physical workplaces leveraging mobile, analytics, social media, IoT and other technologies.	Build autonomous business
6	In Smart Inn, business processes are not digitalized to make it nearly autonomous.	Build autonomous business

Source: Author

Table 10.7: Deriving Digital Business Strategic Themes

In summary, the four digital business strategic themes that can help Smart Inn achieve its digital dream state are given in Figure 10.10. These are the pillars of excellence that will ensure that Smart Inn will reach its dream digital state.

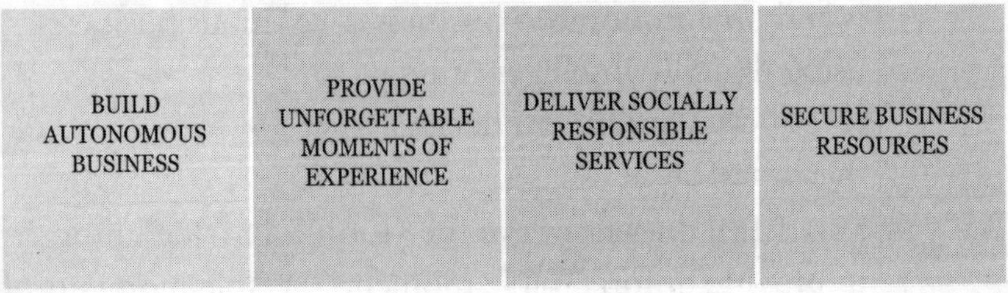

BUILD AUTONOMOUS BUSINESS	PROVIDE UNFORGETTABLE MOMENTS OF EXPERIENCE	DELIVER SOCIALLY RESPONSIBLE SERVICES	SECURE BUSINESS RESOURCES

Source: Author

Figure 10.10: Digital Business Strategic Themes

David and Peter met with Bryan, updated him about the progress and shared a clear roadmap for the next 90 days. Bryan asked whether they needed any additional support. David requested Bryan to hire an anthropologist and a psychologist to carry out a detailed analysis of the persona of different segments of guests. Bryan approved the request immediately. Based on the digital business strategic themes identified, David and Peter introduced a digital hospitality programme to prepare Smart Inn for the digital era.

Preparing the Core Team

Bryan arranged a meeting on 15 October 2015 with the core team. He congratulated the team for finalizing the digital strategic themes, which was the basis for creating an excellent digital business strategy for Smart Inn.

Surprisingly, the general managers raised a concern with Bryan that they have been spending a lot of time in these meetings and this was disturbing their day-to-day guest service operations. Bryan asked them how much time they had been spending on the digital hospitality initiative. They said that it took up almost 15 hours per week. Bryan immediately approved recruitments within different departments to take care of this additional load. Bryan spent quality time separately with David and Peter and advised them to focus on completing the digital business strategy map immediately.

On 31 October 2015, Rao conducted a workshop with the core team on how to create a great digital business strategy for Smart Inn. See Figures 10.11 and 10.12 for two of the slides Rao used in the workshop.

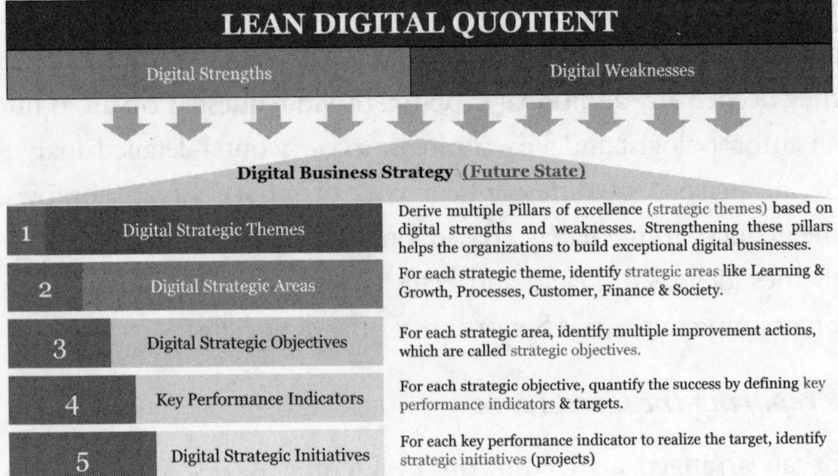

Source: Author

Figure 10.11: High-Level Framework for Digital Business Strategy Map

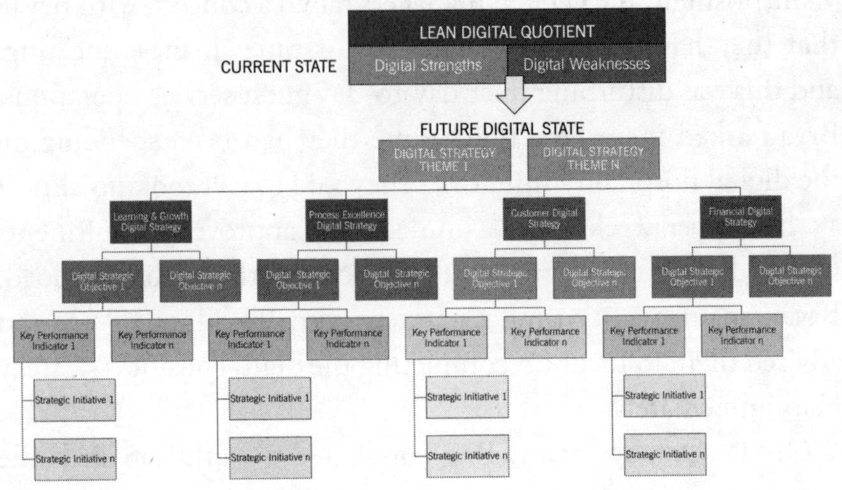

Source: Author

Figure 10.12: Detailed Framework for Digital Business Strategy Map

Identifying Digital Business Strategic Objectives

The next big challenge for David and Peter was to identify the improvement actions against each strategic theme. They distributed the following template to each core team member and asked them to submit their inputs by 25 November 2015.

Digital Business Strategy Objectives Framework				
Strategic Themes	Learning and Growth	Process Excellence	Customer Delight	Financial Excellence
Build autonomous business	Improvement Action 1 Improvement Action n	Improvement Action 1 Improvement Action n	Improvement Action 1 Improvement Action n	Improvement Action 1 Improvement Action n
Provide unforgettable moments of experience	Improvement Action 1 Improvement Action n	Improvement Action 1 Improvement Action n	Improvement Action 1 Improvement Action n	Improvement Action 1 Improvement Action n
Deliver socially responsible services	Improvement Action 1 Improvement Action n	Improvement Action 1 Improvement Action n	Improvement Action 1 Improvement Action n	Improvement Action 1 Improvement Action n
Secure business resources	Improvement Action 1 Improvement Action n	Improvement Action 1 Improvement Action n	Improvement Action 1 Improvement Action n	Improvement Action 1 Improvement Action n

Source: Author

Table 10.8: Digital Business Strategy Objectives Framework Template

David and Peter got an overwhelming response from the core team. Many improvement actions were proposed by the core team against each strategic theme. On 4 January 2016, David and Peter arranged a workshop with the team and filtered the improvement actions as listed in Table 10.9. They baselined these improvement actions as **strategic objectives.**

David and Peter appreciated the efforts of the core team and their passionate involvement in identifying 26 strategic objectives, which was a great achievement.

S. No.	Strategic Theme	Learning and Growth Strategic Objectives	Process Excellence Strategic Objectives	Customer Delight Strategic Objectives	Finance Excellence Strategic Objectives
1	Build autonomous business	• Introduce new digital business roles • Institutionalize digital culture and policies • Implement digital systems	• Implement human-machine interaction processes • Implement data management processes • Implement lean processes	• Offer personalized services • Reach long-tail customers	• Improve profits • Reduce operational costs • Enable asymmetric growth
2	Deliver socially responsible services	• Improve lean and agile capabilities	• Build environment-friendly processes	• Offer socially responsible services	• Create positive social impact
3	Provide unforgettable moments of experience	• Improve experience design capabilities	• Build experience management processes	• Deliver personalized services • Deliver unforgettable moments of experience • Offer faster, better and affordable services	• Increase revenue from repeat business • Increase revenue and impact of new digital offerings
4	Secure business resources	• Improve cybersecurity capabilities	• Build cybersecurity management processes	• Safeguard the safety, security and privacy of guests	• Minimize revenue loss due to security breaches

Source: Author

Table 10.9: Identification of Sample Digital Business Strategic Objectives

Creating a Digital Business Strategy Map

David and Peter used a balanced scorecard framework and created a digital business strategy map. According to it:

1. **Financial performance** will improve only upon delivering high-value proposition, personalized services and unforgettable moments of experience to customers.
2. **Customers get high-value proposition**, personalized services and unforgettable moments of experience only when the processes are improved and digitalized.
3. **Internal processes** can be improved or digitalized only upon improving internal skills, policies and systems.

The balanced scorecard framework creates a significant cascading effect across multiple pillars of the balanced scorecard. If the learning and growth perspective is strengthened, then the organization can execute the business processes exceptionally well. With the process of excellence, an organization can deliver excellent customer services. With exceptional customer services, organization get more customers and revenue.

On 2 February 2016, the core team reviewed the digital business strategy map given in Figure 10.13 and approved the same.

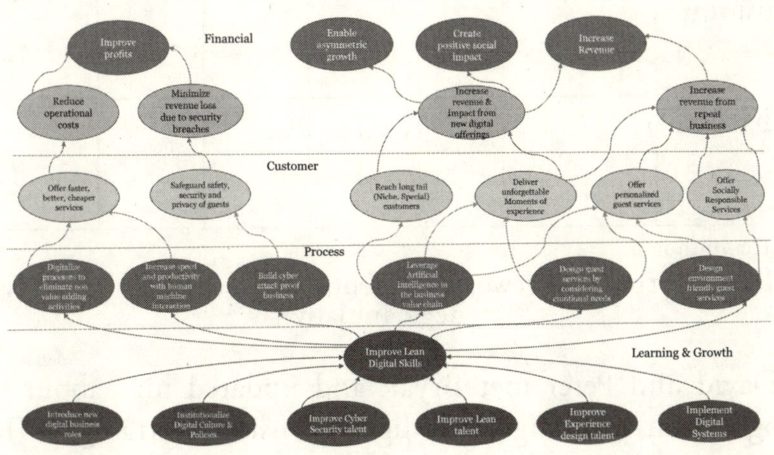

Source: Author

Figure 10.13: Sample Digital Business Strategy Map

David and Peter conducted an orientation session on how to identify digital business strategic initiatives from the digital business strategy map. The core team actively participated in the meeting and geared up to identify game-changing strategic initiatives, which would help Smart Inn prepare for its digital journey. The template given in Table 10.10 was used to orient the core team towards the process of identifying strategic initiatives. **Strategic objectives, key performance indicators (KPIs) and targets reflect the future state of the organization.**

Strategic Areas	Strategic Objectives	KPIs	FY'16 Target	FY'17 Target	FY'18 Target	Strategic Initiatives to Realize Targets
Learning and growth						
Internal processes						
Customer						
Financial						

Source: Author

Table 10.10: Framework for Identification of Digital Business Strategic Initiatives

David and Peter met Bryan and updated him about the progress in creating the digital business strategy. They expressed their concern that considerable budget and resources will be required to implement the digital business strategy. They promised Bryan that within 45 days, they would release the portfolio of the strategic initiatives, priority and budget needed to drive the digital hospitality transformation programme.

After David and Peter left, Bryan was deep in his thoughts, concerned about the significant budget required to drive the digital hospitality programme. He knew that the budget available was minimal but the work ahead was humongous.

Identifying Digital Business Strategic Initiatives

On 20 February 2016, David arranged a workshop with the core team to brainstorm and identify the strategic initiatives based on the template designed. Peter guided the core team members to form groups based on the skill mix of the core team members. They successfully collated a significant number of strategic initiatives mentioned in Table 10.11 with the help of the core team. They were pretty much satisfied with the outcome of the workshop and were confident about kick-starting the digital hospitality programme very soon.

Strategic Area	Digital Business Strategic Objectives	KPIs	FY'17 Targets*	Digital Business Strategic Initiatives
Learning and growth	**LO1:** Introduce new digital business roles	No. of new roles introduced to drive digital business	4	Design and implement digital business organization structure
	LO2: Institutionalize digital culture and policies	No. of new digital policies introduced	3	Design and implement lean digital culture, attitude and behaviour within the organization
		No. of new digital values and principles introduced	4	
	LO3: Implement digital systems	Percentage of the systems of record, the systems of engagement and the systems of things built to support digital businesses	Systems of records: 70% Systems of engagement: 20% Systems of things: 10%	Build digital technology ecosystem
	LO4: Improve lean and agile capabilities	No. of lean, agile methods and specialists inducted	5	Implement lean, agile and digital methods
	LO5: Improve experience design capabilities	No. of guest experience design methods and specialists inducted	3	Establish a lean digital academy
	LO6: Improve cybersecurity capabilities	No. of cybersecurity methods and specialists inducted	3	

Process	**PO1:** Increase value, speed and productivity with human-machine interaction	No. of business processes automated by integrating IoT	5	Establish lean digital business process centre of excellence
		No. of business processes automated using chatbots and virtual assistants	2	
		No. of business processes automated using blockchain smart contracts	1	
	PO2: Leverage data and AI in the business value chain	Percentage of unstructured and semi-structured data used for business intelligence and predictability	80%	Build "systems of intelligence" ecosystem
	PO3: Digitalize processes to eliminate non-value-added activities	Percentage of reduction in the cost due to the elimination of non-value-added activities	8%	Establish a lean centre of excellence
	PO4: Design environment-friendly guest services	No. of guest services designed by factoring environmental impact	2	Establish digital services innovation centre of excellence
	PO5: Design guest services by considering emotional needs	No. of guest services implemented by considering the emotional needs of the guests throughout their journey	3	
	PO6: Build cyberattack proof business	No. of cybersecurity processes implemented	2	Implement cybersecurity policies, governance and technologies
		Percentage of cybersecurity attacks prevented	90%	

Customer	**CO1:** Offer personalized guest services	No. of guest services introduced based on guest category (Gen Z, Gen Y, Gen X, Baby boomers)	2	Establish digital services innovation centre of excellence
	CO2: Reach long-tail (niche) customers	Percentage of long-tail guests reached due to digital services	20%	Implement digital platforms to leapfrog Airbnb and similar competitors
	CO3: Offer socially responsible services	No. of environment-friendly guest services introduced	1	Establish digital services innovation centre of excellence
	CO4: Deliver unforgettable moments of experience	No. of guest services that meets emotional needs throughout the journey	2	
	CO5: Offer faster, better, affordable services	Percentage of price reduction in guest services due to digitalization	5%	
	CO6: Safeguard safety, security and privacy of guests	Percentage of security and privacy incidents/breaches addressed	100%	Implement cybersecurity policies, governance, and technologies

Finance	**FO1:** Improve profits	Percentage of increase in profit	5%	Establish digital channels to have low touch sales and marketing
	FO2: Reduce operational costs	Percentage of cost reduction in acquiring a new guest	10%	
	FO3: Enable asymmetric growth	No. of guest services that enable asymmetric growth	2	Create complementary businesses for asymmetric growth
	FO4: Create positive social impact	Sustainability index of the organization (1–7) 1: Poor 7: Excellent	4	Develop processes, policies and software application to collate the data and derive sustainability index
	FO5: Increase revenue from repeat business	Percentage of revenue from repeat business	80%	Introduce innovative pricing models for repeat guests
	FO6: Increase revenue and impact from new digital offerings	Percentage of revenue from new digital guest services	20%	Implement a new budget system for digital guest services
	FO7: Minimize revenue loss due to security breaches	Percentage of revenue loss due to cyberattacks	0%	Implement cybersecurity policies, governance and technologies

Source: Author

Note: * It is easy to set the targets but identifying data sources, collating data, filling the data gaps and building systems to mine the data are the challenges to be addressed by the organizations.

Table 10.11: Digital Business Strategic Initiatives for Smart Inn

Finalizing Digital Business Strategies

On 2 March 2016, David and Peter presented the list of digital business strategic initiatives to Bryan to get his approval for creating a charter and budget for each strategic initiative with priorities. Bryan reviewed the list of initiatives and expressed his happiness about the team effort and his confidence that they were now clear on what is to be done to kick-start the digital journey of Smart Inn. Bryan instructed David and Peter to identify the priority strategies immediately.

Digital Processes and Services Innovation	Digital Methodologies and Talent	Sales and Growth	Organization Structure and Digital Culture	Digital Technology Ecosystem
• Establish a digital services innovation centre of excellence • Establish a lean centre of excellence • Establish a lean digital business process centre of excellence	• Establish a lean digital academy • Implement cybersecurity policies, governance and technologies • Implement lean, agile and digital methods	• Create complementary businesses for asymmetric growth • Establish digital channels to have low touch sales and marketing • Implement a new budget system for digital guest services • Introduce innovative pricing models for repeat guests	• Design and implement digital business organization structure • Design and implement lean digital culture, attitude and behaviour within the organization	• Build "systems of intelligence" ecosystem • Build a digital technology ecosystem • Implement cybersecurity policies, governance and technologies • Implement digital platforms to leapfrog Airbnb and similar competitors

Source: Author

Table 10.12: Final List of Digital Business Strategies for Smart Inn

In this section, the process of identification of digital business strategies or strategic initiatives was discussed in a detailed manner. The dates mentioned for each meeting and workshop are indicative.

The strategic initiatives identified in this chapter are based on a few indicative digital weaknesses. They are not based on the actual digital strengths and weakness report as produced by each organization. Hence, the list of strategic initiatives identified here is in no way comprehensive, it is only indicative.

Each organization is expected to define its business-specific strategic themes (pillars of excellence, strategic objectives, KPIs, digital business strategies, etc.). By using the LDQ finder, organizations can obtain the list of digital strengths and weaknesses, which forms an excellent basis to derive the right digital business strategies. Creating digital business strategies is not a one-time activity but a continuous task and that must be done iteratively.

Building Block 4: Business Outcomes

Dream Bank CEO Jeffrey shared his candid view on current key performance indicators (KPIs) at the monthly senior leadership meeting. The KPIs failed to provide the current picture of their digital business outcomes. They had invested considerably in building new digital systems to meet the needs and wants of their digital age customers. They had also started seeing some good business outcomes but the current KPIs were not reflecting the digital business outcomes.

He requested CSO Martin to design a sophisticated business outcome management system with the right and relevant performance indicators. Martin formed a small team representing various business units and technology units. He remembered the way their organization had used **lean digital thinking** principles to build its digital business ecosystem. He wanted to apply the same principles for designing the **business outcome management system.**

Achieve business outcomes by performing various tasks under each business building block of an organization.

Martin further added that any organization aims to achieve business outcomes in two dimensions:

1. **Business outcomes to prepare for the digital future**: How are we making our business building blocks to prepare for a digital future so that our organization becomes relevant and leapfrogs the new competition?
2. **Business outcomes to excel today:** How our digital business building blocks are performing today to gain profits, revenue, growth and customer happiness?

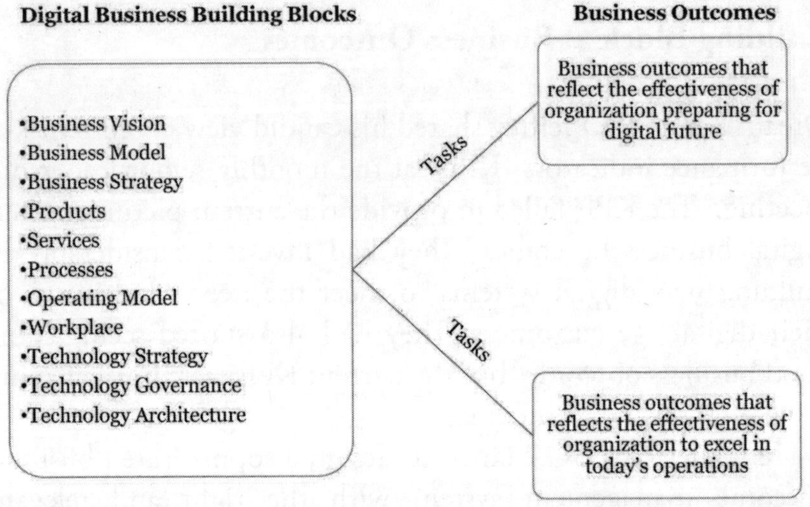

Source: Author

Figure 10.14: Business Building Blocks and Digital Business Outcomes

Business Outcome Management

Rashmi, one of the members of the core team, asked Martin how to effectively manage digital business outcomes. She also expressed her confusion about defining the right KPIs for both strategic and operational business outcomes. Digital business is different from traditional business and indeed the KPIs of traditional business must be refined. A few new KPIs must be added to drive digital businesses.

Martin shared a framework to manage the business outcomes of digital businesses effectively.

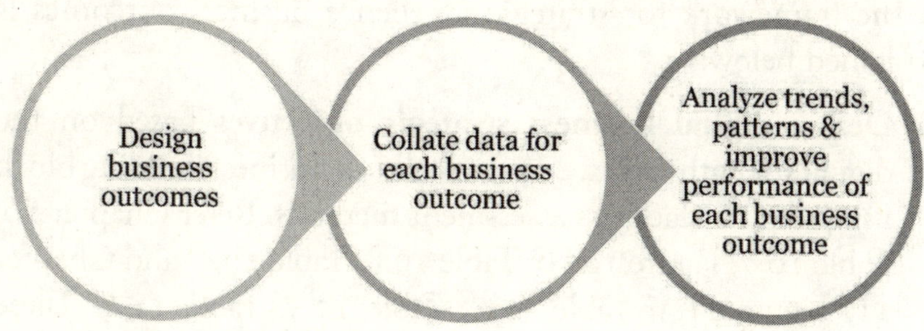

Source: Author

Figure 10.15: Business Outcome Management Framework

Designing Business Outcomes

The business outcomes are designed under two categories to prepare for a better future and excel in today's business:

1. Design business outcomes for strategic excellence in digital business
2. Design business outcomes for operational excellence in digital business

Designing Business Outcomes for Strategic Excellence

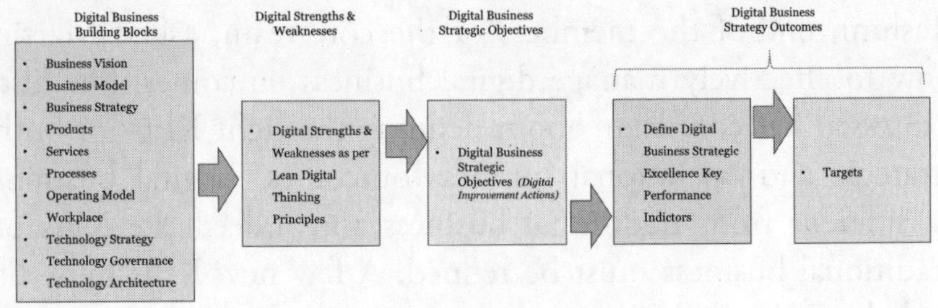

Source: Author

Figure 10.16: Business Outcomes Framework for Strategic Excellence

The framework for strategic excellence business outcomes is explained below:

1. **Define digital business strategic objectives** based on the digital strengths and weaknesses of each business building block using digital readiness assessment methods. Refer Chapter 10, Table 10.7, Figure 10.10, Table 10.8, Table 10.9 and Chapter 11, Figure 11.3, Table 11.4, Table 11.5, Table 11.6. These tables cover both business and technology strategic objectives.
2. **Quantify the strategic objectives into KPIs**. Digital business strategy map and digital technology strategy map would form the basis to define the strategic key performance indicators. Refer Chapter 10, Table 10.11 and Chapter 11, Table 11.8 for more details. .
3. There may be around **25 KPIs** to measure the effectiveness of strategies. Table 10.13 has details on five examples of strategic excellence KPIs. These KPIs, targets and industry benchmarks are given only for illustrative purpose.

S. No.	Strategic Excellence KPIs	Target	Industry Benchmark
1	No. of customer experience design methods implemented	4	Not available
2	No. of cybersecurity policies and methods implemented	5	10

3	Percentage of customer services introduced by considering the emotional needs of the customers throughout their journey	50%	70%
4	Time taken to introduce and market new digital services and products	4 months	3 months
5	Percentage of reduction in the total cost of ownership (TCO) after digitalization of technology operations	10%	20%

Source: Author

Note: * Industry benchmarks may not be available for all the KPIs. Professional bodies of respective industry sectors usually maintain the benchmark information.

Table 10.13: Sample Business Outcomes for Strategic Excellence

Designing Business Outcomes for Operational Excellence

Lean digital thinking principles help in deriving operational parameters. These parameters become the basis for designing business outcomes for digital business operational excellence.

S. No.	Lean Digital Thinking Principles	Digital Business Operational Parameters for Outcomes
1	Create a better today and better tomorrow	Bimodal business operational excellence
2	Embrace constant and instant change	Agility and time to market
3	Nurture digital business talent	Good digital skills
4	Identify and eliminate non-value-added activities	Minimal waste
5	Predict, recommend and improve overall business performance	High predictability

6	Optimize human interventions	Hyperautomation
7	Establish real-time and contextual communication and collaboration	Real-time collaboration
8	Safeguard privacy, safety and security of business resources	High security, safety and privacy
9	Establish a technology start-up ecosystem	Agility to respond to changing business needs
10	Create socially responsible products, services and value chains	Social responsibility
11	Design and deliver unforgettable moments of experience to stakeholders	Super customer experience
12	Build a human-centric digital culture	Human-centric culture

Source: Author

Table 10.14: Applying Lean Digital Thinking to Define Operational Excellence Outcomes

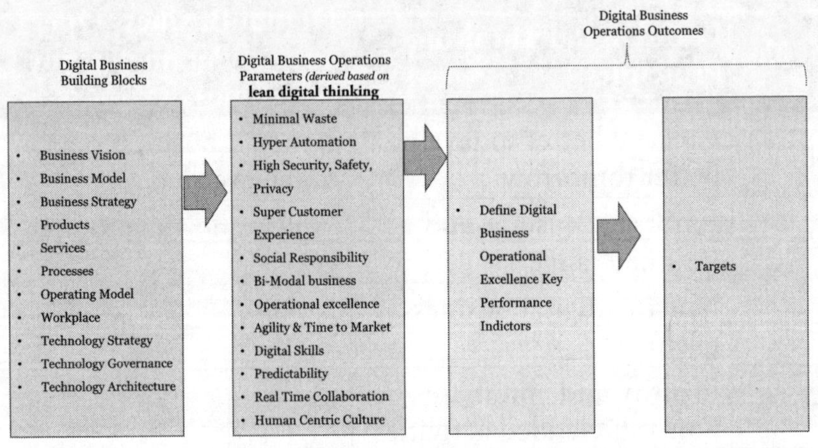

Source: Author

Figure 10.17: Business Outcomes Framework for Operational Excellence

The framework for operational excellence outcomes is explained below:

1. **Analyze each business building block** against the digital business operational parameters for outcomes derived from **lean digital thinking** (refer to Table 10.14).
2. **Identify KPIs** for digital business operational excellence against each business building block against the digital business operational parameters for outcomes derived using **lean digital thinking**. Refer to Table 10.15.
3. There may be around **50–75 KPIs** to measure the effectiveness of digital business operations. Table 10.15 has details on seven examples of operational excellence KPIs. These KPIs, targets and industry benchmarks are given only for illustrative purpose.

S. No.	Operational Excellence KPIs	Target	Industry Benchmark*
1	No. of cybersecurity incidents	3/year	1/year
2	Percentage of employees trained on new digital skills	50%	75%
3	Cost of acquiring a new customer	US$10	US$8
4	Revenue from digital products and services	35%	50%
5	Customer retention rate	65%	80%
6	NPS	7	9
7	Profit from digital products and services	6%	8%

Source: Author

Note: * Industry benchmarks may not be available for all the KPIs. Professional bodies of respective industry sectors usually maintain the benchmark information.

Table 10.15: Sample Business Outcomes for Operational Excellence

Streamlining Key Performance Indictors

Martin proposed a two-step process to streamline business outcomes.

1. Prioritize KPIs and remove redundant and unimportant KPIs.
2. Remove contradicting KPIs. For example, increasing the scope and product features while reducing the number of skilled resources may not be possible. Conflicting KPIs must be streamlined.

At the organization level, there may be around 100 KPIs covering both strategic and operational excellence.

Collating Data for Each Business Outcome

Martin created a simple framework to identify the sources and availability of data for collation at regular intervals. Without collating the data, KPIs will be of no use. Hence, it is essential to identify the source from where the data for the KPI could be collected and also identify the KPIs for which there is no or partial data available.

Martin emphasized the importance of creating a plan, building new software applications and refining existing software applications to fill the data gaps so that the organization is ready to collate the required data against each KPI in no time.

S. No.	Business Outcomes/KPIs	Data Source	Data Availability
1	No. of customer experience design methods implemented	Manual	No data available
2	No. of cybersecurity policies and methods implemented	Manual	Partial
3	Percentage of customer services introduced by considering the emotional needs of the customers throughout their journey	Manual	No data available

4	Time taken to introduce and market new digital services and products	Product life cycle management system	Full
5	Percentage of reduction in the TCO after digitalization of technology operations	Finance system and manual	Partial
6	No. of cybersecurity incidents	Incident management system	Full
7	Percentage of employees trained in new digital skills	Learning management system	Full
8	Cost of acquiring a new customer	Finance system	Partial
9	Revenue from digital products and services	Finance system	Partial
10	Customer retention rate	CRM system	Full
11	NPS	Social media system	Full
12	Profit from digital products and services	Finance system	Full

Source: Author

Table 10.16: Sources of Data Collation for Business Outcomes of Digital Businesses

Analyzing Trends and Patterns to Improve Business Outcome Performance

IT Director Gibson asked Martin a fundamental question, "Why there is no correlation between the data collected against the KPIs of business outcomes and the actual performance?"

For example, their customer net promoter score (NPS) is always good, which means many customers recommend company products and services to others. But the fact is that there is no corresponding improvement in sales. Martin replied that the most common reason for this could be that the data collated against each KPI may not be accurate and does not reflect the actual business performance.

Automation is the key to collate and analyze data. Manual dependency and using excel sheets does not provide a single source of truth and real-time insight on business outcomes. With the emergence of AI and machine learning (ML) algorithms built as part of software applications, it is now easy to establish a good correlation between the KPIs. Analyzing trends and patterns against each performance indicator must be made in real time using data and analytics.

With the help of data, AI, ML and analytics, for each KPI (business outcome), we get a 360-degree view as given in Figure 10.18.

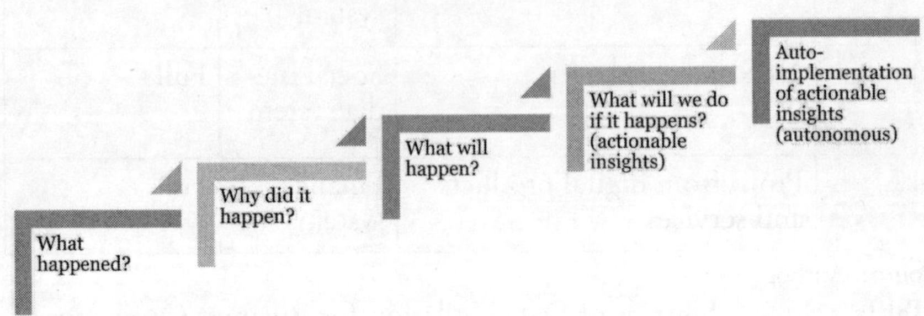

Source: Author

Figure 10.18: Five-Level Analysis of Business Outcomes/KPIs

Five-Level Analysis of Key Performance Indicator

Business Outcome: Customer Retention Rate

S. No.	Level	Hindsight/ Insight/Foresight	Action
1	What happened?	Customer attrition has gone up by 25%.	—
2	Why did it happen?	Irrelevant products and services; inability to meet the needs of Gen Z customer segment.	Perform root-cause analysis based on various trends, patterns (bar charts, pie charts and others) available in the dashboards.
3	What will happen?	Software system predicts customer attrition rate based on the data collected from various sources and by using predictive models and ML algorithms retention rate in future.	Continuous refinement of predictive models and relevant algorithms based on data sets is the key to improving predictability.
4	What will we do if it happens? (actionable insights)	Automatic system recommendations by software based on the knowledge acquired by the system from both internal and external sources. The system recommends actionable insights to improve customer retention.	Design and develop recommendation engines as part of software applications. The recommendation engine is a system that recommends products, services, information based on the analysis of users data.
5	Auto-implementation of actionable insights (autonomous)	In a few cases, the software system automatically implements the actionable insights proposed by it. For example, personalized ads to customers based on their Facebook profile can be implemented automatically without human intervention.	Management must decide in which situation and scenarios, auto-implementation of actionable insights can be applied.

Source: Author

Table 10.17: Example of Five-Level Analysis of Business Outcome—Customer Retention Rate

Similarly, for each business outcome/KPI, the five-level analysis framework could be used by the concerned manager/executive to improve performance.

Within three months, Martin and his team completed the business outcomes system design. Martin presented the same to CEO Jeffrey and other senior leaders. With minor modifications, he got approval to implement the designed system.

Jeffrey concluded his meeting with Martin by saying that he and all the managers must monitor the business performance in real time anywhere and anytime using any device. He wanted Martin to make sure that he worked with the CIO and other business leaders. Together, they were to make the intelligent system ready so that the five-level analysis for critical business outcomes was available automatically to each manager/employee in the company.

Martin committed to Jeffrey that he will work with all the key stakeholders and make the system ready in two phases within 12 months.

CHAPTER 11
Digitalization Cluster 2: Digitalize Technology Foundation

Building Block 5: Technology Strategy

The chief information officer (CIO) of Smart Inn hotel, Richards, who had attended the sessions on **lean digital thinking** and digital business strategy, had been thinking about preparing the Smart Inn technology organization for the digital future. In general, within his technology organization, there was much confusion about the following:

1. What is a digital strategy? Is it digital business strategy or digital technology strategy or both?
2. What is the difference between a digital business strategy and a digital technology strategy?
3. Do we need both strategies or only one?

In Smart Inn, till now, a digital strategy meant a digital marketing strategy that was driven by the chief marketing officer (CMO). In reality, digital marketing strategy is the tip of the iceberg in the overall digital business strategy.

Richards knew that digital business strategy and digital technology strategy were different, and the digital technology strategy must be aligned with the digital business strategy.

Establishing the Core Technology Team

The first thing Richards did was form a core team representing different technology and domain disciplines. He was very keen on devising a digital technology strategy that would help Smart Inn to leapfrog the competition and provide enhanced value to hotel guests. Richards selected the best team available to him.

Richards identified the core technology team representing the following areas:

- Enterprise architecture
- Mobile technology
- Cloud and edge computing
- Guest relationship management
- Wearable devices
- Microprocessors and embedded software
- Hotel business processes (domain)
- Augmented reality (AR), virtual reality (VR) and mixed reality (MR)
- Sensors
- Blockchain smart contracts
- Artificial intelligence (AI)
- Short-range communication networks
- Data and analytics
- Long-range communication networks
- Gamification

A few days later, Richards spoke with Rao, the lean digital specialist, and shared his plan to form a core technology team to start working on a digital technology strategy. He requested Rao to conduct a workshop for the core team on digital technology strategy and **lean digital thinking** so that he could collaboratively work with the team to create their digital technology strategy.

Preparing the Core Technology Team

On 2 April 2016, Rao visited Richards in his office and spent some time discussing the workshop agenda. Both proceeded to the conference room to kick-start the workshop with the core team.

Rao asked the team to share their thoughts on technology strategy. A few members responded that it was a selection of technologies, governance and architectural guidelines. There was no common understanding among the core team members. The first thing Rao cleared to them was that technology strategy was not governance and architecture.

Rao explained to them that digital technology strategy means well-planned long-term technology initiatives that help in moving a technology organization from the current state to the future digital state to support the digital business strategy.

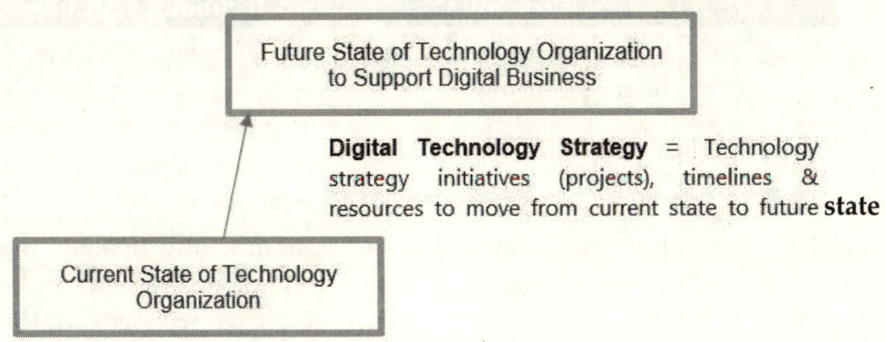

Source: Author

Figure 11.1: Moving from Current State to Future State of Technology Organization

Rao shared that the biggest challenge for the core team was to imagine the future state of Smart Inn. He identified a list of strategic initiatives that would help in transitioning from the current state to the future state.

Sai, the enterprise architect, said that it was quite interesting and inquired about the availability of any framework that would help the core team to start thinking about the future state. Rao suggested the 12 principles of **lean digital thinking** which would provide an excellent framework for them to effectively devise their digital technology strategy.

Applying Lean Digital Thinking to Design Digital Technology Strategy

After the tea break, Rao started sharing information about **lean digital thinking** as mentioned in Table 11.1. He guided the team to identify improvement actions in relation to the 12 principles. This would direct the core team towards kick-starting the digital technology strategy.

S. No.	Lean Digital Thinking Principles	Changing the Anatomy of Technology Strategy
1	Create a better today and better tomorrow	Create a digital technology strategy that seamlessly integrates with the existing technologies and evolve incrementally.
2	Embrace constant and instant change	Continuously analyze the technology trends, shifts and data received on the performance of the software applications, infrastructure, networks and devices. Conduct root-cause analysis of the deficiencies and improve evolutionarily.
3	Nurture digital business talent	Nurture digital technology leadership skills regularly to refine the digital technology strategy as per the changing needs.

4	Identify and eliminate non-value-added activities	Identify improvement actions that help in eliminating non-value-added activities in the technology department using digital technologies.
5	Predict, recommend and improve overall business performance	Predict the digital technology strategy performance by using various simulation tools and implement actionable recommendations to get the best out of digital technology strategies.
6	Optimize human interventions	Identify automation opportunities that help in designing, developing, releasing, deploying and maintaining the digital software applications and providing elastic infrastructure within a shorter period of time.
7	Establish real-time and contextual communication and collaboration	Establish a real-time communication and collaboration system to virtually involve business executives, employees, partners, suppliers and other stakeholders anywhere and anytime to innovate digital technology strategies.
8	Safeguard privacy, safety and security of business resources	Identify improvement actions that safeguard software applications, infrastructure, networks, devices and intelligent things from cyberattacks.

9	Establish a technology start-up ecosystem	Build an entrepreneurial mindset to devise unique technology strategies.
10	Create socially responsible products, services and value chains	Identify improvement actions that make sure that the software applications and infrastructure are built with a low impact on the environment and society.
11	Design and deliver unforgettable moments of experience to stakeholders	Identify improvement actions that provide unforgettable moments of experience to employees, business executives and partners throughout their journey while getting services from internal technology department driven by the CIO.
12	Build a human-centric digital culture	Establish a digital culture and set of values that promote continuous digital technology strategy innovation.

Source: Author

Table 11.1: Applying Lean Digital Thinking Principles to Design Digital Technology Strategy

IT Director Ken turned to Rao and said, "We are already digital. We have been developing mobile, social and analytics apps for the last couple of years. Why are we talking about this digital technology strategy today?"

Ken continued the conversation with Rao, "We have established private cloud infrastructure centres, and recently we also started working on edge computing. We are on top of the technologies as and when they get into the market."

He proudly shared that one of his young, passionate team members had recently written an ML algorithm to understand the guest persona based on different data sources.

Ken asked Rao whether they were unnecessarily spending time on these workshops as they were already working on a few digital technology initiatives. With one look at the expressions on the faces of the other core team members, Rao understood that they also had similar questions in their minds. With a deep breath, Rao answered:

"Yes, you have already begun to implement digital technology initiatives. Still, they may not be strategic as you have not taken digital business strategy as an input for your digital technology initiatives. Hence, any initiative, which is not aligned with a digital business strategy may not yield the required business outcomes.

"At present, as per my discussion with Richards, there is a think tank within your technology organization. A think tank gets ideas and solutions from technology teams to identify the initiatives, which again may not align with what the business needs.

"ML algorithms written by your team will be useful and will add value to guests. But, are these algorithms enough to provide unforgettable moments of experience as designed by your business? Does your team have the list of algorithms to be written to offer a superior experience to guests?"

The answer from Ken was a "no".

Rao concluded by saying that he was not challenging what they have been doing. "What I have been emphasizing is that you must have a long-term digital technology strategy at least for a 1,000 days. The initiatives must be as per the digital business strategy and according to the digital strengths and weaknesses of your organization. In that way, you will have a high return on investment (ROI) and business value."

The core team had been slowly digesting the multiple dimensions of digital technology strategy based on **lean digital thinking** principles and realized that devising a digital technology strategy was not an easy task.

Methodology to Create Digital Technology Strategy

Rao shared a simple methodology with the core team to devise the digital technology strategy.

1. Smart Inn had already created digital business strategic objectives which would be one of the important inputs to derive digital technology strategic objectives (improvement actions). Refer to Chapter 10, Table 10.9.
2. In addition to this, Rao also explained and gave a demo of the lean digital quotient (LDQ) finder tool, which helps in assessing the current technology strategy maturity to find digital technology strategy strengths and weaknesses. This helps derive digital technology strategic objectives.

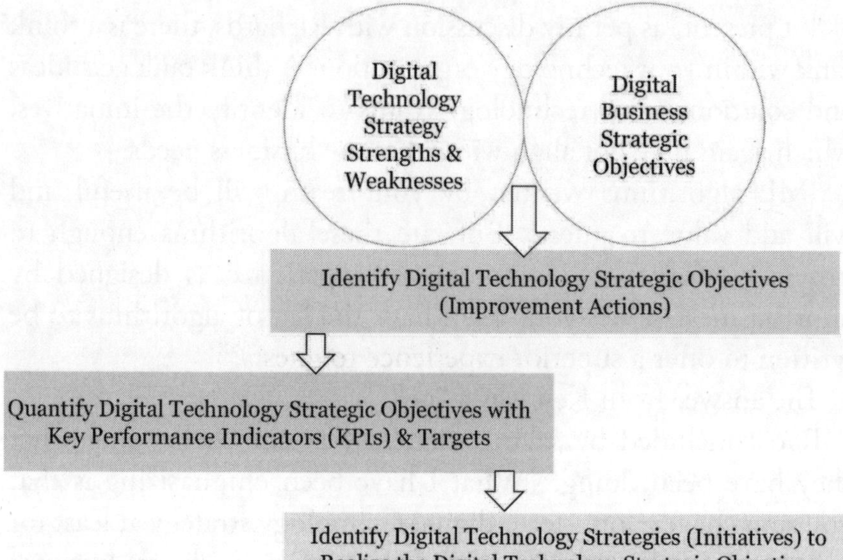

Source: Author

Figure 11.2: Methodology for Creating Digital Technology Strategy

Rao concluded the workshop with two key messages:

1. Technology brains must put their focus on identifying game-changing strategic initiatives that help in realizing Smart Inn's digital business strategy faster, better and cost-effectively.
2. Technology organizations must transform themselves to become digital business innovators and strategic advisors to drive digital business effectively.

Richards ended the workshop by announcing the date for the next meeting which was 25 April 2016. He requested all the core team members to go through the material provided by Rao during the workshop. Everyone was asked to complete their responses to the assessment questions regarding technology strategy in the LDQ tool. He also asked everyone to come prepared for the next meeting to work on digital technology strategic objectives (improvement actions).

Richards asked Tom to drive this initiative like a project and finalized the timeline given in Table 11.2 with the core team.

S. No.	Activity	Target Date of Completion
1	Generate technology strategy strength and weakness report using the LDQ finder tool	20 April 2016
2	Define digital technology strategic themes	25 April 2016
3	Define digital technology strategic objectives	25 April 2016
4	Create a digital technology strategy map	15 June 2016
5	Identify digital technology strategic initiatives (strategies)	31 July 2016

Source: Author

Table 11.2: Indicative Schedule for Designing Digital Technology Strategy

Identifying Digital Technology Strategic Objectives

By 20 April 2016, Richards and the team had identified the digital strengths and digital weaknesses report of technology strategy using the LDQ finder tool. This report was to be the basis for the digital technology strategies identified for Smart Inn.

Before the workshop on 25 April 2016, Richards asked the core team to go through the digital strength and weakness report to identify digital technology strategic objectives. He sent a detailed mail with the agenda items for the upcoming workshop.

Identifying Digital Technology Strategic Objectives Based on Digital Strengths and Weaknesses

The core team finalized **digital technology strategic themes** (pillars of excellence) given in Table 11.3 after a detailed brainstorming session held on 25 April 2016. These themes were derived from the **digital technology strengths and weaknesses** identified using the LDQ finder tool.

S. No.	Digital Weaknesses	Technology Strategic Themes
1	Smart Inn does not identify the negative impact on their business due to digital technologies.	Future-proof and future-ready technology ecosystem
2	Smart Inn does not create long-term plans to build new digital applications and digital infrastructure that generate new business opportunities.	Technology and business alignment
3	Smart Inn does not create long-term plans to build new digital applications and digital infrastructure that improves business operations excellence.	Technology and business alignment
4	Smart Inn's internal technology department does not create a long-term plan for hyperautomation of applications development.	Hyperautomation of the software life cycle

5	Smart Inn technology department does not have a long-term plan to maintain emotional bonding with its digital technology ecosystem partners.	Superior partner relationship
6	Smart Inn technology department does not have a long-term plan to assess the changes required in our technology infrastructure to support digital businesses	Future-proof and future-ready technology ecosystem

Source: Author

Table 11.3: Deriving Digital Technology Strategic Themes for Smart Inn

FUTURE-PROOF & FUTURE-READY TECHNOLOGY ECOSYSTEM	TECHNOLOGY & BUSINESS ALIGNMENT	HYPER-AUTOMATION OF SOFTWARE LIFE CYCLE	SUPERIOR PARTNER RELATIONSHIP

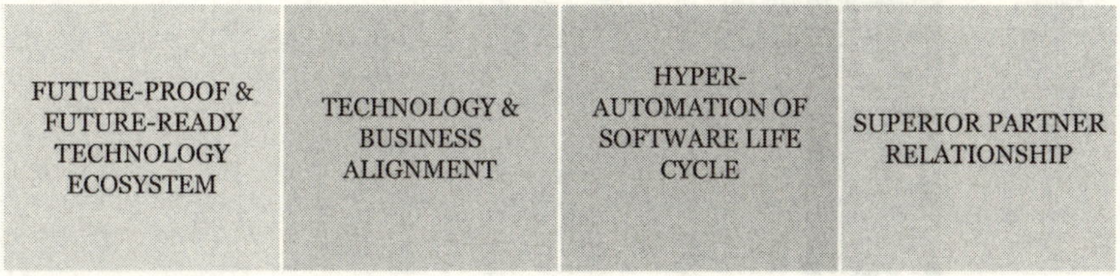

Source: Author

Figure 11.3: Digital Technology Strategic Themes for Smart Inn

Richards and the team agreed to identify the improvement actions required for each strategic theme. Improvement actions are nothing but strategic objectives that help in improving the digital capabilities, processes, customer value proposition and financial outcomes of an organization. Richards requested the team to identify the improvement actions in the following areas for Smart Inn:

1. Learning and growth strategic objectives
2. Process excellence strategic objectives
3. Customer delight strategic objectives
4. Finance excellence strategic objectives

The core team had argued, discussed, brainstormed and finalized the list of the digital technology strategic objectives (improvement actions), as given in Table 11.4.

S. No.	Digital Technology Strategic Theme	Learning and Growth Strategic Objectives	Process Excellence Strategic Objectives	Customer delight Strategic Objectives	Finance Excellence Strategic Objectives
1	Future-proof and future-ready technology ecosystem	• Establish digital technology centre of excellence	• Accelerate the digitalization of existing software applications and infrastructure	• Prepare business executives to leapfrog the competition	• Increase revenue from new digital services
2	Technology and business alignment	• Improve business and domain knowledge of technology executives	• Improve interaction among business and technology executives using unified communications	• Empower business executives to predict and improve business performance	• Create new business opportunities leveraging digital technologies • Improve business operational excellence with digital technologies

3	Hyperautomation of the software life cycle	• Develop agile, DevOps, DevSecOps skills • Introduce automation tools for software and infrastructure management	• Digitalize software management process • Digitalize infrastructure management process	• Introduce new software that enhances customer services	• Reduce the cost of technology operations
4	Superior partner relationship	• Acquire new digital skills from partners	• Enhance real-time collaboration and communication with partners using unified communications		• Create new business opportunities leveraging digital technologies

Source: Author

Table 11.4: Digital Technology Strategic Objectives Based on Digital Weaknesses

Richards was quite happy with the active participation of his team and outcomes. Edwin, the blockchain smart contract specialist, was very excited about the LDQ finder tool. He mentioned that the LDQ was a critical game changer and helped them observe the as-is situation of the digital technology strategy. Richards emphasized that the 18 digital technology strategic objectives identified by them were not exhaustive. Some more objectives derived from the strategic business objectives must be added to make a more comprehensive list.

Identify Digital Technology Strategic Objectives Based on Digital Business Strategic Objectives

After lunch, the whole team assembled and started working on deriving improvement actions based on the digital business strategic objectives of Smart Inn. To realize the digital business strategic objectives, corresponding digital technology strategic objectives must be defined as well. The core team went through all the digital business strategic objectives and derived a comprehensive list of digital technology strategic objectives as given in Table 11.5.

Strategic Areas	Digital Business Strategic Objectives	Digital Technology Strategic Objectives
Learning and growth	**LO1:** Introduce new digital business roles	Introduce new digital technology roles
	LO2: Institutionalize digital culture	Institutionalize digital culture
	LO3: Implement digital systems	Acquire capabilities to manage digital software applications
	LO4: Improve lean and agile capabilities	Improve agile, DevOps and DevSecOps capabilities
	LO5: Improve experience design capabilities	Acquire capabilities to manage digital software applications
	LO6: Improve cybersecurity capabilities	Improve cybersecurity capabilities

Process	**PO1**: Increase value, speed and productivity with human-machine interaction	Optimize business processes with digitalization
	PO2: Leverage data and AI in the business value chain	Build predictable business processes with digitalization
	PO3: Digitalize processes to eliminate non-value-added activities	Accelerate software development with agile and DevOps methods
	PO4: Design environment-friendly guest services	Design environment-friendly software applications, computing resources and data centres
	PO5: Design guest services by considering emotional needs	Analyze human emotions for better experience
	PO6: Build cyberattack-proof business	Improve security with DevSecOps methods
Customer	**CO1**: Offer personalized guest services	Empower business executives with technology to offer personalized services
	CO2: Reach long-tail (niche) customers	Not applicable
	CO3: Offer socially responsible services	Minimize travel and other overheads with unified communications
	CO4: Deliver unforgettable moments of experience	Not applicable
	CO5: Offer faster, better and cheaper services	Respond to business needs faster, better and cost-effectively
	CO6: Safeguard safety, security and privacy of guests	Safeguard safety, security and privacy of employees, partners and suppliers

Finance	**FO1**: Improve profits	Not applicable
	FO2: Reduce operational costs	Reduce the cost of technology operations
	FO3: Enable asymmetric growth	Not applicable
	FO4: Create positive social impact	Create positive social impact
	FO5: Increase revenue from repeat business	Not applicable
	FO6: Increase revenue and impact from new digital offerings	Not applicable
	FO7: Minimize revenue loss due to security breaches	Minimize revenue loss due to security breaches

Source: Author

Table 11.5: Digital Technology Strategic Objectives Based on Digital Business Strategic Objectives

Consolidated Digital Technology Strategic Objectives

After identifying the digital technology strategic objectives based on digital technology strategy strengths and weaknesses as well as digital business strategic objectives, Richards and the core team consolidated the following 33 digital technology strategic objectives as shown in Table 11.6. This implies that 33 improvement actions were identified to transform the technology department of Smart Inn.

Learning and Growth Strategic Objectives	Process Excellence Strategic Objectives	Customer Delight Strategic Objectives	Finance Excellence Strategic Objectives
• Establish digital technology centre of excellence	• Accelerate the digitalization of existing software applications and infrastructure	• Prepare business executives to leapfrog the competition	• Increase revenue from new digital services
• Improve business and domain knowledge of technology executives	• Improve interaction among business and technology executives using unified communications	• Empower business executives to predict and improve business performance	• Create new business opportunities leveraging digital technologies
• Introduce automation tools for software and infrastructure management	• Digitalize software management process	• Introduce new software that enhances customer services	• Improve business operational excellence with digital technologies
• Acquire new digital skills from partners	• Digitalize infrastructure management process	• Empower business executives with technology to offer personalized services	• Reduce the cost of technology operations
• Introduce new digital technology roles	• Enhance real-time collaboration and communication with partners using unified communications	• Minimize travel and other overheads with unified communications	• Create positive social impact

• Institutionalize digital culture	• Optimize business processes with digitalization	• Respond to business needs faster, better and cost-effectively	• Minimize revenue loss due to security breaches
• Acquire capabilities to manage digital software applications	• Build predictable business processes with digitalization	• Safeguard safety, security and privacy of employees, partners and suppliers	
• Improve agile, DevOps, DevSecOps capabilities	• Accelerate software development with agile and DevOps methods		
• Improve cybersecurity capabilities	• Design environment-friendly software applications, computing resources and data centres		
	• Analyze human emotions for better experience		
	• Improve security with DevSecOps methods		

Source: Author

Table 11.6: Consolidated List of Digital Technology Strategic Objectives

Creating Digital Technology Strategy Map

On 15 June 2016, Richards and the core team used a balanced scorecard framework and created a digital technology strategy map. This meant:

1. **Financial performance** will improve only upon delivering high-value proposition, personalized services and unforgettable moments of experience to internal business executives and customers.
2. **Internal business executives and customers get high-value proposition**, personalized services and unforgettable moments of experience only when the processes are improved and digitalized.
3. **Internal processes** can be improved or digitalized only upon improving the internal skills, policies and systems.

The balanced scorecard framework creates a significant cascading effect across multiple pillars of the balanced scorecard. If the learning and growth perspective is strengthened, then the organization can execute the internal technology management processes exceptionally well. With the process excellence, an organization can deliver excellent technology services to business units. With exceptional services to business units (internal customers), the organization gets more customers, revenue and profit.

Once the interrelationships and cascading effect between the 33 strategic objectives were identified, the team came up with the technology strategy map given in Figure 11.4.

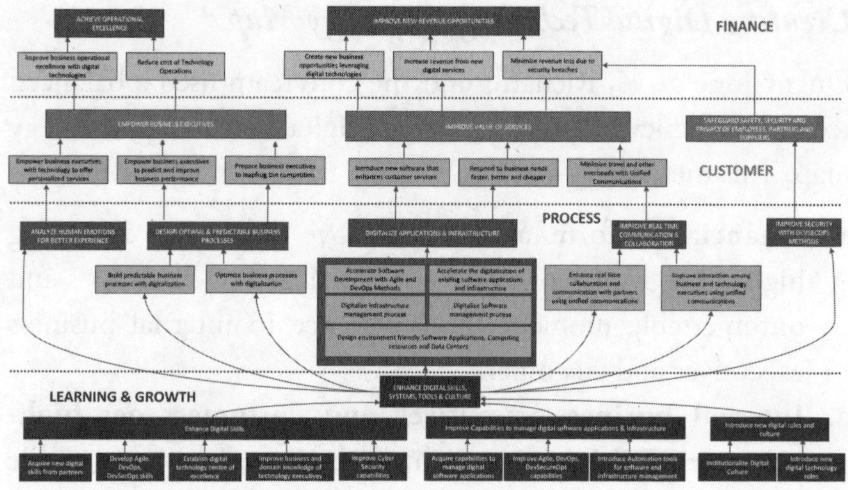

Source: Author

Figure 11.4: Digital Technology Strategy Map for Smart Inn

Identifying Digital Technology Strategic Initiatives

Richards conducted an orientation session for the core team on how to identify digital technology strategic initiatives from the digital technology strategy map. The core team actively participated in the meeting and was geared up to identify game-changing digital technology strategic initiatives, which would help Smart Inn prepare for the digital business journey. The following template was used to orient the core team on the process of identifying digital technology strategic initiatives. **Strategic objectives, key performance indicators (KPIs) and targets reflect the future state of the Smart Inn's technology department.**

Digital Technology Strategic Areas	Digital Technology Strategic Objectives	KPIs	FY'16 Target	FY'17 Target	FY'18 Target	Digital Technology Strategic Initiatives to Realize Targets
Learning and growth						

Internal processes						
Customer						
Financial						

Source: Author

Table 11.7: Framework for Identification of Digital Technology Strategic Initiatives

Once they had identified the digital technology strategic initiatives based on the designed template, Richards guided the core team to form the groups based on the skill mix of the core team members. They successfully collated a significant number of digital technology strategic initiatives as mentioned in Table 11.8. They were pretty much satisfied with the outcome of the workshop and were confident about kick-starting digital technology programmes very soon.

Strategic Areas	Digital Technology Strategic Objectives	KPIs	FY'17 Targets*	Digital Technology Strategic Initiatives
Learning and Growth	**LO1:** Establish digital technology centre of excellence	• No. of digital solution proofs of concept built	10	• Build digital technology centre of excellence • Establish digital technology learning academy • Establish digital software management methods and automation tools development centre • Re-engineer technology organization and policies
	LO2: Improve business and domain knowledge of technology executives	• Percentage of technology executives trained in the digital business domain	50%	
	LO3: Develop agile, DevOps and DevSecOps skills	• Percentage of technology executives trained in agile, DevOps and DevSecOps	75%	
	LO4: Introduce automation tools for software and infrastructure management	• No. of automation tools built/acquired to manage digital software and infrastructure	15	
	LO5: Acquire new digital skills from partners	• No. of new digital skills acquired from partner collaboration	4	

	LO6: Introduce new digital technology roles	• No. of new digital roles introduced in the organization	4	
	LO7: Institutionalize digital culture	• No. of events and sessions conducted on digital culture	12	
	LO8: Acquire capabilities to manage digital software applications	• No. of new methods, automation tools and policies implemented for just-in-time software apps development	New methods = 3 Automation tools = 4 Policies = 4	
	LO9: Improve agile, DevOps and DevSecOps capabilities	• No. of new methods, automation tools, policies implemented for just-in-time software apps development		
	LO10: Improve cybersecurity capabilities	• Percentage of technology executives trained in cybersecurity	40%	

Process	**Po1:** Accelerate the digitalization of existing software applications and infrastructure	• Percentage of existing software applications digitalized	40%	• Implement unified communications infrastructure • Manage digital business process portfolio • Introduce RPA bots • Manage digital software applications portfolio • Digitalize technology infrastructure • Create digital software management process handbook • Secure business resources
	Po2: Improve interaction among business and technology executives using unified communications	• Percentage of reduction in communication and collaboration issues between business and technology executives	60%	
	Po3: Digitalize software management process	• Percentage of effort reduction in releasing digital software applications	25%	
	Po4: Digitalize infrastructure management process	• Percentage of time reduction in providing the technology infrastructure	70%	
	Po5: Enhance real-time collaboration and communication with partners using unified communications	• Percentage of reduction in communication and collaboration issues with partners and suppliers	50%	

	Po6: Optimize business processes with digitalization	• Percentage of business processes digitalized	30%	
	Po7: Build predictable business processes with digitalization	• Percentage of business processes digitalized		
	Po8: Accelerate software development with agile and DevOps Methods	• Percentage of effort reduction in releasing digital software applications	30%	
	Po9: Design environment-friendly software applications, computing resources and data centres	• Percentage of digital resources that are environment-friendly	80%	
	Po10: Analyze human emotions for better experience	• No. of business processes designed with integrated customer experience (emotional needs)	5	
	Po11: Improve security with DevSecOps methods	• Percentage of digital resources secured against cyberattacks and privacy issues	50%	

Customer	**Co1:** Prepare business executives to leapfrog the competition	• Percentage of digital offerings for which the revenue is higher than the competition	20%	• Introduce innovative digital offerings ahead of the competition • Establish predictive business analytics system • Establish customer experience management centre • Establish AI-driven omnichannel-based customer service centre • Establish customer security and privacy management centre
	Co2: Empower business executives to predict and improve business performance	• No. of proactive decisions and actions taken by business executives leveraging digital resources	40%	
	Co3: Introduce new software that enhances customer services	• Percentage of new software releases that offer better customer services in a timely manner	75%	
	Co4: Empower business executives with technology to offer personalized services	• Rate of retention of customers after introducing personalized services	90%	
	Co5: Minimize travel and other overheads with unified communications	• Percentage of reduction in business travel cost due to unified communications	35%	
	Co6: Respond to business needs faster, better and cost-effectively	• Time taken to introduce new digital services and products	30–90 days	
	Co7: Safeguard safety, security and privacy of employees, partners and suppliers	• Percentage of cybersecurity and privacy incidents in the organization	20%	

Finance	**F01:** Increase revenue from new digital services	• Percentage of revenue generated from digital offerings against overall revenue	30%	• Create long-tail customer reach strategy using digital resources • Identify new customer segments and revenue potential due to digitalization • Establish technology cost optimization centre
	F02: Create new business opportunities by leveraging digital technologies	• No. of new customer segments reached due to digital offerings	3	
	F03: Improve business operational excellence with digital technologies	• Percentage of reduction in the cost of overall business operations after digitalization	20%	
	F04: Reduce cost of technology operations	• Percentage of reduction in the TCO after digitalization of technology operations	30%	
	F05: Minimize revenue loss due to security breaches	• Percentage of revenue loss due to cybersecurity breaches	<1%	

Source: Author

Note: *It is easy to set the targets but identifying data sources, collating data, filling the data gaps and building systems to mine the data are the challenges to be addressed by the organizations.

Table 11.8: Digital Technology Strategic Initiatives for Smart Inn

Finalizing Digital Technology Strategies

On 31 July 2016, Richards presented the list of digital technology strategic initiatives to David and Peter for their consent on creating a charter and budget for each digital technology strategic initiative with priorities. David and Peter reviewed the list of initiatives (digital technology strategies). They agreed with Richards to set up a digital technology strategic programme management office to drive the portfolio of digital technology strategic initiatives that would help in preparing Smart Inn for the digital future.

Richards shared a list of 18 strategies identified as per the previous section and grouped them into five groups. He believed that the execution of these strategies would prepare Smart Inn for a digital journey adequately. He cautioned the team that these strategies were not enough, there were many others to be identified, but it was a good starting point.

One of the core team members challenged Richards on the need to follow such a complicated process to define these digital technology strategies. She also emphasized that they had already identified a few digital technology strategies.

Richards answered that following a balanced scorecard approach ensured that they identify the digital technology strategies, which align with digital business vision and objectives. The balanced scorecard approach also ensured high ROI from technology initiatives.

Customer Engagement and Value-Added Strategies	Digital Infrastructure Management Strategies	Digital Software Management Strategies	Digital Solution Innovation and Learning Strategies	Technology Organization Management Strategies
• Create long-tail customer reach strategy using digital resources • Establish AI-driven omnichannel-based customer service centre • Establish customer experience management centre • Establish customer security and privacy management centre • Identify new customer segments and revenue potential due to digitalization • Introduce innovative digital offerings ahead of the competition	• Digitalize technology infrastructure • Implement unified communications infrastructure • Secure business resources	• Create digital software management process handbook • Establish digital software management methods and automation tools development centre • Manage digital business process portfolio • Introduce RPA bots • Manage digital software application portfolio	• Build digital technology centre of excellence • Establish digital technology learning academy	• Establish predictive business analytics system • Establish technology cost optimization centre • Re-engineer technology organization and policies

Source: Author

Table 11.9: Final List of Digital Technology Strategies for Smart Inn

Building Block 6: Technology Governance

Auto Gear is a manufacturing company that has embarked on a digital journey and is in the process of defining digital business strategies, business models, products and services.

Paul, the CIO of Auto Gear, had established reasonably good IT governance in the last five years. They had been able to support a conventional "lights on" business that included software applications, IT infrastructure management and a few discretionary IT projects for incremental improvement in the business operations.

Unfortunately, Auto Gear's technology department was not geared up to deal with the challenges that emerged with the fusion of various technologies beyond IT. Tom, the CEO, discussed with Paul about the IT department's readiness to support their digital business journey. Paul was not confident and recommended inducting a specialist to help their technology organization prepare for the digital future.

Nancy, a lean digital specialist, was appointed by Auto Gear to support Paul and the team to establish the required technology foundation blocks to support digital business. Nancy requested Paul to form a core team to work with her to design and implement digital technology governance.

Nancy strongly believed that the technology team must have a common understanding of digital technology governance, which was beyond IT governance. Transitioning from IT governance to digital technology governance is not an easy job as it involves transformational changes in many areas.

On 15 January 2016, at 9:00 AM, Paul and the core team assembled in the conference room to kick-start the meeting. Nancy started the session by asking a question: "What is IT governance?" Based on the responses provided by Auto Gear technology leaders, it was quite evident that there was no common understanding of IT governance among the leaders.

Nancy shared her views on the definition of IT governance which she had learned from Rao (lean digital specialist).

Structure, policies, processes, procedures and culture to plan, execute, monitor and control the strategic, tactical and operational objectives and activities of an IT organization.
IT governance helps in mitigating the risks, taking the right decisions at the right time, optimally managing the IT resources like people, software applications, information, IT infrastructure and effectively managing IT investments.
Therefore, IT governance improves business performance, the environment and social responsibility.

Nancy gave Paul and the team some time to understand the difference between IT governance and digital technology governance. She clearly articulated the focus areas of technology governance as mentioned in Figure 11.5.

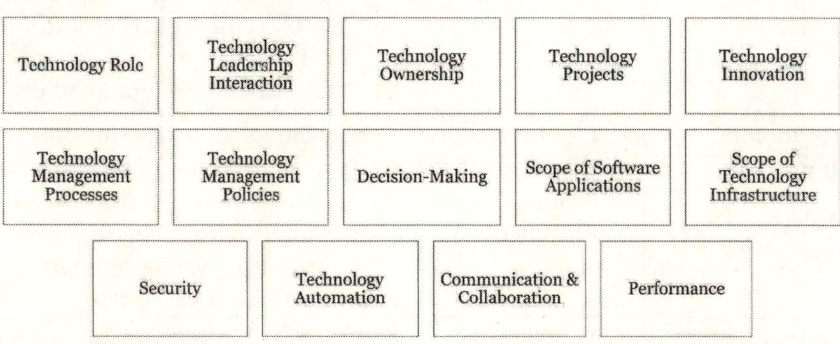

Source: Author

Figure 11.5: Technology Governance Focus Areas

Governance Areas	IT Governance	Digital Technology Governance
Technology role	• Business enabler • System integrator	• Business driver • System integrator • Digital business innovator • Digital business strategy consultant • Facilitator of citizen IT development within business units. (Citizen IT developers = business executives in the business units who build software applications)
Technology leadership interaction	• Interaction with business units, suppliers and partners	• Interaction is wider and is with the board of directors, CXOs, business units, start-ups, suppliers and partners
Technology ownership	• By CIO department	• By the board of directors and CXOs supported by the technology units under the CIO and citizen IT units within the business units
Technology projects	• IT-focused projects	• Projects focusing on the fusion of IT, communication and collaboration technology, operational technology and others
Technology innovation	• IT innovations that support the business to perform better	• Technology fusion innovations that create new business opportunities, business models and growth engines which help in achieving business operational excellence
Technology management processes	• Processes that improve the performance of the technology department that contribute to realizing business objectives and improving the business performance	• Processes that help in creating new business opportunities, new business models, new offerings and also realize the business objectives and improve business performance

Technology management policies	• Internet • Email • Information security • Disaster recovery and business continuity • Change management • Back-up and archival • Vendor selection • Incident response • Media retention • Remote access	• Internet • Email • Information security • Disaster recovery and business continuity • Change management • Back-up and archival • Vendor selection • Incident response • Media retention • Remote access • Data privacy management • Cybersecurity • Bring your own device (BYOD) • Enterprise reputation in social media • Social media usage • Content management • Tech start-ups acquisition • Multi-vendor technologies integration • Others
Decision-making	• Based on the leadership experience and algorithms that provide information, limited to hindsight and insight into technology operations	• Based on the leadership experience and intelligent algorithms that provide hindsight, insight and foresight into technology operations
Software applications	• Systems of record (process automation)	• Systems of record (process automation) • Systems of engagement (emotion and experience management) • Systems of things (products, machines, equipment and things automation) • Systems of intelligence (hindsight, insight and foresight using intelligent algorithms)

Technology infrastructure	• Cloud computing • Networks • Storage management • Relational database • 3G and 4G and fibre-optic networks	• Cloud computing • Edge computing • 6LowPan, ZigBee, LoRaWAN, Bluetooth, 5G and fibre-optic networks • Relational database • Not Only SQL Database (NoSQL) • Wearable devices • Mobile devices • Consoles (game, car, etc.) • Intelligent things/machines/equipment/products
Security	• Information security • Network security • Server security	• Information security • Things security • Device security • Hybrid network security • Server security • Privacy
Technology automation	• Semi automation of software and infrastructure life cycle management	• Hyperautomation of software and infrastructure life cycle management
Communication and collaboration	• Email, telephone-based communication and collaboration among stakeholders	• Real-time and contextual communication and collaboration among stakeholders using mobile, social media and other collaboration tools
Performance management	• Reactive performance management using technology data collated from different sources (business intelligence)	• Predictive, preventive performance management using technology data collated from different sources (leveraging AI, ML with predictive and prescriptive analytics)
Data	• Structured data	• Structured data • Semi-structured data • Unstructured data

Source: Author

Table 11.10: IT Governance vs Digital Technology Governance

Paul and the team found the session an eye-opener for them; they clearly understood the difference between IT governance and digital technology governance. They also realized that they needed to put in a lot of effort to establish a digital technology governance system within Auto Gear. Nancy resumed the session by introducing the objectives of establishing sophisticated digital technology governance.

Digital Technology Governance Objectives

Digital technology governance must effectively manage the fusion of many technologies to realize the following objectives:

1. Prepare the technology organization for two-speed/bimodal business (traditional and digital).
2. Establish an agile technology organization to build a digital business ecosystem in collaboration with suppliers and partners.
3. Align technology outcomes with business outcomes.
4. Remove functional silos and collaborate to work for a common digital business purpose.
5. Nurture the technology department to become a digital business strategic advisor, innovator and consultant to business executives.
6. Introduce predictive governance by leveraging AI and ML and take the right decisions at the right time.
7. Get visibility on technology performance in real time.
8. Integrate citizen IT with central technology organization.
9. Predict and mitigate risks effectively.
10. Establish strong governance for
    a. investment management,
    b. risk management,
    c. applications (complex workflows and many features)/ apps (lightweight and simple features) management,
    d. API management,
    e. public, private, hybrid and multi-cloud management,
    f. things management,

g. device management,
h. data management,
i. cybersecurity,
j. data privacy and
k. others.

11. Balance the technology investments for exploration, experimentation, transformation and operational excellence.

Nancy emphasized that as the first step, it was essential to establish a common purpose for the Auto Gear technology team. Nancy wanted to use the **lean digital thinking** principles to design the technology governance for Auto Gear. She discussed with Paul and the team to plan for a workshop immediately. They agreed to conduct the workshop on 25 January 2016.

Applying Lean Digital Thinking to Digitalize Technology Governance

On 25 January 2016, Nancy conducted an orientation session on the 12 principles of **lean digital thinking**. She explained each principle in the context of digital technology governance as mentioned in Table 11.11.

S. No.	Lean Digital Thinking Principles	Changing the Anatomy of Technology Governance	Digital Technology Governance Elements
1	Create a better today and better tomorrow	Establish governance in such a way that the technology organization supports both traditional businesses and digital businesses at the same time	• Bimodal/two-speed business

2	Embrace constant and instant change	Continuously collate and analyze the changing trends and shifts in technologies, customer needs, wants and bring improvements to the software applications, infrastructure, devices, technology processes, policies, etc.	• Technology change management
3	Nurture digital business talent	Nurture digital technology leadership skills regularly to prepare the technology organization for the digital business future	• Digital talent management
4	Identify and eliminate non-value-added activities	Design lean processes for technology management and implement lean methods, tools and techniques for eliminating non-value-added activities within the technology organization	• Lean execution
5	Predict, recommend and improve overall business performance	Predict the technology organization performance by collating and mining the data received from various resources like software applications, infrastructure, devices, things and implement actionable recommendations to enable the technology organization to improve the business performance	• Predictable technology performance management • Governance and compliance management
6	Optimize human interventions	Automate software applications and infrastructure management life cycles for optimal utilization of resources	• Hyperautomation of technology management processes, machines and assets

7	Establish real-time and contextual communication and collaboration	Establish a real-time communication and collaboration system (collaboration tools, chatbots, virtual assistants, crowdsourcing tools, etc.) to deliver faster and better technology systems and services	• Real-time communication and collaboration
8	Safeguard privacy, safety and security of business resources	Define and implement effective governance for cybersecurity and risk management	• Cybersecurity management • Risk management
9	Establish a technology start-up ecosystem	Build an entrepreneurial mindset and drive technology initiatives and transformation programmes as start-ups or acquire technology niche start-ups to enhance new technology skills	• Internal start-ups
10	Create socially responsible products, services and value chains	Design technology management processes, services, software and infrastructure by complying with various environment safeguarding standards and methods	• Social responsibility
11	Design and deliver unforgettable moments of experience to stakeholders	Design and introduce technology services that help in improving employee and leader productivity, effectiveness, business operations excellence end consumer delight and experience	• Technology service management
12	Build a human-centric digital culture	Establish a digital culture and set of values that enable the technology teams to adopt a business-first attitude and work towards achieving business objectives	• Digital culture

Source: Author

Table 11.11: Applying Lean Digital Thinking Principles to Design Digital Technology Governance

Defining Digital Technology Governance Framework

Paul and the team were very happy that they got excellent clarity on the different elements of digital technology governance that were designed on the basis of **lean digital thinking** principles. They grouped the digital technology governance elements under the categories given in Figure 11.6 with the help of Nancy.

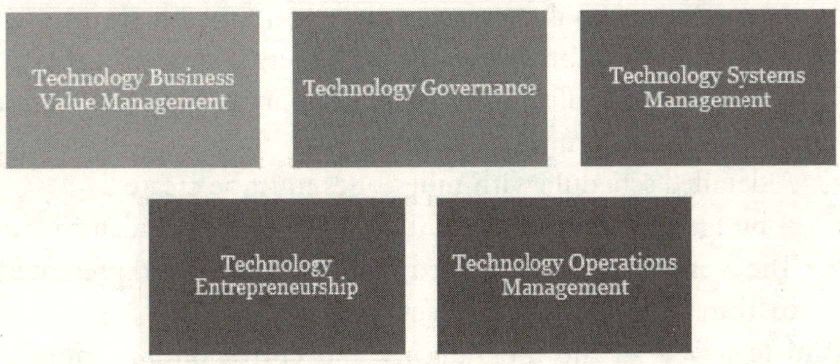

Source: Author

Figure 11.6: Digital Technology Governance Framework

Paul and the team elaborated on the foundational blocks of the digital technology governance framework as mentioned in Figure 11.7.

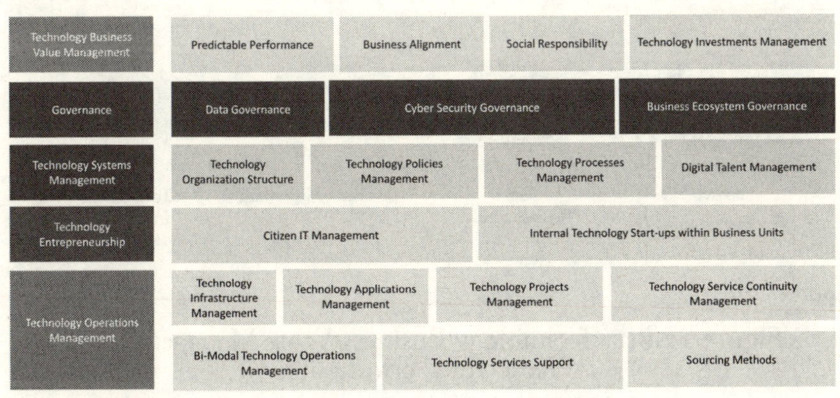

Source: Author

Figure 11.7: Foundation Blocks of Digital Technology Governance

Paul and the core team agreed that defining digital technology governance is a very complex and time-consuming task. Hence, different groups must be formed to drive it as a significant initiative with a charter, plan, budget and resources. They agreed upon the following guidelines:

1. Each leader must form a cross-functional team to execute their work.
2. A common purpose must be established for each group.
3. Regular interaction with business leaders is very critical.
4. Baselining a detailed problem and opportunity statement is a must before executing the work.
5. A detailed schedule with milestones must be created.
6. A budget must be created and resources must be identified.
7. The work must be completed within 100 days and presented to Tom and other senior executives.

Paul reviewed all the programmes and corresponding internal projects on a regular basis and provided guidance to the team to complete the required deliverables.

Technology Business Value Management

Predictable Performance

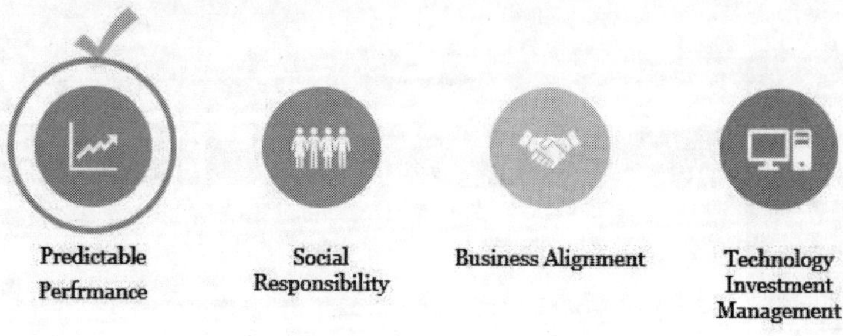

Source: Author

Figure 11.8: Technology Business Value Management—Predictable Performance

Paul opened the session by saying that digital technology governance must deal with the opportunities, challenges and operations that were being driven by the fusion of various technologies beyond IT. This meant the performance indicators of the technology organization must be revisited and reimagined.

Paul also shared that we were under pressure as Tom had been asking them to establish the digital technology foundation blocks for quite some time. They realized that the current KPIs designed for IT governance may not be enough to prepare for the digital journey of Auto Gear.

Paul emphasized the fact that one of the **lean digital thinking** principles which they had learned from Nancy was: **predict, recommend and improve overall business performance**. The team developed a method to predict, recommend and improve technology performance.

Paul asked Rohit, the head of software applications of Auto Gear, to share his experience. Rohit responded by saying that due to high automation in the software and infrastructure life cycles, humongous data was available within the technology organization. He shared his experience of applying AI on this data to predict the schedule overrun, cost overrun, defect density and application performance well in advance so that corrective actions could be taken effectively. He also shared his plan to introduce a recommendations engine so that the system itself provides actionable insights for the successful delivery and deployment of software applications.

Paul shared that they had taken KPIs from two sources and finalized the best suitable KPIs to improve the effectiveness of digital technology governance.

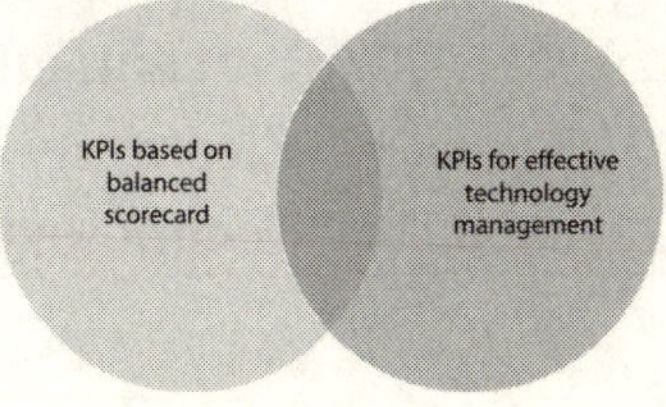

Source: Author

Figure 11.9: Technology Business Value—KPIs

Paul and the team reviewed the KPIs defined as part of the digital technology strategy map of Auto Gear. They created the digital technology strategy map using balanced scorecard in the past with the help of Nancy. Paul and the team also identified the KPIs for the areas which required special attention by the technology organization. After consolidating the KPIs from these two sources, they finalized the following KPIs to measure the performance of digital technology governance.

Performance Category	KPIs
Business value	• No. of new business models proposed due to digital technologies • No. of new product features or products introduced • No. of new customer services introduced due to digital technologies • No. of new revenue streams created due to digital technologies • Percentage of business operational cost reduced due to digital technologies
Environment, security and compliance	• No. of cybersecurity incidents • Percentage of non-conformance items as against the compliance standards • Percentage of CO2 emission • Electricity consumption • Technology risk score

Technology operations	• No. of technology management processes digitalized • Percentage of availability of mission-critical systems, services, assets and applications • Percentage of employees trained in digital technologies • Percentage of systems of engagement, systems of things and systems of intelligence • Percentage of travel cost reduced due to unified communications • Average time taken to provision the cloud infrastructure • Average time taken to deploy new mobile and digital apps • No. of citizen development team members nurtured (within business units) • Cost of waste (eight non-value-added activities) • No. of successors nurtured for all technology leadership positions • On-time delivery performance (in days) of suppliers
Technology projects execution	• No. of proofs of concept completed in digital technologies • Percentage of projects selected without business justification • Percentage of projects with < 5% cost overrun • Percentage of projects with < 5% schedule overrun • Percentage of projects with zero defect density

Technology budget	• Budget mix for bimodal/two-speed technologies management • Technology budget percentage of operations cost • Technology budget percentage of revenue

Source: Author

Table 11.12: Sample KPIs for Technology Performance

Tom expressed that while these KPIs were excellent, he was concerned that they didn't have the required sources to collate the data and measure the performance. Paul addressed this by explaining that it was quite a common situation and shared that they had already started implementing a few initiatives to get all the required data from different sources to measure the digital technology governance performance in real time.

Paul also highlighted the importance of using predictive analytics on the data collected by the technology organization to improve the effectiveness of digital technology governance.

Paul promised that they will institutionalize the new KPIs from August onwards after completing a few pilot projects in collaboration with business executives.

Social Responsibility

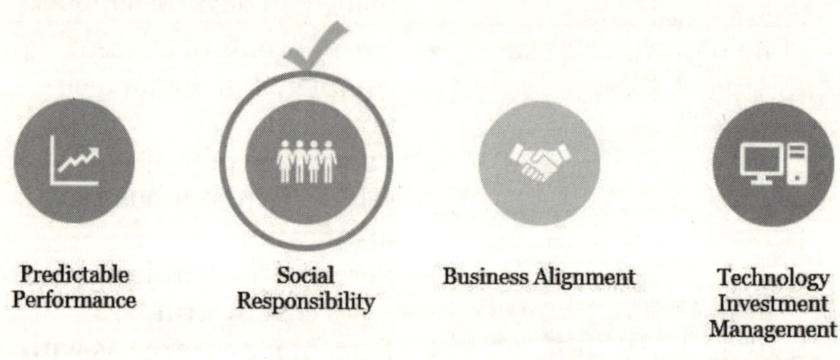

Source: Author

Figure 11.10: Technology Business Value Management—Social Responsibility

Paul reiterated Auto Gear's policy of safeguarding the environment and creating value for society while managing a profitable business. He shared that they had conducted a detailed analysis of new technologies and their impact on the environment and society and come across the challenges listed in Figure 11.11.

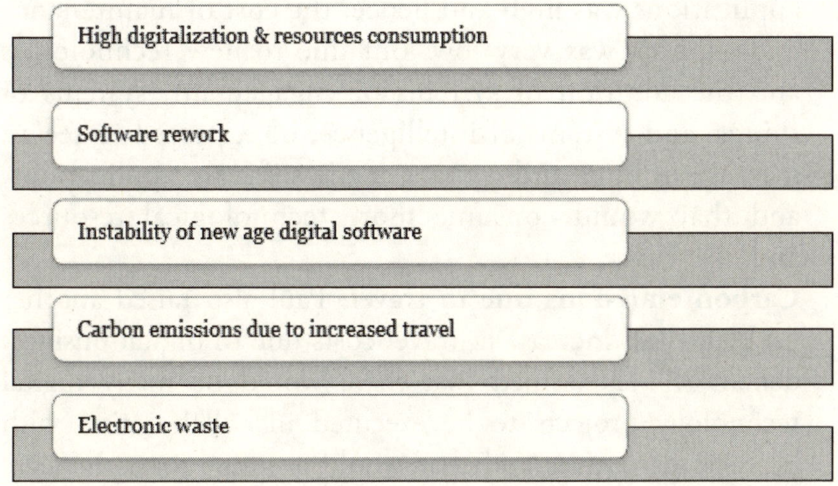

Source: Author

Figure 11.11: Impact of Digital Technologies on the Environment

Paul explained the impact of digital technologies on the environment.

1. **High digitalization and resources consumption:** In the next three years, there will be a 10% increase in the electricity consumption of Auto Gear due to digitalization. This would be mainly due to the automation of the shop floor and other assets with the internet of things (IoT), high usage of applications, devices and the use of high computing power in the data centres.
2. **Rework:** Currently, the cost of software products which includes quality control, quality assurance and fixing of defects at Auto Gear is around 38% as against the overall

software development cost. That means Auto Gear has been spending a lot of time to identify the quality issues in a reactive mode and then putting a lot of effort and other resources to fix them. Hence, rework must be reduced.

3. **Instability of new-age digital software applications:** Paul proudly stated that the stability of their current software applications was high and hence, the cost of maintenance and support was very low. But due to new technologies and the addition of systems of engagement, systems of things, and systems of intelligence, he expected to see an increase in bugs and defects in the software applications, and that would consume more technological resources and electricity.
4. **Carbon emissions due to travel:** Paul also raised another problem—an increase in travel costs due to digital business demands. He clarified that there would be many digital technology projects to be executed in collaboration with business executives and that would amount to an increase of 10% of the current travel costs. Increasing travel means a higher contribution to carbon emission levels.
5. **Electronic Waste:** Tom enquired about the electronic waste patterns within Auto Gear. Paul answered that the increase in the use of mobile devices, electronic devices, machines and computers have been generating electronic waste. It may go up to 20% from the current level.

Keeping all this in mind, Paul shared the plan of action and a few recommendations to be implemented within 18 months to safeguard the environment and to become a socially responsible technology organization. All the recommendations mentioned in Table 11.13 were to be implemented in a full-fledged manner. Tom and the business executives appreciated the efforts put in by Paul and his team.

S. No.	Measures for Safeguarding Environment	Recommendations
1	Reduce electric consumption due to digitalization	• Implement green data centre, elastic computing, elastic storage, elastic bandwidth and build new-age applications that consume fewer resources
2	Reduce rework and cost of software quality	• Implement preventive and proactive defect/failure management standards, processes and practices to reduce rework and consumption of technological resource • For example, root-cause analysis using Pareto analysis and fishbone diagram for defects and incidents • Use low-code platforms to improve quality
3	Increase software applications stability for effective utilization of technology resources	• Establish agile digital technology architecture • Use low-code platforms Build microservices-based modular software applications Develop and implement high-quality software applications to enhance stability during the software maintenance phase and reduce project life cycle costs
4	Recycle electronic waste to lessen the environmental impact	• Establish partnerships with OEMs and make electronic waste management a part of the contractual agreement

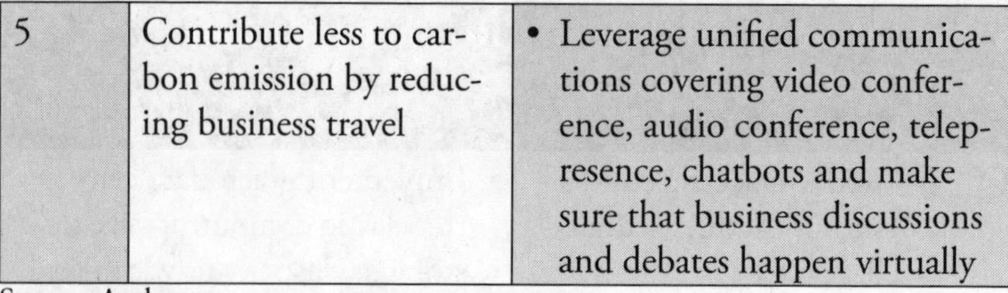

5	Contribute less to carbon emission by reducing business travel	• Leverage unified communications covering video conference, audio conference, telepresence, chatbots and make sure that business discussions and debates happen virtually

Source: Author

Table 11.13: Digital Technologies for Safeguarding the Environment

Business Alignment

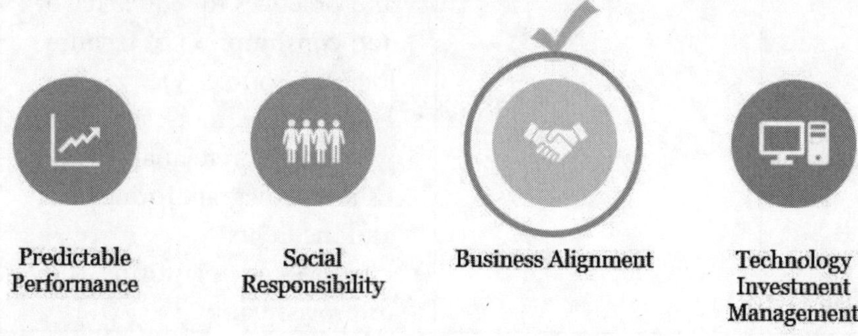

Source: Author

Figure 11.12: Technology Business Value Management—Business Alignment

Paul highlighted the point that in the era of IT governance, business and IT alignment was focusing on process automation, internet-based customer service delivery, productivity improvement, communication and collaboration requirements of businesses. Accordingly, the applications, infrastructure, information and other resources were planned to support the companies.

To support the digital business of Auto Gear, Paul emphasized that the alignment between business and technology must be beyond processes, services, productivity and communication. He suggested that they work on the recommendations given in Figure 11.13 which will be implemented in the following 10–12 months.

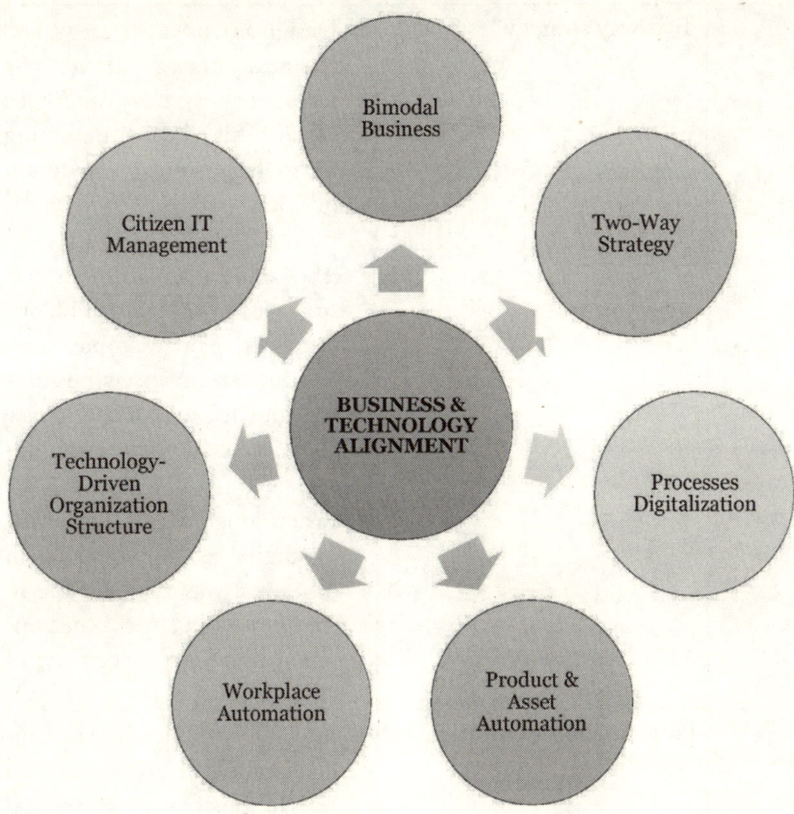

Source: Author

Figure 11.13: Business and Technology Alignment Elements

S. No.	Business and Technology Alignment Elements	Recommendations
1	Bimodal business	• Bimodal business management by business executives means managing traditional and digital businesses at the same time. • Prepare the technology organization and its resources like people, applications, infrastructure, information, intelligence and other resources to support both traditional and digital businesses at the same time.

2	Two-way strategy	• Design a two-way strategy, i.e., business strategy driven by technology strategy and vice versa. This means designing technology strategy based on business strategy (which is normal). • Also, devise unique business strategies based on technology strategies. For example, new revenue streams, new customer segments, new services, new customer channels, new products, etc. • Reach long-tail customers using digital technologies, which otherwise was not possible in traditional business (personalized bikes for a segment of customers).
3	Processes automation	• Integrate mobile, social, unified communications, IoT and AI with business processes and workflow to make them intelligent business processes that help in offering personalized customer services.
4	Product and asset automation	• Prepare the technology organization to automate products like cars, bikes, trucks and assets like machines, equipment, etc. using IoT, analytics and other technologies. In the past, IT organizations have not worked in this space. Hence, new capabilities and resources must be introduced.

5	Workplace automation	• Prepare the technology organization to support the employees to work from anywhere, anytime and use any device. • Introduce technology solutions that help employees get automatic alerts about their wellness. For example, BYOD, working on office tasks at Starbucks, IoT-enabled chairs to send alerts to employees to take a break or stretch/walk for some time. • Establish ATMMs to remotely connect with doctors from the office.
6	Technology-driven organization structure	• Leverage unified communications and other collaborative platforms that help the organization improve productivity. • Introduce technology policies that help Gen X, Gen Y and Gen Z to align and work together without a hierarchical structure. For example, an employee can directly have a chat with the vice president to get any clarification or direction.
7	Citizen IT management	• A technology organization that is driven by the CIO must build new processes, policies and standards for preparing business executives to implement/develop their software applications required for business units with zero code platforms and other tools available today. Make sure that the citizen IT developers completely align with the central technology organization policies.

Source: Author

Table 11.14: Description of Business and Technology Alignment Elements

Technology Investment Management

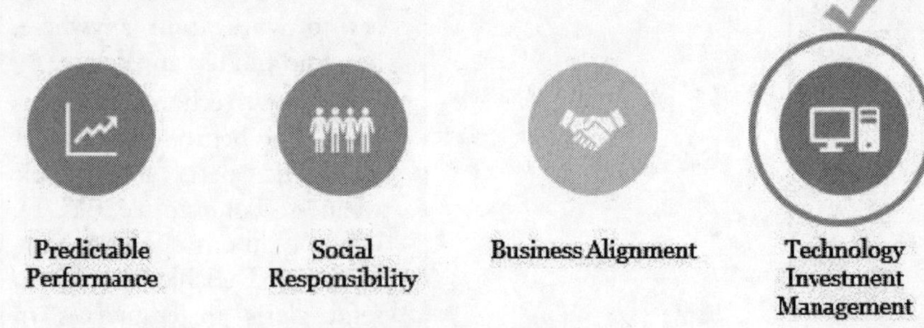

Source: Author

Figure 11.14: Technology Business Value Management—Technology Investment Management

Tom knew that the technology department/organization had not been getting the required budget for the last two years to prepare for the digital business future. Considering the importance of technology as a strategic driver, Tom had recently changed the reporting structure. Now, the CIO directly reported to the CEO. With this change, Tom sent a clear message that technology is his topmost priority.

Paul was quite happy with the change in organization structure, but CFO Greg and the previous boss of Paul were quite unhappy with this change in the reporting structure. Paul wanted to use the opportunity of his direct reporting relationship with the CEO. He used his time judiciously and came out with an excellent framework for technology budget.

Paul realized that, until now, they had been spending around 4% of their revenue on IT. But to transition to digital business, they needed to invest in other technologies like communication and collaboration technologies and operational technology. Hence, the scope of the work of the technology organization must be extended. Now, due to the emergence of citizen ITs, they must also factor that in the budget. Accordingly, Paul proposed 7-8% of the revenue as the technology budget.

He justified the need for a higher technology budget to support the digital business by highlighting the changes needed in the existing IT elements and the new elements added due to the adoption of digital technologies.

Technology Elements	Information Technology	Digital Technology	Remarks	Approximate Technology Budget Share* (%)
Strategy	IT strategy	Digital technology strategy (strategic initiatives)	**Change**	5%
Governance	IT governance	Digital technology governance	**Change**	5%
Architecture	IT architecture	Digital technology architecture, centres of excellence	**Change**	10%
Intelligent products and assets	Not applicable	Digital products (software-defined cars) Digital assets (software-defined machines, equipment, physical infrastructure)	**New**	30%
Applications	Software applications, services, systems of record	Software applications, systems of records, systems of engagement, systems of things, systems of intelligence, digital platforms and digital marketplace	**Change**	30%

Infrastructure/data centre	IT infrastructure and services that include cloud computing, web, desktop, 2G, 3G, fibre optics	Digital technology infrastructure that includes edge computing, cloud computing, web, desktop, mobile, wearable devices, consoles, 6LowPan, ZigBee, LoRaWAN, 4G, 5G and fibre optics	**Change**	20%
Technology	IT	IT + communication technology + operational technology + others	**Change**	Total = 100%

Source: Author

Note: *Percentage distribution of technology budget is only indicative and it varies based on the organization's vision and digital journey roadmap.

Table 11.15: Technology Budget Distribution

Greg was livid and challenged Paul about doubling the technology budget and declared that it was impossible. Paul asked Greg the following questions:

1. Do you agree that we must now work on technologies beyond IT?
2. Do you agree that we need to invest in technologies to build intelligent products and shop floor that includes machines, equipment, logistics infrastructure?
3. Do you agree that we need to invest in high computing power and storage as we get data from applications, social media, intelligent machines and assets?
4. Do you agree that we need to spend on new devices like wearable devices, mobile devices and others?

5. Do you agree that we need to invest in new capabilities either by acquiring new-age technology companies or services from costly specialist firms?

Greg reluctantly agreed with Paul. On Tom's instructions, Greg decided to listen to what Paul had to say before moving on to optimizing the budget. Paul said that they needed to spend a considerable budget on the following:

1. Technology strategic initiatives, which they must invest in immediately to prepare for the next three years. They must no longer work in a reactive mode. They must understand market dynamics, changing customer needs and the competition.
2. Technology centres of excellence to build proofs of concept in collaboration with partners and service providers.
3. The technology team must closely work with the product R&D team while designing the products to understand technology interventions. Going forward, the products and assets will have millions of lines of software code.

Technology governance designed in the IT era does not work in the digital age. Digital technology governance will be complicated due to cybersecurity, big data, extended business ecosystems, intelligent products, assets and new-age software applications and infrastructure.

Paul suggested the distribution of the budget with respect to the budget line items as well.

S. No.	Budget Line Items	Distribution (%)
1	Prepare for digital business	25
2	Grow current business	20
3	Improve business operational excellence	45
4	Manage risks	10

Source: Author

Table 11.16: Technology Budget Distribution for Auto Gear

Tom asked Greg and Paul to meet him after they both come to a consensus about the budget before he gives final approval.

Paul concluded the discussion, saying that he was confident about

1. introducing new measures to improve technology performance with effective digital technology governance that helps driving digital businesses,
2. improving social responsibility,
3. aligning technology organization with business units that have been working to prepare for the digital era and
4. increasing the technology budget to factor in the digitalization of various business building blocks beyond business processes.

Technology Governance

Data Governance

Source: Author

Figure 11.15: Digital Governance—Data Governance

Paul arranged a meeting with the CEO, senior executives as well as his cybersecurity and data management specialists, Martin and Alan. He started the session with a few examples of how other organizations have been leveraging the organization's data using AI, ML and analytics to improve the business outcomes. In the past, the business world used to leverage only 15%, of the data available to it. This was structured data from enterprise resource planning (ERP), transactional systems and others. Due to technology fusion, the remaining 85%, comprising unstructured and semi-structured data emerging from different data sources (social media, IoT devices, web content and others), was available for obtaining business intelligence holistically.

Due to lack of data governance, a manufacturing company which had a footprint in different locations lost price advantage. The procurement systems being used by various departments in separate areas used a different supplier ID. Hence, even though each department received material from a common supplier, the organization lost the volume discount which it deserved.

Paul also shared the example of an automotive organization that maintained different legacy systems for customer order management and customer care service management. Hence, there was no single source of truth about the customer who interacted with various departments of the organization. Therefore, they missed out on offering a better experience to their customers.

The final example was of a manufacturing organization in which the citizen IT team in a business unit started using a cloud-based customer relationship management (CRM) system (on rental basis), without even letting the technology department/ organization know about it. There was no integration of this new CRM system with the existing systems of the organization. This forced the customers to share the same information repeatedly as they interacted with each department which upset them very much.

Paul summarized that without data governance or with poor data governance in the era of data storm, Auto Gear will have to face some serious issues as given in Figure 11.16.

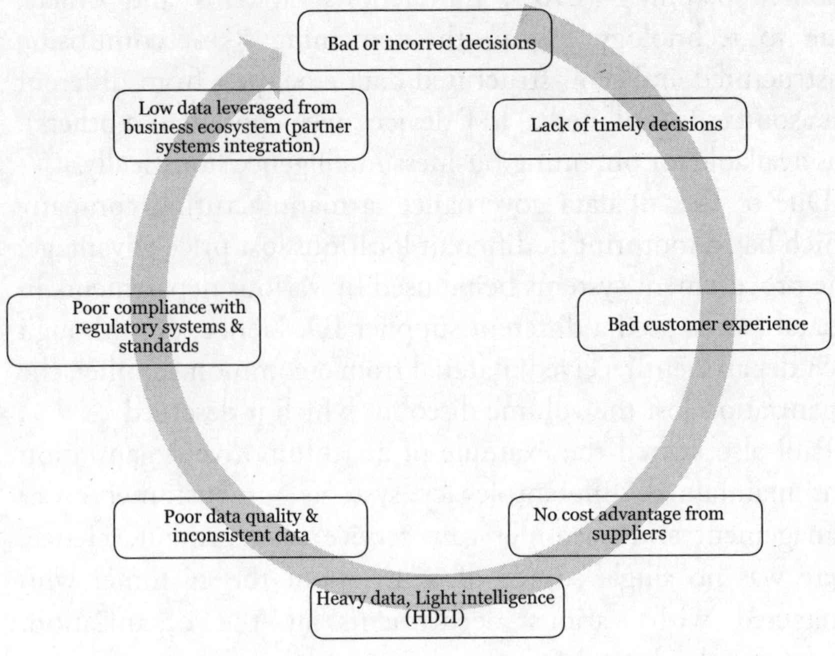

Source: Author

Figure 11.16: Impact of Poor Data Governance

The Troika of Data Management

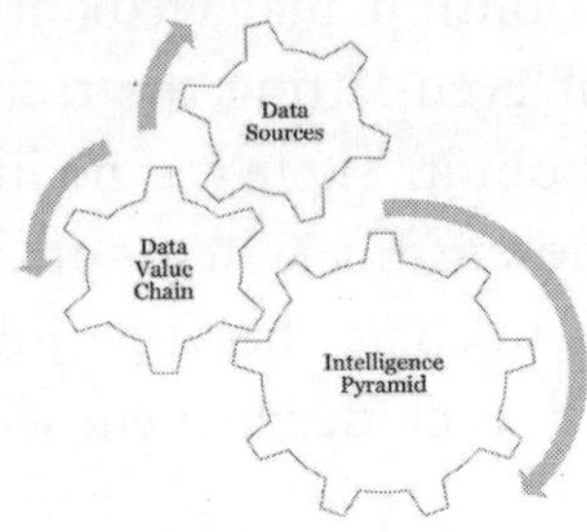

Source: Author

Figure 11.17: The Troika of Data Management

Paul wanted to give a bit more clarity to all the senior executives on how the data sources and data value chain had changed in the last few years due to technology fusion. He introduced the troika of data management that includes data sources, data value chain and intelligence pyramid. Paul specifically highlighted one of the critical elements of data management troika—the intelligence pyramid. He believed that Auto Gear could improve business predictability and business performance with the use of intelligence pyramid.

Data Sources

Auto Gear needed to deal with different types of data sets received from various sources. According to Paul, the key challenge ahead of Auto Gear was to determine how this data could be leveraged to

1. create new business opportunities,
2. create new revenue streams,
3. improve customer delight and
4. improve operational excellence.

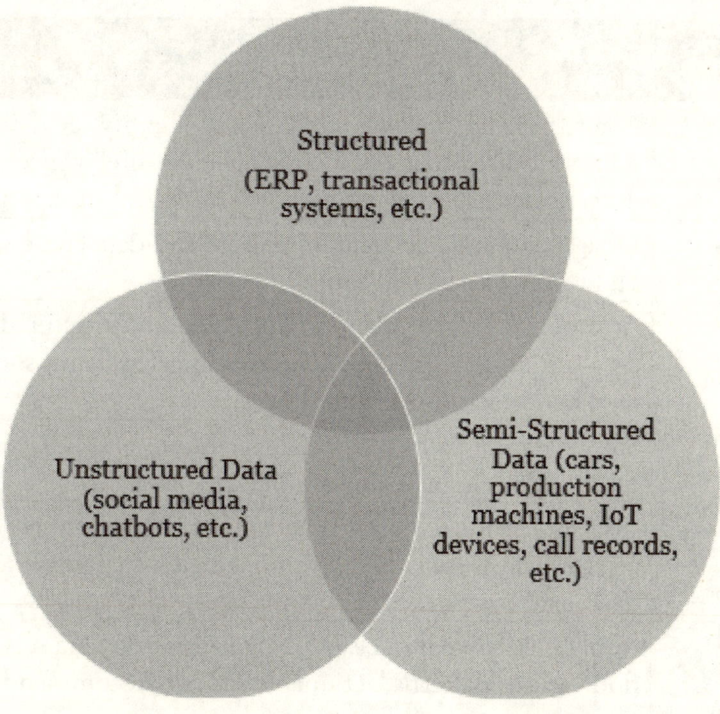

Source: Author

Figure 11.18: Data Sources

Data Value Chain

Figure 11.19 depicts the data value chain defined by Paul's team in collaboration with business stakeholders. Auto Gear did not have adequate data governance to manage the data value chain. He emphasized the importance of bringing a common understanding of the data value chain of Auto Gear among all stakeholders.

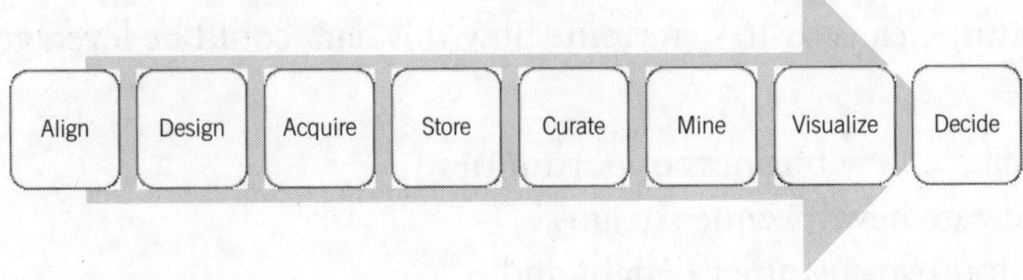

Source: Author

Figure 11.19: Data Value Chain of Auto Gear

Value Chain	Description
Align	Analyze the business value chain and identify the data points required to get real-time business intelligence to different stakeholders as per the context and situation
Design	Design data models, data warehouse and data lake based on business needs and wants
Acquire	Acquire structured, semi-structured and unstructured data from different sources (transactional systems, social media, devices, things and others)
Store	Store data in relational databases, not only SQL databases (NoSQL), lightweight mobile databases
Curate	Inspect, cleanse, transform and integrate the data received from various sources into a data warehouse and data lake
Mine	Apply predictive models, ML algorithms on data sets to find various patterns, actionable insights/recommendations

Visualize	View the intelligence (hindsight, insight and foresight) on mobile devices, wearable devices, desktops, consoles and others in the form of graphs, trends and auto recommendations
Decide	Take the right decisions, corrective actions and preventive actions as per the context and situation

Source: Author

Table 11.17: Description of Data Value Chain of Auto Gear

Intelligence Pyramid

Paul and his core team worked very hard to design an intelligence pyramid (Figure 11.20) that would help Auto Gear improve its current business performance, design business future and create new business opportunities. Only those organizations which leverage data, AI and ML will become leaders in the digital era. They were committed to implementing the intelligence pyramid in 18 months to take advantage of new digital opportunities.

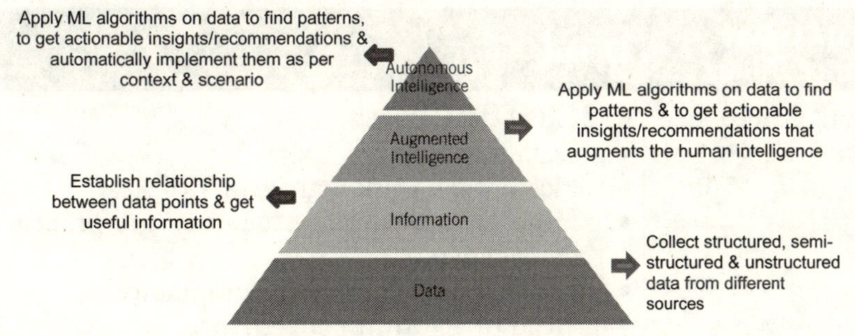

Source: Author

Figure 11.20: Intelligence Pyramid

Sophisticated data governance will be required to effectively manage data sources, data value chain and intelligence pyramid at Auto Gear. Paul and the team spent considerable time in designing Auto Gear's data governance model, which Paul shared with the CEO, the CFO and other senior executives of Auto Gear.

360-Degree View of Data Governance

Data governance is the effective management of data with the help of appropriate structure, policies, processes and tools. It includes planning, monitoring and controlling the data throughout the data value chain, and helps in translating the data into value and money.

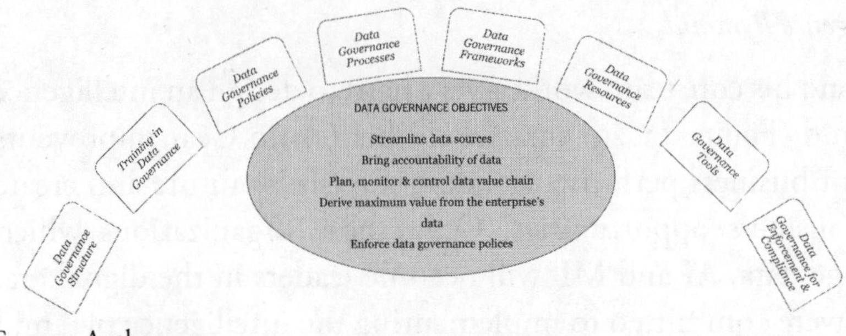

Source: Author

Figure 11.21: 360-Degree View of Data Governance

Data Governance Pillars	Details
Data governance structure	• Chief data officer • Data steward • Head of data risk management • Head of data governance policy and process management • Head of data regulatory compliance • Head of algorithm governance • Head of data value management
Training in data governance	• Training in data governance policies • Training in cybersecurity processes • Training in best practices of data governance • Training in data governance tools and techniques • Training in data sources • Training in crisis management • Training in data value chain • Training in intelligence pyramid

Data governance policies	• Data availability policy • Metadata management policy • Data authentication and authorization policy • Data security policy • Regulatory compliance policy • Data integration policy • Data integrity policy • Data innovation policy • Data privacy policy • Data monetization policy • Data value policy • Data audit policy • Algorithms development policy
Data governance processes	• Enterprise data strategy management process • Data value management process (data steward) • Data source management process • Lean data management process • Data change management process • Data risk management process • Data audit management process • Data planning process • Data monitoring and controlling process • ML algorithm management process • Data regulatory compliance process • Data management tools and technologies selection process • Auto-implementation management process of system recommendations
Data governance frameworks	• Stanford maturity model • EDM Council—DCAM (data capability assessment model) • ARMA (information governance maturity model) • DAMA-DMBOK (data management body of knowledge)
Data governance resources	• Data governance budget • Data governance specialists and team • Data governance tools and technologies • Physical infrastructure

Data governance tools	• Collibra • Truedat • Oveledge • Io-Tahoe • Talend • Informatica • IBM Data Governance
Data governance for enforcement and compliance	• Data risk analysis, mitigation and contingency planning • Data governance effectiveness—internal project management • Senior leadership oversight and commitment • Data management non-conformances reporting and prevention • Data governance key performance indicators and accountability • Internal data governance compliance audits

Source: Author

Table 11.18: Description of Data Governance Pillars

Paul concluded the session with the decision that they will create a detailed data governance strategy for the next 3 years and a comprehensive plan for the next 12 months. Tom was pleased with the progress made by Paul and the team, and gave the go-ahead to begin the work on data governance.

Cybersecurity Governance

Source: Author

Figure 11.22: Digital Governance—Cybersecurity Governance

In the next session, Paul shared his knowledge of cybersecurity governance. He began with examples of cybersecurity incidents in the automotive sector.[2]

1. A disgruntled former employee of a Texas auto center hacked into the company's computers and remotely activated the vehicle immobilization system and disabled more than hundred vehicles.
2. In a separate incident, hackers disabled the brakes of a car moving at 65 km/h by plugging in software from the laptop into OBD II port. They remotely controlled the car's door locks, headlights, wipers and horn while the car was in motion.
3. Japanese automotive company Honda was forced to halt production at one of its manufacturing plants when its computer network was infected by WannaCry ransomware.
4. Electric charge stations are nothing but minicomputers, and hackers can hack the identity information of the drivers who charge their vehicles at these stations. Later, hackers can perform transactions using the identity details of the driver without their knowledge.

Tom and Greg were deeply concerned about these security issues related to connected cars and smart factories, which are built with the fusion of many technologies. They were very keen to listen to Paul about Auto Gear's game plan to safeguard various software-enabled assets of the organization. Paul was confident that with sophisticated cybersecurity governance, they could prevent many cybersecurity threats and attacks. He cautioned that considerable budget needed to be factored in to integrate security throughout the business life cycle of Auto Gear.

Paul emphasized that, till now, Auto Gear had focused on securing information, software applications, servers and networks. Now, due to digitalization, almost all assets of Auto

2 Upstream Security Ltd. Upstream Security Global Automotive Cybersecurity Report 2019. Available at: https://industrytoday.com/wp-content/uploads/2018/12/Upstream-Security-Global-Automotive-Cybersecurity-Report-2019.pdf

Gear will be connected to the internet and will become highly vulnerable to cyberattacks. It was imperative to secure the digital assets of Auto Gear, given in Figure 11.23.

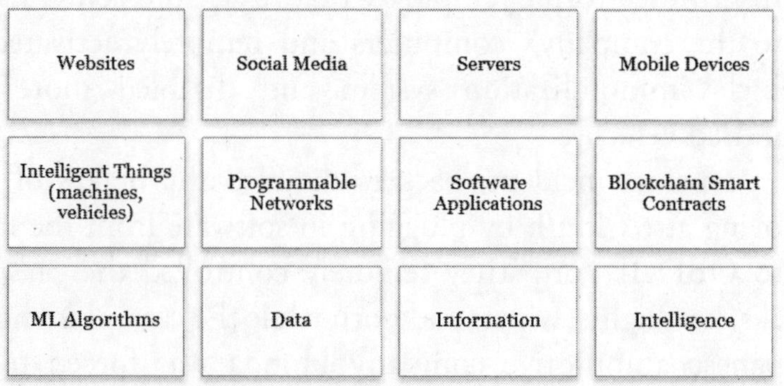

Source: Author

Figure 11.23: Digital Assets to Be Secured

Paul explained the work done by his team while designing a sophisticated cybersecurity governance framework. He had asked them to design a comprehensive cybersecurity governance framework. It would be challenging to implement all the elements of cybersecurity governance at the same time. Hence, Paul proposed that they establish cybersecurity governance evolutionarily.

Paul comforted Tom, Greg and other business executives that they will start small with prioritization based on business needs, budget and resources. Hence, there was no need to implement all improvements and initiatives at the same time.

360-Degree View of Cybersecurity Governance

Cybersecurity governance is the effective management of security with the help of appropriate structure, policies, processes and tools. It includes planning, monitoring and controlling various cybersecurity incidents to help safeguard the organizations' assets, people and brand.

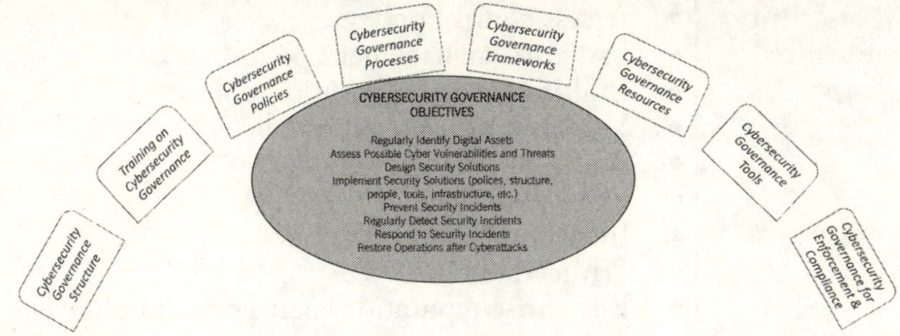

Source: Author

Figure 11.24: 360-Degree View of Cybersecurity Governance

Cybersecurity Governance Pillars	Details
Cybersecurity governance structure	• Chief cybersecurity officer • Head of identify and access management • Head of cybersecurity programme management • Head of things and assets security governance • Head of product engineering cybersecurity governance • Head of cybersecurity policies, processes and compliance • Head of cyber risks and threats prevention • Head of incident prevention and response acceleration • Head of applications security governance • Head of networks security governance • Head of digital infrastructure security governance • Head of data security and privacy governance • Head of cybersecurity training
Training in cybersecurity governance	• Training in cybersecurity policies • Training in cybersecurity processes • Training in best practices of cybersecurity governance • Training in cybersecurity tools and techniques • Training in cybersecurity technologies • Training in business continuity and disaster recovery management • Training in crisis management

Cybersecurity governance policies	• Access control policy • Password management policy • Multifactor authentication policy • Vulnerability detection policy • Audit policy • Firewall policy • Intrusion prevention policy • Privacy policy • Enterprise reputation management policy (social media) • Device control policy • Data protection policy • Remote access policy • Network management policy • BYOD policy • Incident response policy • Disciplinary policy • Encryption policy • Software code development policy
Cybersecurity governance processes	• Asset catalogue management • Asset protection management • Asset life cycle vulnerability detection management • Cybersecurity risk management • Cyberattacks detection management • Incident response management • Communication management • Disaster recovery management • Business continuity management • Identify and access management
Cybersecurity governance frameworks	• PCI DSS (Payment Card Industry Data Security Standard) • ISO 27001/27002 • CIS Critical Security Controls • NIST Cybersecurity Framework
Cybersecurity governance resources	• Cybersecurity budget • Cybersecurity specialists • Cybersecurity tools • Physical infrastructure

Cybersecurity governance tools	**Application security** • Arxan Application Protection • Black Duck Software • Burp Suite • Veracode • Immuniweb® Mobilesuite • Kiuwan **Files and URL Scanning** • Virus Total **IoT Security** • AWS IoT Device Defender • Azure IoT hub • Google Cloud IoT Core • FortiNac **Network Security** • Wireshark • Air Cack • Metasploit • Snort • Cain and Abel **Data Security** • Oracle Audit Vault and Database Firewall • IBM Guardium • MSSQL Data Mask • Scuba • AppDetectivePro **Anti Virus** • McAfee • TotalAV • Norton • BullGuard
Cybersecurity governance for enforcement and compliance	• Cybersecurity operations excellence • Risk analysis, mitigation and contingency planning • Cyberattacks prevention tools implementation planning • Cybersecurity excellence—internal project management • Business continuity and disaster recovery planning • Senior leadership oversight and commitment • Security non-conformances reporting and prevention • Disinfecting schedule for cleansing the potential virus affected devices and things • KPIs and accountability • Internal security compliance audits

Source: Author

Table 11.19: Description of Cybersecurity Governance Pillars

Paul concluded the session with the assurance that his team will submit a detailed cybersecurity governance strategy for the next 3 years and a comprehensive plan for the next 12 months. They decided to begin work next month. Tom was pleased with the progress made by Paul and his team. But Tom was anxious about his team's preparedness for robust security to safeguard Auto Gear's assets, customer privacy and safety because of the recent cybersecurity crimes in the automotive sector.

Business Ecosystem Governance

Source: Author

Figure 11.25: Digital Governance—Business Ecosystem Governance

Once the group had convened for the next session, Paul took a deep breath and said that automotive companies may not survive without building a business ecosystem powered by digital technologies. Auto Gear had not yet started any work on this. But in the last three months, Paul and his team had put in considerable effort in coming up with a framework to nurture the Auto Gear business ecosystem.

Greg interrupted Paul and said that after listening to Paul for the last few days, he felt as if they will be out of business in a few years. Then, he asked a question, "Are we thinking too much about digital? Is it only apprehension or reality?" Paul smiled and replied that six months ago they used to think the same, but after spending a considerable amount of time on the way the auto market was moving, they found eye-opening facts. Hence, they had been working to completely reimagine and re-engineer the entire technology organization of Auto Gear to enable new business models and discover new opportunities.

Paul shared that new-age customer behaviour and expectations were different. Businesses built only around their own ecosystem limited to particular processes, policies, products and services as well as software systems will not survive for a long time. It was critical that they understand how to create an extended ecosystem with various suppliers and partners.

The team had realized that business ecosystems today constitute suppliers, consumers, start-ups, advertisers, government agencies and even the competition. This meant the organization must be agile enough to work with new ecosystem partners as per the changing needs of the business.

Tom concurred with Paul and shared his view that just selling cars will not make it a competitive automotive company. An ecosystem in collaboration with other non-automotive products and services for value exchange would help them create new opportunities and reach new customers.

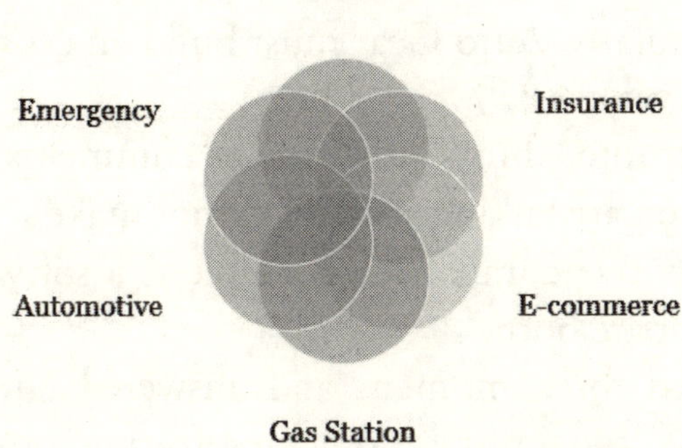

Source: Author

Figure 11.26: Example of an Automotive Business Ecosystem

Greg admitted that he was overwhelmed by the information shared by Paul and the team. To implement the initiatives proposed by their CIO, it will take considerable amount of resources and budget, and he was still not sure about Auto Gear's capabilities to prepare for the digital future. He wanted to know whether there will be any value for money.

While Tom agreed with Greg's views and said that there was no choice except investing in digital capabilities if they wanted to be in business at least 10 years from now. Tom asked Paul to go ahead and share their game plan in building Auto Gear's digital business ecosystem.

Paul started with a few examples of how digital natives like Apple, Microsoft and Google had built a business ecosystem around their products. The lifespan of a product is very short, but the lifespan of an ecosystem is longer. Similarly, Auto Gear, which was not a digital native also needed to build an ecosystem in the context of the automotive industry.

For example, Apple iPhone and iPad are the products. Around these products, Apple has built an ecosystem with the help of millions of developers who can develop many apps that can be accessed by millions of iPhone and iPad users. Hence, it is tough for the users to come out of the experience they get through many apps running on their iPad/iPhone. This means Apple started offering services and experience on top of their products. Similarly, Auto Gear must build an ecosystem around its products (vehicles).

Greg interrupted him with cynical comments, "It's enough Paul. What you are talking about does not make sense to me. We are an automotive company and Apple is a software company. We can't compare both."

Paul paused for a moment and answered Greg in a single sentence, "We too must become a software company, else we will sunset very soon." He requested Greg to listen to him to understand the importance of an automotive business ecosystem driven by software.

Greg sat down impatiently. Paul continued to explain the concept of the business ecosystem using the following example. There are producers and consumers whom they needed to bring on to their digital platform. Till now, they had been used to selling cars/vehicles as products, and they did not put any effort into providing superior services and experiences to their customers around their products, which was the need of the hour in the digital era.

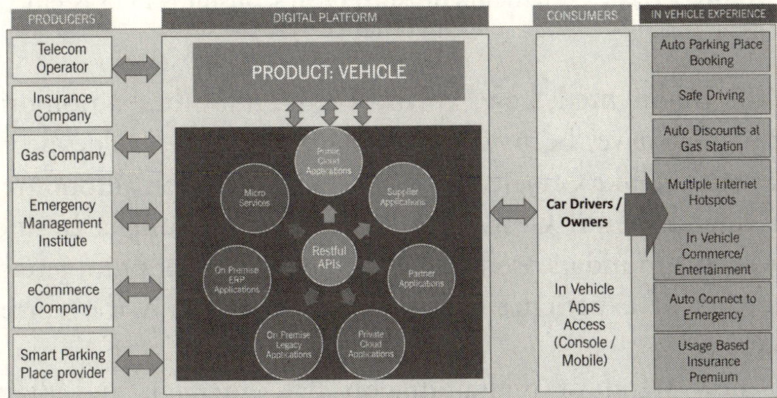

Source: Author

Figure 11.27: Digital Business Ecosystem Elements of Auto Gear

Ecosystem	Description
Producers	• **Telecom provider:** Provides 6–7 internet hotspots within the vehicle so that passengers can connect to the internet within the vehicle. • **Insurance company:** Automatically monitors the driving style of the driver, provides safe driving instructions and calculates the insurance premium based on the driving history. • **Gas company:** While filling the vehicle at the gas station, provides discounts automatically. • **Emergency management institute:** In case of an accident or any emergency, allows the driver to make a call from the car console. In case the air-bags open due to crashing, automatically makes a call to an emergency number. • **E-commerce company:** Allows the passengers from the car to order any products/services using a mobile phone or the in-vehicle console app. • **Smart parking place provider**: Allows the driver to book a parking place in advance.

Digital platform	The software platform that integrates the systems of the **producers** (third-party service providers) with the systems of the **digital platform** and **product owner**.
Consumers	Car owners and drivers.

Source: Author

Table 11.20: Elements of Auto Gear's Business Ecosystem Governance

Paul highlighted how Volvo, Ford and other automotive companies have been transforming themselves. Unexpected competition, like Google, has also been working on autonomous cars. Hence, Auto Gear must strategize and build a business ecosystem around its products immediately to offer unforgettable moments of experience to its customers beyond the current products.

A new trend of digital dragon has emerged, and digital immigrants have been facing serious challenges from these digital dragons. Gartner defines digital dragons as organizations with the following characteristics[3]:

1. Invest massively in R&D of new technologies, which others can't beat
2. Game-changing AI/ML algorithms
3. Extraordinary supply chain and logistics infrastructure and system
4. Offer exceptional customer experience
5. Provide cloud infrastructure as a service (IaaS)
6. Provide software as a service (SaaS)

Amazon acquired Zoox, a California-based company which designed autonomous ride-hailing vehicles. This means it has

3 Available at: https://www.gartner.com/en/doc/367758-winning-in-a-world-of-digital-dragons#:~:text=Winning%20in%20a%20World%20of%20Digital%20Dragons%20Digital,dragons%20are%20a%20critical%20piece%20of%20that%20story

entered the automotive segment. Google is already into banking and auto segment.

Today's digital dragons are Alibaba, Amazon, Baidu, Google and Tencent and a few others may emerge in future. These digital dragons possess unmatched capabilities in digital, supply chain and customer experience capabilities, and they could get into any industry segment easily.

Digital immigrants must create a long-term strategy to partner with digital dragons to enhance their digital capabilities to introduce new business models, products and services.

In a few cases, digital immigrants may need to compete with digital dragons. Hence, digital dragons are called frenemies (business friend + business enemy to digital immigrants).

360-Degree View of Business Ecosystem Governance

Business ecosystem governance is the effective use of structure, policies, processes and tools to expand an organization's capabilities and services. This is done by enrolling partners (producers and service providers) and consumers into one digital platform to offer extended services and unforgettable moments of experience to customers apart from current products and services.

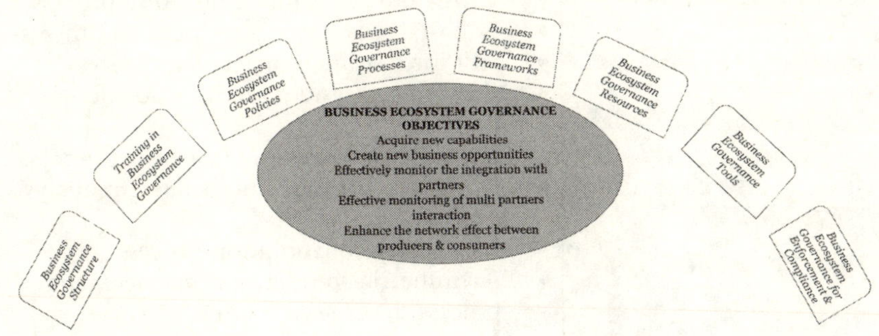

Source: Author

Figure 11.28: 360-Degree View of Business Ecosystem Governance

Business Ecosystem Governance Pillars	Details
Business ecosystem governance structure	• Chief business ecosystem officer • Head of producer and consumer feedback governance • Head of consumer behaviour governance • Head of producer curation governance • Head of API governance • Head of producer and consumer interaction governance
Training in business ecosystem governance	• Training in business ecosystem governance policies • Training in business ecosystem governance processes • Training in best practices of business ecosystem governance • Training in business ecosystem governance tools and techniques • Training in crisis management • Training in digital platforms features
Business ecosystem governance policies	• API management policy • Consumer and producer acquisition policy • Consumer and producer rewards and recognition policy • Regulatory compliance policy • Multiparty value design policy • Multiparty confidentiality and trust policy • Service disruption by other parties policy
Business ecosystem governance processes	• Multiparty interaction management process • Cross-industry value innovation process • Long-tail customer identification process • API integration management process • Business ecosystem service value design process • Asymmetric growth management process • Consumer attitude and behaviour analysis process • Multiparty collaboration process • Asymmetric competition management process • Matchmaking design process • Multiparty contract management and risk ownership process • Ecosystem parties feedback management process • Ecosystem parties acquisition process

Business ecosystem governance frameworks	• The Platform Manifesto* • Matchmaking algorithms governance • Core value unit governance • Rewards and incentives governance
Business ecosystem governance resources	• Digital platform governance resources budget • Digital platform management tools procurement budget • Physical infrastructure budget • Training budget
Business ecosystem governance tools	• Apigee • 3scale • Akana • IBM
Business ecosystem governance for enforcement and compliance	• Multiparty service dependency risks, mitigation and contingency planning • Business ecosystem improvement—internal project management • Senior leadership oversight and commitment • Business ecosystem non-conformances reporting and prevention • Business ecosystem governance KPIs and accountability • Internal business ecosystem governance compliance audits

Source: Author

Note: *Sangeeta Paul Chowdary describes The Platform Manifesto (pipes to platforms); Source: https://platformed.info/the-platform-manifesto/

Table 11.21: Business Ecosystem Governance Pillars

At the end of the session, Paul had assured the senior leaders that his team will be ready with the 3-year business ecosystem governance strategy for Auto Gear. Furthermore, they will be ready with a comprehensive plan for the next 12 months.

Technology Systems Management

Technology Organization Structure

Source: Author

Figure 11.29: Technology Systems Management—Technology Organization Structure

There had been no changes in the technology organization structure of Auto Gear for the last 10 years. Paul realized that the existing outdated organization structure will not help prepare Auto Gear for its digital journey. Also, due to new technological challenges, compliance issues and security risks, considerable changes needed to be brought in the technology organization. Tom advised Paul to meet Andrew, the chief strategy officer (CSO), to discuss this further.

Andrew was known as a tough guy who asks too many questions until he is satisfied. Paul met Andrew and shared his agenda for the meeting. Andrew asked Paul why he wanted to change the technology organization structure now.

Paul explained that for more than 20 years, their technology organization had been working with the "inside out" approach as a supporting role. Now, in the digital era, technology organizations must function beyond supporting role. Andrew concurred with Paul.

According to Paul, today's CIO must be the **chief intelligence officer** and not the chief information officer. Technology fusion has been creating new business opportunities, and without the technology organization's readiness, traditional businesses will sunset. Paul shared his views on how the scope of technology organizations was changing and it required a new organization structure.

Scope	From	To
Role	Chief information officer	• Chief intelligence officer
Technology operations	Information technology	• **C**ommunication and collaboration technology • **I**nformation technology • **O**perational technology
Automation	Processes	• Processes • Stakeholder's emotions • Things/products/assets
Intelligence	Hindsight	• Hindsight • Insight • Foresight (AI, algorithms, etc.)
Innovations	Automated business processes	• Intelligent business processes • Digital business models • Intelligent products • Intelligent customer services • Intelligent workplaces

Source: Author

Table 11.22: Scope of Technology Organizations

Andrew smiled at Paul and remarked, "It seems you want to take my role as the CSO." Paul replied, "It's not taking your role, Andrew. It is complimenting your efforts to bring the technology innovations that help you and other business leaders in creating unique business strategies, business models, products and services." Andrew was quite happy with the justification given by Paul.

Guiding Principles for Technology Organization Design

Paul further emphasized that they needed to introduce new roles, behaviours, processes and policies to transform the technology organization. Andrew asked how could he be of help. Paul wanted to share the indicative organization structure which he and his team had designed for the technology organization. But before that, Andrew asked Paul to share the guiding principles used by Paul while designing the structure. The guiding principles used by Paul and his team are listed below:

1. Play the role of strategic innovator, consultant and system integrator to create new business opportunities in the digital era.
2. Prepare the whole organization for digital business by enhancing digital skills, behaviours and responsibilities.
3. Empower citizen IT teams (technology teams within business units) to work as per organization standards and policies.
4. Build agility into the structure to respond to rapid changes in technologies.
5. Strengthen technology research and innovations.
6. Effectively manage new risks and threats that emerge in the digital era.
7. Comply with new global standards in line with digital businesses.
8. Deliver technology services faster, better and **consistently**.

Building Blocks of Technology Organization Structure

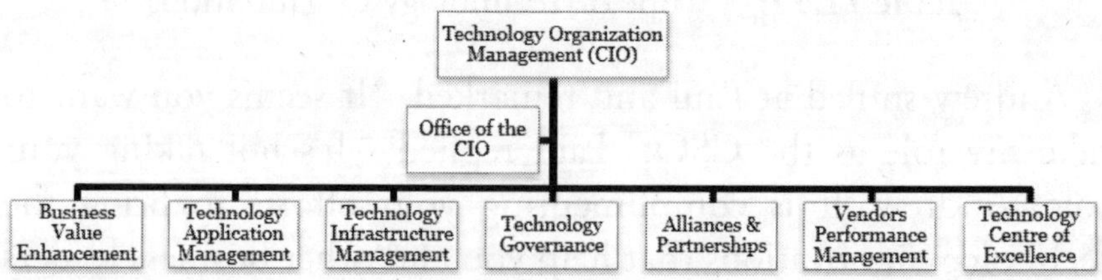

Source: Author

Figure 11.30: Building Blocks of Technology Organization Structure

S. No.	Technology Organization Structure Building Block	Role and Focus	Execution Model
1	Technology organization management	**Chief Intelligence Officer** (it's no more chief information officer) The convergence of IT, communication and collaboration technology and operational technology; it is beyond IT	Internal
2	Business value enhancement	**Head of Digital Business Innovation** (a role like the chief digital officer) Technology-driven business strategies, business model innovations, intelligent products, customer service innovations, business relationship management and technology strategy	Internal and with support from external specialists
3	Technology application management	**Head of Software Applications** Application life cycle management (fusion of IT, communication and collaboration technology and operational technology)	Internal, outsourcing, crowdsourcing
4	Technology infrastructure management	**Head of Technology Infrastructure** Infrastructure life cycle management	Internal, outsourcing, crowdsourcing
5	Technology governance	**Head of Technology Governance** Policies, processes, culture, talent, business continuity, data governance and security governance.	Internal and with support from external specialists

6	Alliances and partnerships	**Head of Strategic Alliances** Start-ups, technology vendors and specialist firms	Internal
7	Vendors performance management	**Head of Vendor Performance** Supplier/vendor contract performance	Internal
8	Technology centre of excellence	**Head of Technology Centre of Excellence** Technology architecture, R&D and proof of concepts	Internal and with support from external specialists

Source: Author

Table 11.23: Building Blocks of Technology Organization Structure

Andrew commented that the proposed structure was huge and needed a lot of investment and leadership bandwidth. Paul clarified that it was a holistic blueprint of the technology organization. They will not be focusing on all the building blocks at the same time. It all depended upon the business plan and budget availability. Paul emphasized that it was very clear that they must shape their technology organization in the next 18–24 months to survive and excel in the digital era. He candidly shared that at Auto Gear, the technology organization usually focused on the initiatives to be completed in next year and did not focus on long-term plans. Hence, the organizational structure needed to be refined with a focus on long-term strategy and initiatives too.

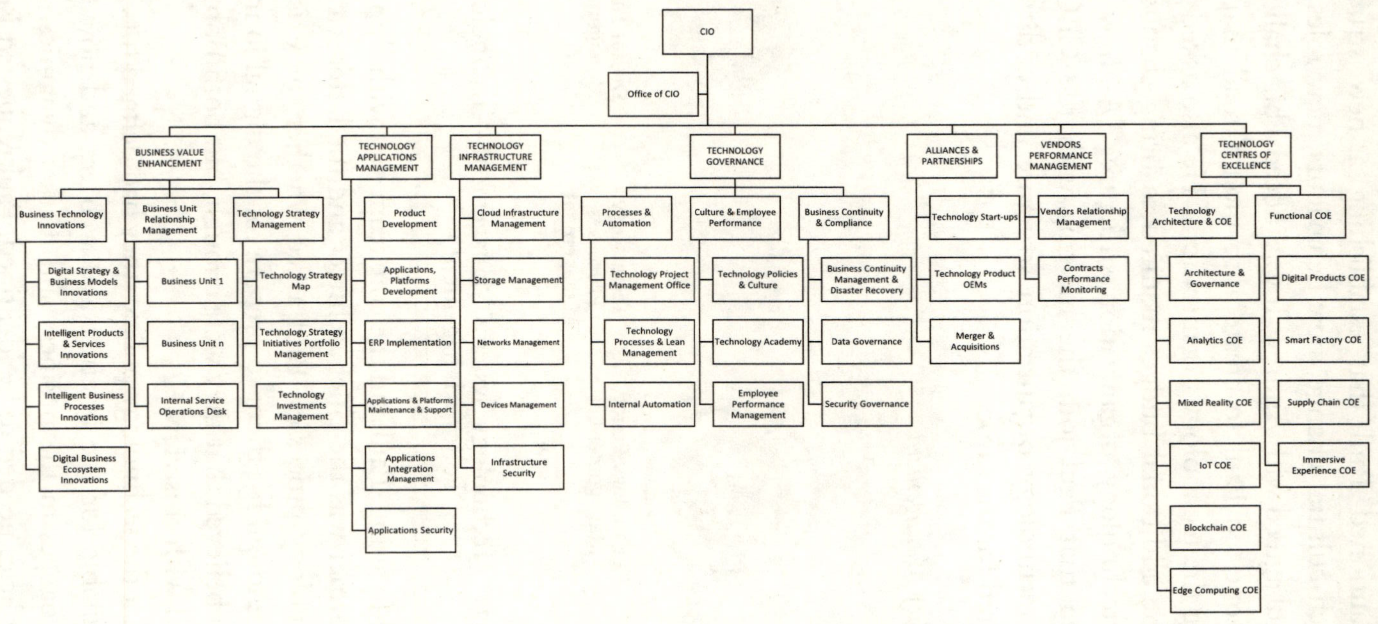

Source: Author

Figure 11.31: Expanded View of Technology Organization Structure

Andrew cautioned Paul that establishing a new structure involved much cultural change, new behaviours, new roles and new skills. Paul agreed with Andrew. It was indeed a big challenge.

Andrew requested Paul to create a 12-month road map for strengthening the technology organization. He also asked Paul to provide the additional budget required to bring changes in the current technology organization structure. Andrew also recommended that Paul join the meeting with the CEO to discuss new technology organization structure and additional budget required for the same.

Technology Policy Management

Source: Author

Figure 11.32: Technology Systems Management—Technology Policy Management

At the next meeting, Paul and the core team met with Bob and Roger. The former was the chief architect and the latter was the head of technology processes. Paul asked whether the existing policies were enough for them to prepare for the digital journey. Most of them believed that they were enough, but both Bob and Roger differed with this view.

Bob said that the technology policies they had defined for Auto Gear were suitable for the internet and web era and now, due to so many changes in the business and market dynamics, new technology policies needed to be identified. Paul requested Roger to share a list of technology policies which Auto Gear may need.

Raj, another member of the core team, sarcastically said, "Paul, enough with these policies, our innovation is dead and many things are getting delayed. Unless we simplify the polices and make them business-friendly, there is no point defining new policies."

Roger concurred with Raj and said, "Policies are like business rules or statements that guide the organization to take the right decisions. With the right and relevant policies, organizations can effectively implement appropriate processes and procedures."

He went on to share a few guiding principles to define the technology policies in the digital era. Technology policies must

1. align with organizational strategy and objectives,
2. reduce redundancy and expedite the decision-making process,
3. empower the leaders to take decisions without depending on the higher-ups,
4. must be unambiguous and unbiased,
5. agile and flexible, to align with changing needs, and
6. promote a positive culture and work environment.

He also proposed a few new technology policies that he felt were essential for Auto Gear's readiness for digital business. Roger added that these were only representative policies and new policies must be added continuously.

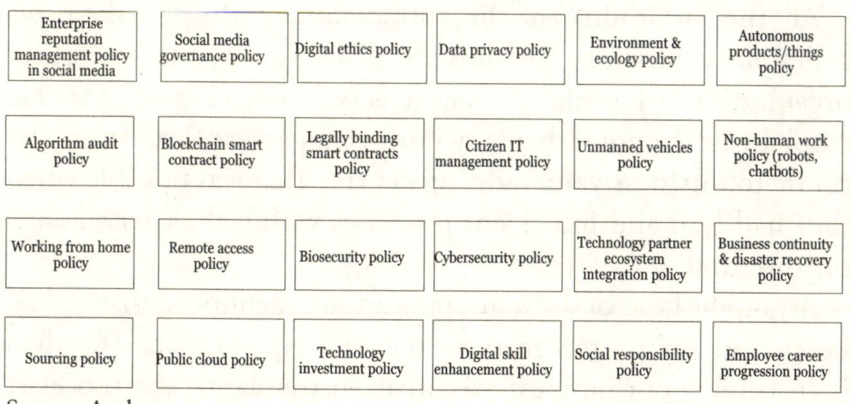

Source: Author

Figure 11.33: New Technology Policies for Auto Gear

Since most of these policies were new, Roger suggested that once the detailed policies were released, training should be conducted for the employees.

Technology Process Management

Source: Author

Figure 11.34: Technology Systems Management—Technology Process Management

Roger was very familiar with the end-to-end processes of the technology organization of Auto Gear. He firmly believed that technology processes were very important business processes as every business in the digital era is considered to be a technology business. Paul requested Roger to present his thoughts to the senior executives of the technology organization.

At the beginning of his presentation, Roger asked his colleagues a question, "What is the value chain of our technology organization?" He did not get a very convincing answer. He emphasized the point that they must change their focus from cost centre to **business value-adding centre**. That was possible when they had lean and intelligent processes within their technology organization.

Arjun, the head of software applications, mentioned that he had never seen any process map for their IT organization. They had been helping out business executives in translating the processes of their line of businesses into software applications. He felt that it would be quite interesting to automate their technology management processes as they did for the business units.

Roger shared six core processes and seven enabling processes to deliver exceptional value to business units and to manage technology operations effectively and efficiently.

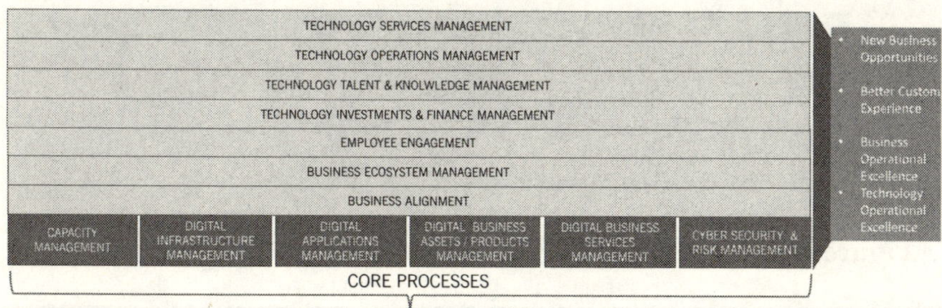

Source: Author

Figure 11.35: Core and Enabling Processes for Technology Process Management

Roger continued to explain each process in detail to provide better clarity to his colleagues. He advised Paul and the other leaders about the need to digitalize processes using mobile, social, analytics, AI, ML and other technologies to make the technology organization agile and predictable to support the digital business of Auto Gear. Arjun asked Paul whether there was any separate budget for digitalizing the technology process management, which was more internally focused.

Paul said that he had already added that line item in the new technology budget template. He asked Arjun to closely work with Roger to finalize the budget for digitalization of technology process management.

The process map, which Roger had shared with Paul and the team, gave an exceptional insight to the whole team. They realized that by building capable and mature technology management processes, the technology organization will be geared up to support the digital business journey of Auto Gear effectively.

BUSINESS ALIGNMENT	EMPLOYEE ENGAGEMENT	TALENT & KNOWLEDGE MANAGEMEENT	TECHNOLOGY SERVICES MANAGEMENT	BUSINESS ECOSYSTEM MANAGEMENT	TECHNOLOGY OPERATIONS MANAGEMENT	INVESTMENTS & FINANCE MANAGEMENT
Citizen IT Management	Employee Performance Management	Digital Technology Talent Management	Digital Technology Service Portfolio Management	Digital Technology Alliances Management	Hyper Automation Management	Digital Technology Strategy Management
Legal, Regulatory, Environment Compliance	Employee Career Progression Management	Digital Technology Knowledge Management	Digital Technology Services Continuity Management	Digital Technology Start-ups Management	Digital Technology Policies Management	Technology Investments Management
Digital Technology Talent & Knowledge Management					Digital Technology Suppliers Management	Finance Management
Social Responsibility Management					Technology Service Level Management	
Technology Change Management					Business Relationship Management	

Source: Author

Figure 11.36: Enabling Processes for Technology Management

S. No.	Technology Process Management	Details
1	Capacity management process	Capacity management is a very critical process that covers the following essential tasks: • **Identify technology services** to translate business needs and demand into new technology services or change current technology services. • **Refine service-level agreements (SLAs)** as per the changes in technology services. • **Identify resources/technology assets** to be built or procured to support the new technology services or changes in the existing technology services. • **Monitor service performance** to continuously monitor the performance of the technology services and refine the capacity as per the changing needs of the business.
2	Digital infrastructure management process	• Infrastructure sizing management • Application deployment management • Cloud infrastructure life cycle management

3	Digital applications management process	• Application portfolio management • Application life cycle management • Right sourcing management (outsourcing, insourcing, crowdsourcing or hybrid sourcing)
4	Digital business assets/products management process	• Software-defined product/asset life cycle management (e.g., intelligent car/machine)
5	Digital business services management process	• Software-defined customer service life cycle management (e.g., immersive experience with MR technology for customer inquiries)
6	Cybersecurity and risk assessment process	• Identity and access management • Security vulnerability assessment • Security threat prevention management • Security threat detection management • Operations restoration management
7	Business alignment process	• Technology management within business units (citizen IT) • Legal, regulatory, environment policies and rules compliance management • Social responsibility management • Technology change management
8	Business ecosystem management process	• Design business ecosystem • Identify interaction/interface points with partners and suppliers • API life cycle management (internal and external) • Integration management with partners using APIs
9	Employee engagement process	• Employee performance management • Career progression management • Rewards and recognition management

10	Technology investment and financial management process	• Technology service prioritization • Estimate the technology asset and resource requirements • Return on investment and business case management • Create technology budget covering strategic, operational and R&D initiatives • Manage technology expenditure • Track the budget against actuals
11	Technology talent and knowledge management process	• Learning strategy management • Technology service portfolio and skill identification management • Training partners management • Knowledge discovery, acquisition and usage management • Portal management for knowledge lifecycle management
12	Technology operations management process	• Technology process automation management • Technology policy management • Technology supplier relationship management • Technology partners/start-up relationship management • Service level management • Internal business relationship management
13	Technology service management process	• Technology service portfolio management • Technology service continuity management

Source: Author

Table 11.24: Core and Enabling Processes for Technology Management

Paul firmly believed that the elimination of (non-value-adding activities) in technology organization was imperative, and each employee must contribute to the same. He requested Roger to make sure that while designing processes for technology management, lean techniques and tools must be integrated. Roger shared the eight non-value-added activities, mentioned in Table 11.25, which the technology organization must address.

S. No.	Waste	Description
1	Waiting time	• Poor automation of technology management processes and high manual interventions • Poor performance of software applications and infrastructure • Delays in getting approvals instead of depending on business rules-based software (e.g., blockchain smart contracts)
2	Overproduction (features)	• Developing applications, reports and dashboards beyond the requirement • Establishing infrastructure beyond the requirement • Digitalizing processes and assets without prioritization
3	Rework	• Negligence on defect prevention practices, automated development and testing tools leads to poor quality and high rework
4	Excessive travel	• Too many people travelling to resolve business requests • Too much travel to resolve hardware/software instead of using remote monitoring and controlling system
5	Overprocessing	• No automation and non-alignment of technology processes output to improve the business performance
6	Inventory	• Low or too much usage of computing power (servers, clouds) • Too many unused mobile devices, desktops • Too many suppliers • Too much data collection and poor usage • Too many AI/ML algorithms and poor usage • Too many dashboards and poor usage

7	Energy consumption	• High electric consumption with poorly designed data centre, infrastructure and applications
8	Unutilized talent	• Poor digital innovation • No digital knowledge management • Reinventing the wheel without leveraging existing best practices and lessons learned

Source: Author

Table 11.25: Non-Value-Added Activities of Technology Process Management

Digital Talent Management

Source: Author

Figure 11.37: Technology Systems Management—Digital Talent Management

Sriram, the head of technology academy, had done extensive work to identify the skills required for roles at operational, tactical and strategic levels. Sriram reiterated that to be successful in the digital era, new roles, skills and behaviours must be introduced within the technology organization.

The Talent Building Pyramid in Figure 11.38 depicts the various skills required for employees at operational, tactical and strategic levels to perform better in the digital era.

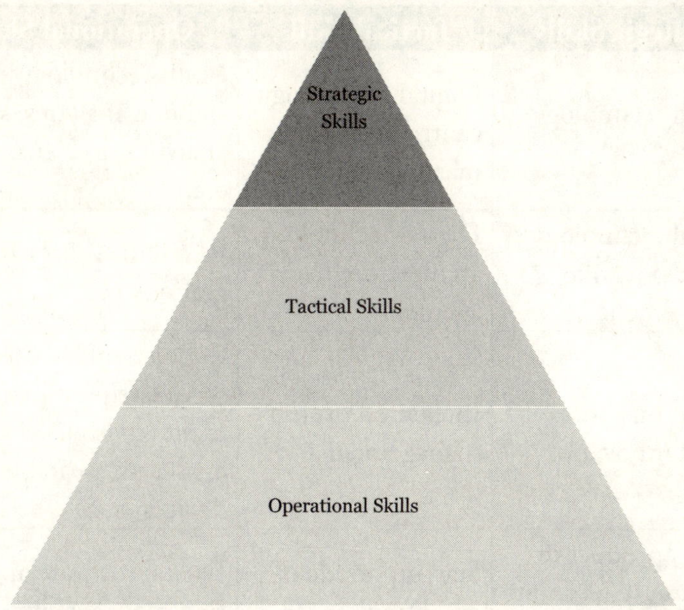

Source: Author

Figure 11.38: Digital Talent Building Pyramid

1. **Strategic skills:** Skills that are required to design the future that includes structure, new roles, new products, new services, new value proposition and long-term plans to achieve the strategic objectives.
2. **Tactical skills:** Skills required to design, create and manage the policies, processes, procedures, methods and plans to meet the strategic objectives within the stipulated period (365 days focus).
3. **Operational skills:** Skills required to perform daily activities by adhering to the policies, processes and procedures so that the technology organization delivers technology services faster, better and steadily (daily focus).

Sriram also shared a few illustrative skills required within the technology organization.

Strategic Skills	Tactical Skills	Operational Skills
Digital technology strategy	Digital technology centre of excellence management	Multi-technology skills (mobile analytics, social analytics, IoT analytics, etc.)
Digital technology service portfolio management	Digital technology architecture management	RESTful API for integrations
Digital business model innovation	Integrated project management	Omnichannel-based service request management (chatbots, virtual assistants, wearable devices, etc.)
Digital product innovation (e.g., smart cars, smart bikes)	Start-up mode of execution	Digital software application management
Digital service innovation (e.g., omnichannel order taking)	Data, information and intelligence management	Digital technology infrastructure management

Source: Author

Table 11.26: Digital Skills

Most in-demand skills for 2022 as per the World Economic Forum[4] (WEF)

- Analytical thinking and innovation
- Active learning and learning strategies
- Creativity, originality and initiative
- Technology design and programming
- Critical thinking and analysis
- Complex problem-solving
- Leadership and social influence
- Emotional intelligence
- Reasoning, problem-solving and ideation
- Systems analysis and evaluation.

4 Available at: https://www.coorpacademy.com/en/blog/learning-innovation-en/world-economic-forum-the-soft-skills-to-prepare-employees-for-the-future-of-work/

Technology Entrepreneurship

Citizen IT Management

Source: Author

Figure 11.39: Technology Entrepreneurship—Citizen IT Management

With the implementation of new technologies, Paul found his inbox flooded with complaints from customers. When CEO Tom came to know of the situation, he called up Paul and expressed his concerns. Paul and his core team carried out a root-cause analysis and found some surprising facts.

Auto Gear was an organization in which business units established **shadow IT** without the knowledge of the technology organization. This meant a few technology-savvy users within the business units had been implementing a few cloud-based applications and had started giving access to their employees, partners and customers.

Challenges with Shadow IT

Paul and his team realized that if this trend was allowed to continue, it will create many silos of technology hubs within Auto Gear business units. Each business unit will follow different standards. Figure 11.40 depicts the challenges posed by shadow IT.

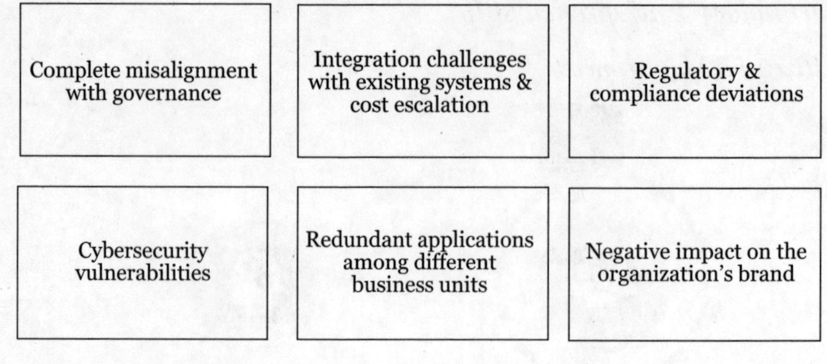

Source: Author

Figure 11.40: Challenges of Shadow IT

Paul requested Bob and Roger to work on streamlining technology management processes for Auto Gear to support business executives and leverage their software development effort to add value to the business.

Business executives familiar with Excel macros, which automate repeated tasks in the Excel sheet, could create useful Excel macros for their work. But they were not comfortable with the responses from the traditional IT organization like "we will take a few weeks to come back to you" or "we can't accept new requirements now due to other priorities".

Today, they have alternate options to get software applications and infrastructure as a service on a pay-as-you-go model. Hence, they started occasionally using external applications and infrastructure without the knowledge of the technology organization.

Now, technology-savvy business executives want to go beyond Excel macros and software as a service (SaaS) applications and develop applications using the latest low-code technology platforms. In addition to this, SaaS apps are readily available for quick deployment.

The Emergence of Citizen IT

Paul and the team realized that there is value in shadow IT if it is managed well with proper governance. Business executives could become an additional resource pool for the technology organization and help develop a few simple software applications. But shadow IT should be driven under the organization umbrella with good governance. They also realized that "shadow" in the name of shadow IT did not resonate with positive vibes. Hence, they renamed shadow IT to citizen IT. This would allow tech-savvy business executives to become citizen IT or software developers.

Bob and Roger also shared a few focus areas, shown in Figure 11.41, which would enable the team to effectively utilize the efforts of citizen IT developers within business units.

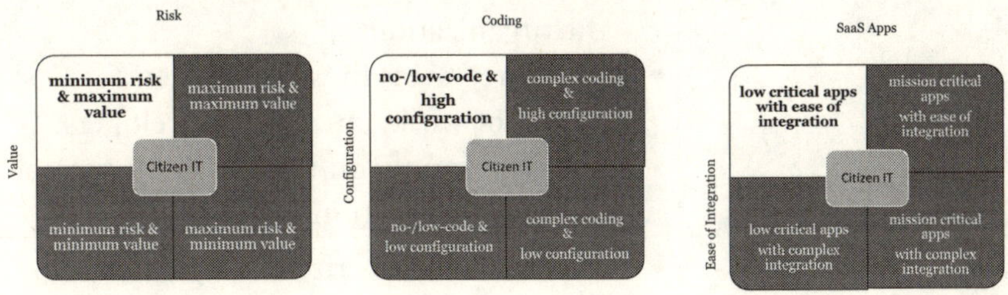

Source: Author

Figure 11.41: Focus Areas of Citizen IT Developers

Citizen IT Focus Areas	Guidelines
Scope of the work	• Low-risk and high-value applications • Applications that can be developed with workflow configuration with no-/low-code • SaaS apps implementation and ease of integration

Governance	• Sofware development life cycle handbook for citizen IT developers • Strict adherence to technology organization's standards, policies and governance • Shadow IT discovery tools to identify siloed initiatives • Two-way communication at scheduled intervals and on-need basis with citizen IT developers by the technology organization • Quality assurance and quality control of apps to be done by the technology organization • Apps must be certified by technology organization before deployment • Promote the apps developed by citizen IT for use in other business units across the organization • Restricted areas in which citizen IT must not work; citizen IT developers must work only on the approved type of software applications
Resources	• Define selection criteria. (e.g., STEM background employees within business units who have excellent domain knowledge to develop apps) • Self-organized teams without overheads • Training in citizen IT governance, standards, policies and guidelines

Source: Author

Table 11.27: Description of Citizen IT Focus Areas

Benefits of Citizen ITs

1. Business and technology alignment—no shadow IT
2. Increased speed of development with reduced development and testing time
3. Reduced load on the technology department

4. Faster translation of business ideas into software apps/ applications
5. Reduced cost of software applications in comparison to traditional methods

Internal Technology Start-ups

Internal Technology Start-ups within Business Units

Source: Author

Figure 11.42: Technology Entrepreneurship—Internal Technology Start-ups Within Business Units

CEO Tom met with his business and technology leaders to discuss the various suggestions and ideas put forward by the core team to prepare for the complex digital journey of Auto Gear. He shared his observations with them:

1. Our customers have been looking for continuous newness and we have not been able to cope with this rapid change.
2. Our competition is releasing new products and services at great speed while we are lagging behind them.
3. The future looks very uncertain due to the impact of technological innovations.
4. We are still not there in terms of leveraging our partners' capabilities to enhance our offerings. We are yet to build an extended digital business ecosystem.
5. We have a lot of inertia in our organization and our current operating model seems to be rigid and a bit bureaucratic.

According to Paul, who was at the meeting as well, most organizations had been driving their line of businesses as technology start-ups. The reason behind this was the uncertainty and huge risks involved in operating companies in the digital era.

Paul emphasized that driving businesses like tech start-ups would be highly beneficial as the new product incubation period would be shorter and results/ROI would be known within a shorter time, unlike other industry start-ups.

He described a scenario in which a business unit wanted to introduce a new product into the market by using tech start-up practices and methods. He cautioned that it requires cultural change and efficient leadership attitude to drive the business like a tech start-up. Figure 11.43 depicts a tried and tested model he shared in the meeting.

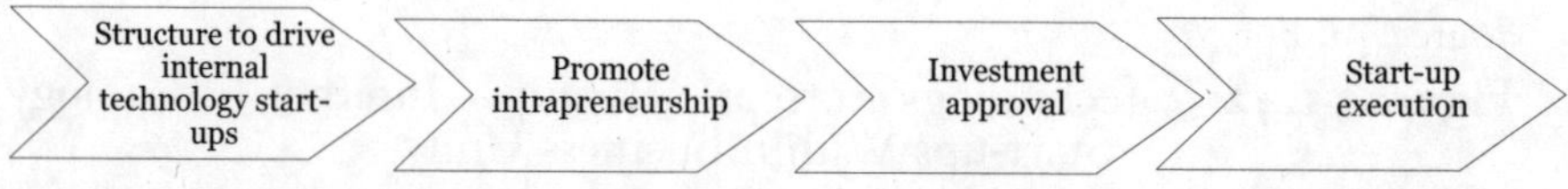

Source: Author

Figure 11.43: Technology Start-up Model in Business Units

Structure to Drive Internal Technology Start-ups

Organization roles can be mapped against start-up roles as follows:

1. Business unit head = Investor (Investor follows the venture capitalist or angel investor principles and guidelines)
2. Cohort group leader = Mini-CEO (employee)
3. Start-up team = Cohort team
4. Board of directors = Mini-CEO, investor, CIO, CSO and chief sales officer

The attributes of the mini-CEO and the cohort team are:

1. Risk tolerance
2. Hope
3. Perseverance
4. Delivering more with less
5. Failure tolerance
6. Uncertainty management

Apart from this, an innovation council could be established at the organizational level for investment approval.

Promote Intrapreneurship

Following are the five steps to successfully establish an intrapreneurship system within an organization:

1. Create intrapreneurship strategy, structure and operational plan.
2. Establish an intrapreneurship fund for each business unit.
3. Develop and institutionalize intrapreneurship culture, policies and skills.
4. Encourage employees to form cohorts to work on business problems.
5. Establish an open innovation platform to receive and assess the best ideas from the cohorts.

Investment Approval

The cohort presents an investment proposal to the innovation council covering the elements listed in Figure 11.44.

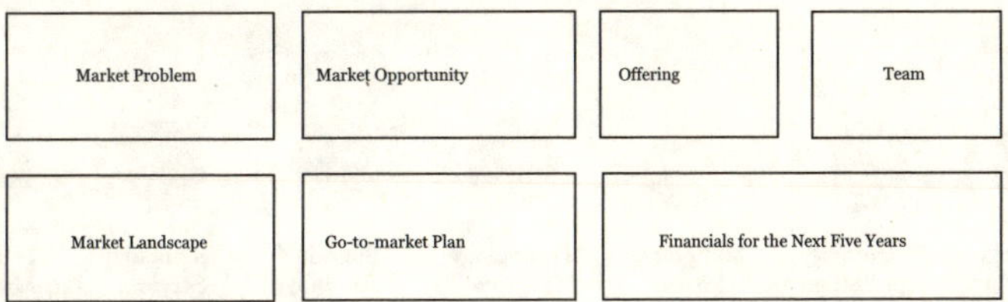

Source: Author

Figure 11.44: Elements of Internal Technology Start-up Investment Proposal

The investment council would decide whether to approve or reject the investment proposal.

Start-up Execution

Following are the seven steps to successfully execute the internal technology start-up within an organization:

1. The start-up team including mini-CEO and the cohort team incubates the minimum viable product (MVP).
2. The internal investor releases the fund based on the progress of the product according to the plan.
3. The start-up team collaborates with the sales team to find the first few customers.
4. The team improves the product with multiple pivots and releases to market.
5. The organization's sales team accelerates product sales and scales up the business.
6. Once the products business reaches a mature state, the start-up team integrates it with the regular business and disassembles the start-up.
7. The organization pays a performance bonus to the internal investor and tech start-up team for the successful execution of the start-up.

In his concluding remarks, Paul focused on the need to strengthen Auto Gear's intrapreneurship system and encouraging many internal start-ups to help prepare them for an uncertain future in the digital era.

Technology Operations Management

Technology Infrastructure Management

Source: Author

Figure 11.45: Technology Operations Management—Technology Infrastructure Management

Technology services are the most vital part of any technology department/organization. Bob, the chief architect, mentioned that they had redefined Auto Gear's technology infrastructure blueprint to support the fusion of operational, communication and collaboration, and information technology. He shared the comprehensive technology infrastructure components, they had defined, based on the business demand as mentioned in Table 11.28.

Infrastructure Components	Details
Computing	• Cloud computing • Edge computing • GPU computing • Multi-cloud life cycle management tools
Hardware	• Switching, routers, gateways, firewalls and servers • Security hardware • Infrastructure management tools
Networks	• Short-range transportation networks (6LowPan, LoRaWAN, Bluetooth, Zigbee, etc.) • Broadband wireless networks like 4G and 5G • Programmable/software-defined networks • Network management tools
Devices	• Mobile devices and other handheld devices • Workstations/laptops • Wearable devices • Consoles (game/car)
System software	• Operating systems • Language translators • Middleware platforms • IoT Platforms (IBM IoT, ThingWorx, Amazon IoT, etc.) • AR/VR development platforms (Vuforia, Google ARCore, etc.) • Blockchain platforms (Ethereum, Hyperledger, MultiChain, etc.) • Cybersecurity software tools (Intruder, Snort, Webroot, etc.)
Storage	• Cloud storage and AI support • Relational database • NoSQL
Things	• Intelligent things/machines/equipment/assets • Electronic microprocessors

Source: Author

Table 11.28: Technology Infrastructure Elements

Bob emphasized that establishing a technology infrastructure to support Auto Gear's digital business was going to be a humongous task. Skill gaps, lack of enough budget and timelines to set up the infrastructure were his major concerns. Bob shared a detailed plan to develop and implement a handbook for technology infrastructure management process. A new set of infrastructure management tools had to be procured (like RackWare, IBM Cloud Orchestrator, Dell EMC storage products, Cisco DNA Centre, Aruba, SolarWinds Server and Application Monitor, IBM MaaS360, etc.).

Bob asked Gilbert and Ivan, two technology infrastructure specialists, how it would be if each of the following activity was done automatically using software program without human interventions.

1. Installation of the database (e.g., MySQL)
2. Creation of user account and password
3. Setting up new databases
4. Verification of the database (whether it is proper or not)
5. Removal of irrelevant/unused databases

Not limiting to databases, infrastructure resources like computing power, storage and networks are managed with software programs with less human intervention. That is called software-defined infrastructure or programmable infrastructure. Simply put, infrastructure as code (IaC). DevOps is built on this concept.

Bob shared a general strategic plan for the next three years and a very detailed plan for next year. CEO Tom and CFO Greg appreciated the plan of action and promised that a considerable budget would be allocated from this year to establish the digital technology infrastructure. Bob thanked Tom for his support.

Technology Application Management

Source: Author

Figure 11.46: Technology Operations Management—Technology Application Management

Bob had closely worked with Arjun, the head of software application, and conducted a detailed analysis of the existing software applications and identified the need for new software applications as per the digital business demand. Bob expressed his concerns that they were not yet ready to develop the full spectrum of software applications required to support Auto Gear's digital business. There were huge skill gaps, lack of development, testing and deployment tools and infrastructure capacity to manage digital age software applications. At present, Auto Gear had around 300 software applications, and Bob shared the distribution according to type, as given in Table 11.29.

Application Type	Focus	Technologies Being Used	Application Mix (%)	Application Mix (Quantity)
Systems of record	Business process automation (custom application and ERP)	IBM, Java, .Net Framework, Opensource, SAP, Oracle, Microsoft, Salesforce	88	264

Systems of engagement	Stakeholder emotion and experience management	Mobile, social, chatbots, virtual assistants, AI, analytics, AR/VR	5	15
Systems of things	Things, asset and product automation	IoT, analytics, mobile, blockchain	5	15
Systems of intelligence	Business hindsight, insight and foresight	Big data, AI, visual analytics	2	6
Total			**100**	**300**

Source: Author

Table 11.29: Auto Gear's Digital Software Applications Distribution

Bob explained how the application life cycle management had changed, as mentioned in Table 11.30. He also emphasized that Auto Gear had to put a lot of effort to adopt the digital-age application life cycle management.

Application Focus Areas	IT Era	Digital Era
Automation	Business process automation and business intelligence based on structured data	Business process automation, products/assets/things automation, stakeholders emotion management, business intelligence based on structured, unstructured and semi-structured data

Technology	IT	Mobile, social, analytics, cloud, IoT, AI, ML, AR, VR, blockchain and others
Requirements	Many functional features and complex workflows	One or two features and simple workflows
Application development period	Months	Days
Application upgrade/releases	Yearly	More than once in a month/quarter
Infrastructure provision	In days/weeks	In hours using cloud
Business ecosystem support	Inward focus and high effort to integrate with partner and supplier ecosystem	Easy to integrate with partner and supplier ecosystems using RESTful APIs

Source: Author

Table 11.30: Focus Areas of Digital Software Applications

Bob and Paul emphasized that Auto Gear's technology organization must re-engineer its application management governance to support the needs of digital business. The new-age applications are built with the combination of technologies like mobile analytics, mobile social, mobile social analytics, mobile IoT analytics, MR, AI, social mobile analytics cloud and others. The innovation lies at the intersection of these technologies.

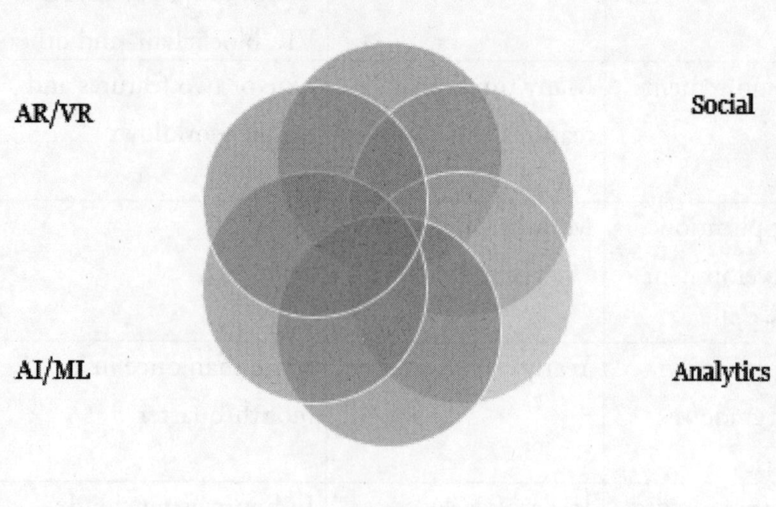

Source: Author

Figure 11.47: Intersection of Digital Technologies

Bob highlighted the application management areas which they wanted to focus on immediately and the gaps that needed to be filled to deliver consistent results.

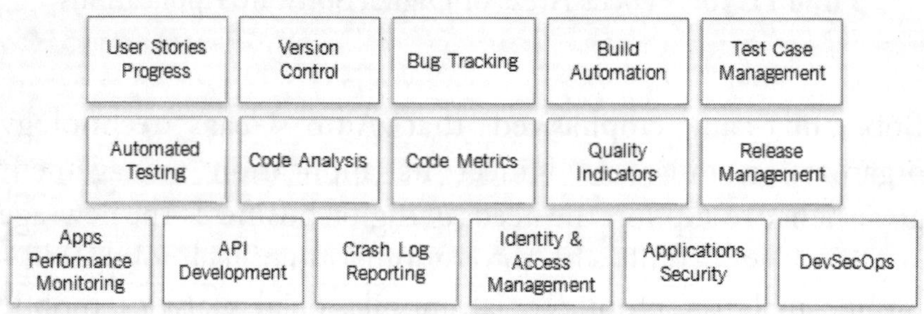

Source: Author

Figure 11.48: Digital Software Application Management Focus Areas

Bob emphasized the importance of identifying various application development life cycles for different types of applications like systems of record, systems of engagement, systems of things and systems of intelligence. One medicine does not fit all. It is crucial to understand the life cycle differences to develop and manage a variety of applications effectively. Digital age applications have the following characteristics:

1. They support high volume, velocity and variety of data.
2. They run on heterogeneous and elastic networks.
3. They use elastic storage.
4. They use elastic cloud computing or edge computing.
5. They are omnipresent (work on multiple channels).
6. They are connected with intelligent things via the internet.

Bob shared a detailed plan to develop an application management handbook and procure or build application management services tools (like Atlassian, Rally, Azure DevOps Server, Kubernetes, like CollabNet, Docker and other open-source tools) to align with the digital era.

Bob also shared the importance of using AI, robotic process automation (RPA), infrastructure as code (IaC), low/zero code platforms in the software development life cycle. This helps in improving the quality of software and accelerate the delivery of software applications at the pace of business expectations.

Organizations need to embrace the best practices being used by Google, Amazon and similar companies. For example, Amazon introduced two pizzas team. That means the individuals in a project should not be larger than what two pizzas can feed. Google has introduced Site Reliability Engineering (SRE) with which software engineering is incorporated to resolve infrastructure and operational problems, which means automating the manual tasks of managing infrastructure and operations using the software.

It is imperative for technology units under the CIO in the organizations to embrace and institutionalize the best practices being used at technology majors like Google, Amazon to be successful in the digital era. That means technology units under CIO should implement software product engineering practices. Bob was known for his futuristic ideas among his colleagues. He shared his view that by the year 2040, there could be nearly autonomous software development platforms (next level of zero code or low-code) that would reduce probably 80–90% of human interventions even for complex business rules and workflow-intensive software applications. He declared, "Robo coders will be the future workforce and they will work closely with human workers."

Robo coders would be intelligent bots with the following features:

1. Robo coders will automatically collate problem statements and create requirements/functional specifications based on the humongous data available from IoT and people.
2. Robo coders will prioritize requirements based on customer, needs and wants.
3. Robo coders will automatically generate user stories and the corresponding software code.
4. Robo coders will test the generated code for security, safety and quality and deploy it automatically.
5. Robo coders will complete all these tasks probably five times faster than a human worker.

CEO Tom and CFO Greg appreciated the plan of action for the next three years put forward by Bob. He had also prepared a detailed plan for the next year. They promised the allocation of a considerable budget for establishing digital software application development and management ecosystem within Auto Gear. Bob thanked them for their support.

Technology Project Management

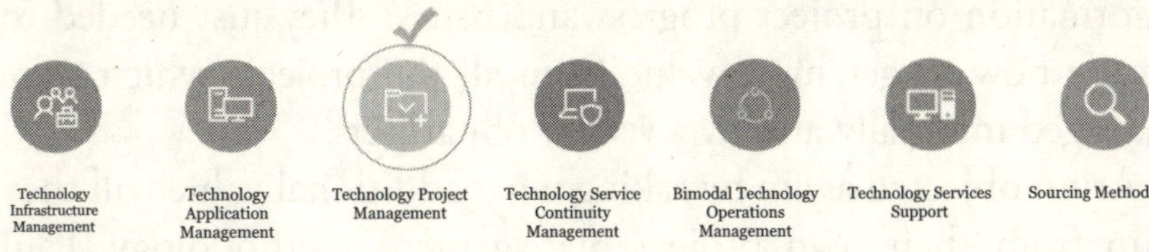

Source: Author

Figure 11.49: Technology Operations Management—Technology Project Management

In the next session with the core team, Paul talked about the project execution challenges faced by Auto Gear from outsourcing partners. They executed projects using their proprietary methods. Hence, the deliverables and status cards were provided in different formats. Therefore, there was no homogeneity in project execution.

Paul reiterated that considering the new capabilities required by Auto Gear to support digital business, they must work with new partners. These partners could be technology start-ups or technology specialist firms in addition to traditional outsourcing partners. This meant that projects would be distributed among many partners.

At present, Auto Gear had been getting updates, information and conducting reviews by outsourcing partners without any standardized project management framework. Hence, most of the times they did not get any comprehensive insight into projects being executed by their outsourcing partners. They needed to develop a digital age technology project management methodology that was lean, unique and agile. Paul believed that providing training to all partner organizations on this method was essential.

The head of software applications, Arjun, argued that it was a waste of money and resources to spend on project management methodology. He felt that it did not add much value except resulting in overheads. Paul explained that at present, they have a project management office that collates project information from internal projects and outsourcing projects to provide inconsistent information on project progress and issues. They just needed to decide how to get more value from all the projects which were managed internally and by a variety of partners.

Arjun objected again by asking what additional value will they gain from their own project management methodology. Paul responded with a smile.

"The project management process is one of the critical business processes of Auto Gear like any other process, be it CRM, supply chain management (SCM) or human capital management (HCM). We never leave the ownership of business processes to partners. Then why do we leave the project management process to our partners? We may take the help of specialists to design the project management process."

He further clarified that Auto Gear must define its unique project management processes as they do for other business processes. Auto Gear's project management processes will help in delivering the digital age software projects faster, better and in a cost-effective manner; not only that, the new project management processes will see them leapfrogging competition too.

Paul shared a few benefits of having their own project management methodology:

1. Project management methodology of Auto Gear will help them respond to business requirements faster than before and also allow them to align with the organization's business objectives.
2. For new/complex projects in the areas of MR, AI and blockchain, even outsourcing partners had not reached the required maturity. Hence, Auto Gear's project management methodology will help them deliver better services.
3. It will be possible to improve the project management process continuously based on changing needs.
4. With homogeneity across all the projects, it will be easy to integrate the deliverables from each partner easily.

"We started seeing the emergence of citizen IT management within business units. Even if they execute smaller and simpler application development projects under overall technology governance, Auto Gear's standardized project management methodology helps them deliver better project outcomes," said Paul.

Paul added that they could also make use of the practices and lessons learned from their partners to improve their project management methodology. He also highlighted that digital age software applications and infrastructure use complex technologies, and considering the volatile market and customer needs and fierce competition, Auto Gear was expected to deliver software applications within shorter cycles and cost-effectively.

He discussed the new trends in project management like agile, scrum, scaled agile framework, DevOps, telepresence, remote virtual walls to connect distributed teams, project analytics and project visibility using mobile devices.

Paul requested Roger to share his work on Auto Gear's proposed project management method that was best suitable for digital age technology projects. Roger came up with various projects, which Auto Gear needed to execute as depicted in Figure 11.50.

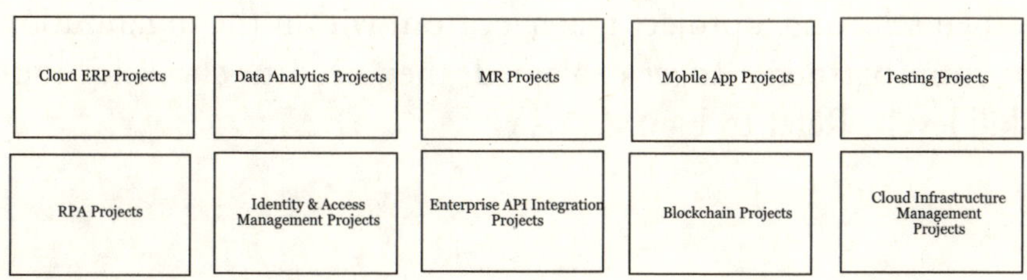

Source: Author

Figure 11.50: Variety of Technology Projects for the Digital Age

Roger reiterated that project management is not a general management domain, it is a specialization. One cannot manage data analytics projects with experience in cloud ERP implementation projects.

The following list provides the constituents of the project management methodology.

1. Processes (activities)
2. Work products (deliverables)
3. Templates
4. Checklists
5. Standards
6. Guidelines
7. Best Practices
8. Lessons learned
9. Automation tools

While there was to be an overarching project management methodology for Auto Gear, it had to be tailored to suit different types of projects like cloud ERP, data analytics, MR, robotic process automation (RPA) and others.

Project Manager Competency—Generalization to Specialization

Roger pointed out that generalization does not work in today's hyperconnected world. Project managers must be specialized. The specialization could be based on project type and domain. Organizations need to nurture specialist project managers. When selecting a project manager from within the organization or from partners, Roger explained, they need to check for four skill levels. Refer to Figure 11.51.

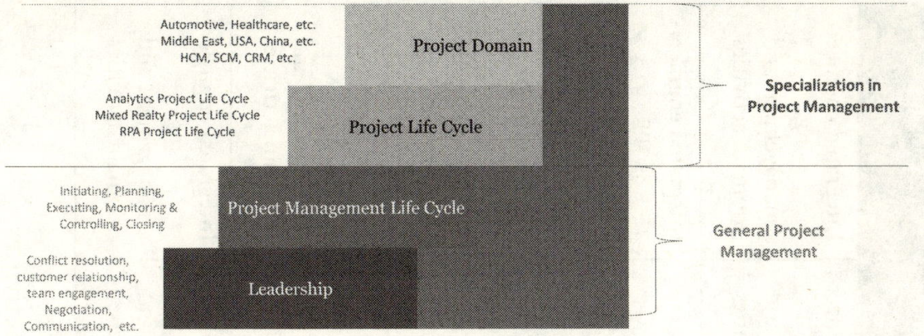

Source: Author

Figure 11.51: Project Manager Maturity Model—Skill Enhancement at Four Levels

Out of the four maturity levels of a project manager, **Level 1 and Level 2 skills are must for a generalist project manager.**

1. **Level 1:** The project manager must possess **leadership** skills. This is the entry criteria for any project manager. If the employees qualify here, they could move to Level 2.
2. **Level 2:** The project manager must possess **project management** skills. If the employees qualify here, they could move to Level 3.
3. **Level 3:** The project manager must possess **project engineering** skills. If the employees qualify here, they could move to Level 4. The specialization starts here.
4. **Level 4:** The project manager must now the domain of the project. At this level project manager is an all-round specialist project manager.

Example of Data Analytics Project Management Methodology

Roger shared two axioms for the development of a project management methodology for data analytics project management:

1. The project life cycle is pretty much different for each project.
2. The project management life cycle is the same for all projects.

The integration of the project life cycle (engineering phases and activities) and project management life cycle (management phases and activities) is essential for delivering projects faster, better and steadily. Table 11.31 is a template for a data analytics project management.

Project Management Life Cycle Phases / Project life Cycle Phases	initiating	Planning	Executing	Monitoring and Controlling	Closing
Define Business Problem	Integrated engineering, management and domain activities	Integrated engineering, management, and domain activities	Integrated engineering, management and domain activities	Integrated engineering, management and domain activities	Integrated engineering, management and domain activities
Identify Data Requirements	Integrated engineering, management and domain activities	Integrated engineering, management and domain activities	Integrated engineering, management, and domain activities	Integrated engineering, management and domain activities	Integrated engineering, management and domain activities
Prepare Data	Integrated engineering, management and domain activities	Integrated engineering, management and domain activities	Integrated engineering, management and domain activities	Integrated engineering, management and domain activities	Integrated engineering, management and domain activities
Formulate Hypothesis and Build Rapid Prototype	Integrated engineering, management and domain activities	Integrated engineering, management and domain activities	Integrated engineering, management and domain activities	Integrated engineering, management, and domain activities	Integrated engineering, management and domain activities
Deploy Production-Ready ML Models	Integrated engineering, management and domain activities	Integrated engineering, management and domain activities	Integrated engineering, management and domain activities	Integrated engineering, management and domain activities	Integrated engineering, management and domain activities

Source: Author

Table 11.31: Phases of Data Analytics Project and Integrated Project Activities

Roger emphasized the point that competent and qualified project managers must possess the skills to manage the integration of engineering, management and domain-specific activities. Hence, project management methodology must unleash the power of the convergence of engineering, management and domain-specific activities.

Brief description of engineering, management and domain activities are given below:

1. **Engineering activities:** Activities that help in building the actual product, asset and benefits.
2. **Management activities:** Activities that help in managing cost, time, scope and quality.
3. **Domain-specific activities:** Activities that help in adding or managing any localization and industry-specific or function-specific activities (e.g., Health Insurance Portability and Accountability Act).

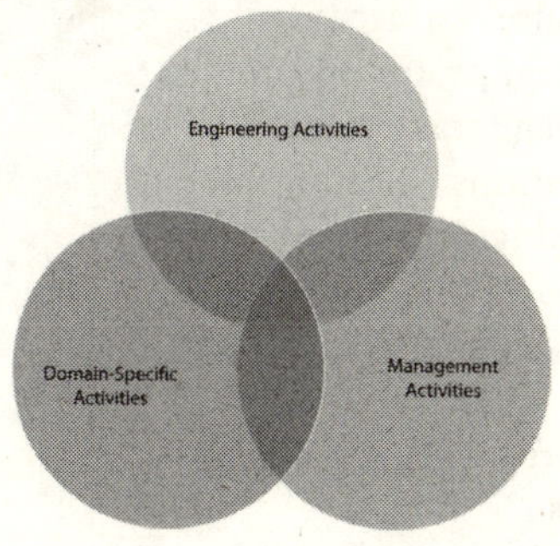

Source: Author

Figure 11.52: Convergence of Project Management Activities

For a typical data analytics application development project, the engineering activities come under the following five phases:

1. Define the business problem
2. Identify data requirements
3. Prepare data
4. Formulate hypothesis and build rapid prototype
5. Deploy production-ready ML models

The project management method described in the tables under the following sections gives an illustrative view of a data analytics project. It's not comprehensive and is only for reference.

Phase 1: Define the Business Problem

	Process	Work Product
Initiating	• Get approval for Phase 1	• Phase 1 approval document
Planning	• Scope planning • Schedule • Risk planning • Quality planning • Resource planning • Communication planning • Budget planning • Procurement planning • Stakeholder expectation management planning	• Detailed plan for business problem definition
Executing	• Acquire resources • Manage team • Manage stakeholders' expectations • Conduct interviews/workshops • Identify the business problem • Identify business expectations • Identify business value and KPIs to be improved • Create user stories and product backlog • Create the initial test strategy	• Business problem, business expectations, KPIs and business value definition document • User stories document • Initial test strategy • Issue log • Risk register
Monitoring and controlling	• Quality monitoring and control • Scope monitoring and control • Cost monitoring and control • Schedule monitoring and control	• Project status report • Project dashboard • Project issue resolution • Project risk mitigation

Closing	• Get the sign-off document from stakeholders • Conduct root-cause analysis for various issues encountered in the business problem definition phase • Create checklists to avoid recurrence of similar issues in the next phase	• Root-cause analysis and preventive action plan • Stakeholder sign-off document for the problem defined

Source: Author

Table 11.32: Data Analytics Project Phase 1—Define the Business Problem

Phase 2: Identify Data Requirements

	Process	Work Product
Initiating	• Get approval for Phase 2	• Phase 2 approval document
Planning	• Scope planning • Schedule Planning • Risk planning • Quality planning • Resource planning • Communication planning • Budget planning • Procurement planning • Stakeholder expectation management planning	• Detailed plan for data requirements identification
Executing	• Acquire resources • Manage team • Manage stakeholders' expectations • Identify the business processes in the context of the business problem • Derive the data requirements	• Business process and data mapping document

Monitoring and controlling	• Quality monitoring and control • Scope monitoring and control • Cost monitoring and control • Schedule monitoring and control	• Project status report • Project dashboard • Project issue resolution • Project risk mitigation
Closing	• Get the sign-off document from stakeholders • Conduct root-cause analysis for various issues encountered • Create checklists to avoid recurrence of similar issues in the next phase	• Root-cause analysis and preventive action plan • Stakeholder sign-off document for the data requirements identified

Source: Author

Table 11.33: Data Analytics Project Phase 2—Identify Data Requirements

Phase 3: Prepare Data

	Process	Work Product
Initiating	• Get approval for Phase 3	• Phase 3 approval document
Planning	• Scope planning • Schedule planning • Risk planning • Quality planning • Resource planning • Communication planning • Budget planning • Procurement planning • Stakeholder expectation management planning	• Detailed plan for the preparation of data

Executing	• Acquire resources • Manage team • Manage stakeholders' expectations • Data cleansing • Data aggregation • Data augmentation • Data sorting • Data formatting	• Cleansed and aggregated data
Monitoring and controlling	• Quality monitoring and control • Scope monitoring and control • Cost monitoring and control • Schedule monitoring and control	• Project status report • Project dashboard • Project issue resolution • Project risk mitigation
Closing	• Get the sign-off document from stakeholders • Conduct root-cause analysis for various issues encountered • Create checklists to avoid recurrence of similar issues in the next phase	• Root-cause analysis and preventive action plan • Stakeholder sign-off document for data analysis

Source: Author

Table 11.34: Data Analytics Project Phase 3—Prepare Data

Phase 4: Formulate Hypothesis and Build Rapid Prototype

	Process	Work Product
Initiating	• Get approval for Phase 4	• Phase 4 approval document.
Planning	• Scope planning • Schedule planning • Risk planning • Quality planning • Resource planning • Communication planning • Budget planning • Procurement planning • Stakeholder expectation management planning	• Detailed plan for building hypothesis and rapid prototype

Executing	• Acquire resources • Manage team • Manage stakeholders' expectations • Convert the business problem into a hypothesis mathematical model • Identify hypothesis test cases for rapid prototyping • Build rapid prototype based on hypothesis and establish the relationship among different variables • Train, iterate and build production-ready ML models	• Hypothesis model • Rapid prototype • Trained ML models
Monitoring and controlling	• Quality monitoring and control • Scope monitoring and control • Cost monitoring and control • Schedule monitoring and control	• Project status report • Project dashboard • Project issue resolution • Project risk mitigation
Closing	• Get the sign-off document from stakeholders • Conduct root-cause analysis for various issues encountered in • Create checklists to avoid recurrence of similar issues in the next phase.	• Root-cause analysis and preventive action plan • Stakeholder sign-off document for the hypothesis and rapid prototype

Source: Author

Table 11.35: Data Analytics Project Phase 4—Build Hypothesis and Rapid Prototype

Phase 5: Deploy Production-Ready ML Models

	Process	Work Product
Initiating	• Get approval for Phase 5	• Phase 5 approval document.

Planning	• Scope planning • Schedule planning • Risk planning • Quality planning • Resource planning • Communication planning • Budget planning • Procurement planning • Stakeholder expectation management planning	• Detailed plan for the deployment of production-ready ML models
Executing	• Acquire resources • Manage team • Manage stakeholder expectations • Deploy the production-ready ML models • Develop visual data (dashboards, reports, trends) • Perform real-world testing • Deliver AI-powered MVP	• Production-ready ML models • Visual dashboards/reports • Test results • AI/ML MVP
Monitoring and controlling	• Quality monitoring and control • Scope monitoring and control • Cost monitoring and control • Schedule monitoring and control	• Project status report • Project dashboard • Project issue resolution • Project risk mitigation
Closing	• Get the sign-off document from stakeholders • Conduct root-cause analysis for various issues encountered • Create checklists to avoid recurrence of similar issues in the next phase.	• Root-cause analysis and preventive action plan • Stakeholder sign-off document for the production-ready ML model and the minimal viable product

Source: Author

Table 11.36: Data Analytics Project Phase 5—Deploy Production-Ready ML Models

In addition to integrated project management phases and activities, other important elements of project management methodology are mentioned in Table 11.37.

Element	Description
Templates	• Project plan template • Business problem definition template • Data requirements template
Checklists	• Checklist for training ML models • Checklist for testing ML models
Standards	• Data sets identification standards • Data management standards
Guidelines	• Hypothesis models guidelines • Custom algorithm development guidelines • Off-the-shelf algorithms selection guidelines
Best practices	• Identify analytics use cases which provide quick wins • Induct data science specialists early in the project • Use the right data and analytics tools • Manage analytics projects as business owned projects not technology owned
Lessons learned	• Involve domain specialists throughout the project • Understand business goals • Timely collaboration among stakeholders • Correct data sets and data integrity
Automation tools	• Project management tool • Darwin • DataRobot • H20.ai • dotData • DORA • DatacCleaner • Missingno • AutoSklearn

Source: Author

Table 11.37: Data Analytics Project—Important Elements of Project Management Methodology

Project Management Framework

In the past, project management was done using waterfall models and iterative models, which will fail in today's uncertain business environment in which business executives themselves do not know the requirements.

There is nothing like frozen requirements; requirements constantly evolve. In the past, the philosophy of project management was to fix the scope. This meant increasing the project budget if there is any change in the scope. However, now the project management philosophy has changed—allow scope change but fix the cost and time. Allowing changes in the requirements within the cost and timelines by adding the prioritized requirements and removing the non-prioritized requirements has become a regular process. This is a paradigm shift and is called **agility**, that is, the ability to absorb evolving changes easily.

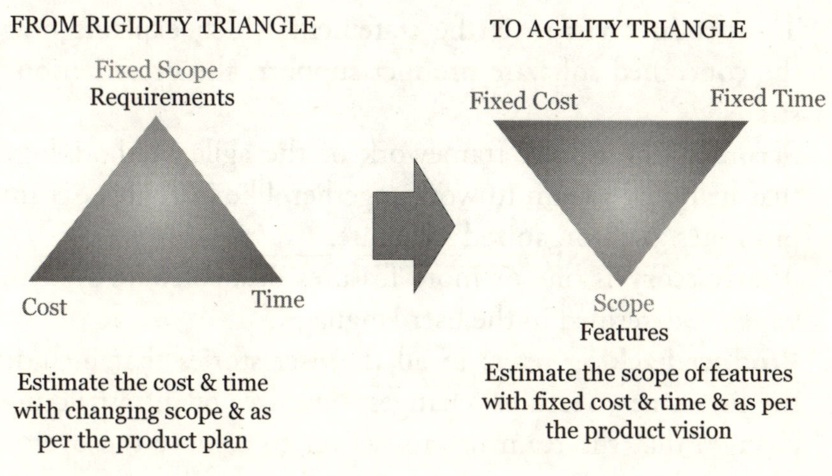

Source: Author

Figure 11.53: Rigidity vs Agility

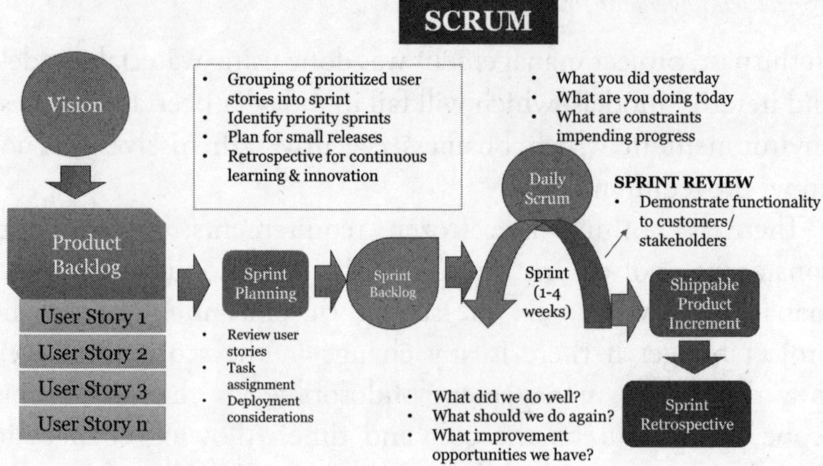

Source: Author

Figure 11.54: Agile Project Management Framework

The various elements of the agile project management framework are:

1. The **product vision** is the statement which explains how the concerned software product supports the organization's strategy.
2. **Scrum** is the overall framework of the agile methodology that helps each team to work together, like in rugby. Scrum promotes a self-organized structure.
3. A **user story** is one or more features of a software system, which is described in the user language.
4. **Product backlog** refers to all the user stories that include the list of new features, changes, bug fixes or infrastructure changes that the team has to deliver to achieve the overall project objectives.
5. **Sprint** is the timeboxed iteration in which the team completes the development of user stories to complete a planned amount of work.
6. **Sprint planning** means identifying the user stories from the product backlog and define the sprints and its timelines.

7. **Sprint backlog means** prioritized user stories that include the list of new features, changes, bug fixes or infrastructure changes that the team has to deliver to achieve a specific objective or outcome.
8. **Daily scrum** is a 15-minute timeboxed meeting, held every day, for the development team to review the progress of work, synchronize activities and create a plan for the next 24 hours.
9. **Shippable product increment** constitutes user stories delivered in each sprint which will be useful to achieve a specific outcome.
10. The **sprint review** is a review meeting held at the end of each sprint with the stakeholders to discuss what was accomplished and not accomplished during the sprint.
11. The **sprint retrospective** is the meeting which happens immediately after the sprint review. Corrective and preventive actions are identified to improve the agile software development process.

Project managers must be acquainted with agile project management methods to deliver the projects within days and weeks unlike in the past.

Project Management Methodology Automation

Implementing project management methodology without automation will become an overhead and may fail as the stakeholders may not get the ROI for their time. Hence, it is important to automate project management processes, templates, checklists and other practices. See Table 11.38 for elements and tools of project management methodology automation.

Project Engineering	Real-Time Collaboration and Communication	Real-Time Intelligence	Omnipresent	Gamification
Automation of software development and machine learning algorithms development activities • Darwin • DataRobot • H20.ai • Others	No/less travel, virtual collaboration and high productivity • Cisco • Telepresence • Zoom • Airmeet • Microsoft Teams • Slack • Box	Project analytics to get insight on cost, time, quality, risks, scope and other aspects of a project • Asana • Jira • VersionOne • Manday.com	Manage projects anywhere, anytime and using any device (mobile, tab, web, wearable devices)	Auto-recognition of star performers based on the project outcomes

Source: Author

Table 11.38: Automation of Project Management Methodology

Technology Service Continuity Management

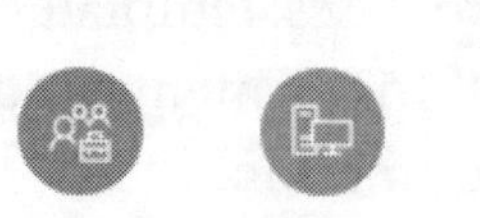

Source: Author

Figure 11.55: Technology Operations Management—Technology Service Continuity Management

CIO Paul and Bob, the chief architect, met with Tom and Greg to discuss business continuity and disaster recovery. Tom was worried because even though they had been spending significantly on business continuity and disaster recovery management activities for the last five years, whenever they encountered disasters, they had been miserably failing. This had been at the cost of brand and revenue.

Tom said that business is highly dependent on technology assets and services, especially in the digital era. Hence, if their technology organization was not ready with sophisticated business continuity and disaster recovery plan, Auto Gear will have to encounter unimaginable disasters.

Paul updated Tom about the progress and informed him that Bob had been working on the **4&16 Model**, which had been designed for digital business continuity. This model was designed based on Auto Gear's experience and after understanding the best practices, lessons learned from other organizations and industries.

Tom and Greg asked them to present the 4&16 Model next week at their office.

Type of Disasters

Technology service continuity is very critical for business continuity. Bob arranged a separate meeting with Paul, Arjun and Roger to discuss how Auto Gear's service continuity management plan, processes and the team to recover from natural and human-made disasters were barely adequate. Such disasters disrupt the business operations in a very significant manner and required highly coordinated efforts and resources to bring the operations back to normalcy.

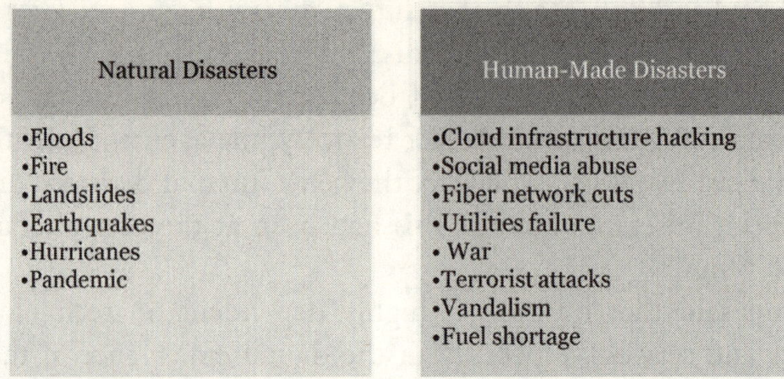

Source: Author

Figure 11.56: Types of Disasters

Bob also expressed a concern that Auto Gear was not clear about the possible disasters that may emerge in the digital era due to so many unknowns, multiple complex technologies and distributed cloud infrastructure.

For Auto Gear to have business continuity and disaster recovery in the digital age, they must change their approach and invest considerably in necessary infrastructure, resources, training, new protocols, policies, processes and procedures.

Bob made it clear that, at Auto Gear, they had never used technologies to predict human-made disasters. By applying ML algorithms on the data available from various technology assets (logs, sensors, databases, social media, etc.), it was possible to predict a few of the human-made disasters like security hacks, fake news, network cuts, etc.

The disaster period could be a few hours, days, weeks or months. It depends on the intensity of the disaster. For instance, during a pandemic, war and severe hurricane restoration may not happen immediately.

4&16 Business Continuity and Disaster Recovery Management Model

Bob and the team held a series of discussions with the business executives, suppliers, partners and employees, and after understanding the digital business and technology strategies, they derived a 4&16 framework for technology services and business continuity. Bob assured Tom and Greg that the four themes they had derived were the real game changers and that they must implement them at the earliest.

Bob said there are 4 themes and 16 pillars for technology services continuity management, which support business continuity

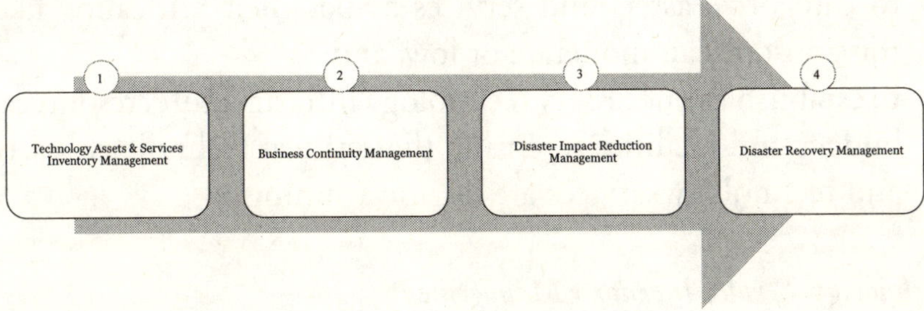

Source: Author

Figure 11.57: Four Themes of Business Continuity and Disaster Recovery Management

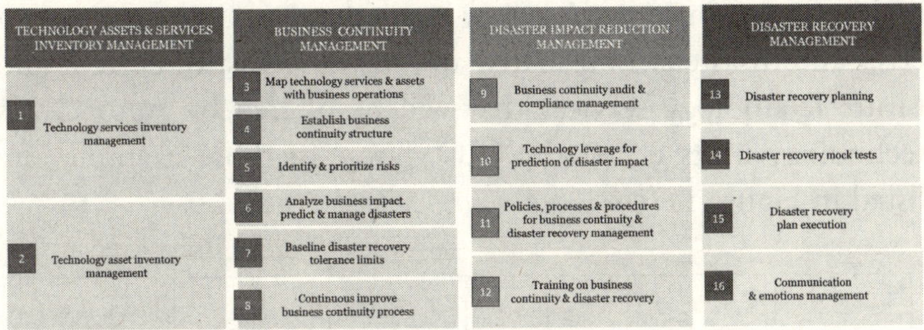

Source: Author

Figure 11.58: 4&16 Business Continuity and Disaster Recovery Management Model

Greg warned Bob, "It's all theory. We have consistently failed to do a good job in business continuity and disaster recovery management. This 4&16 model looks like another slogan." Bob requested Greg to first understand the framework and assured him that this model would bring positive changes. He also emphasized the importance of commitment from Tom, Greg and other senior executives. "Or else, as Greg says, it will be just another slogan."

Theme 1: Technology Assets and Services Inventory Management

Tom asked Bob how would they identify critical technology assets and services. Bob replied that business operations continuity is possible only when the technology assets and services are up and running without disruptions. It is very important

1. to maintain the inventory of assets and services,
2. to categorize assets and services as per their criticality, like mission-critical, moderate or low, and
3. to establish the necessary technology infrastructure, resources, budget and facilities to ensure that critical technology assets and technology services are up and running.

Technology Service Inventory Management

According to Bob, the scope of technology services would widen due to the fusion of operational technology, communication and collaboration technology and IT. Hence, their systems must be improved to continue these new technology services even during crisis and disasters. Bob shared an illustrative list of current and future technology services that will be offered by Auto Gear's technology units under the CIO. Some of those examples are listed in Figure 11.59.

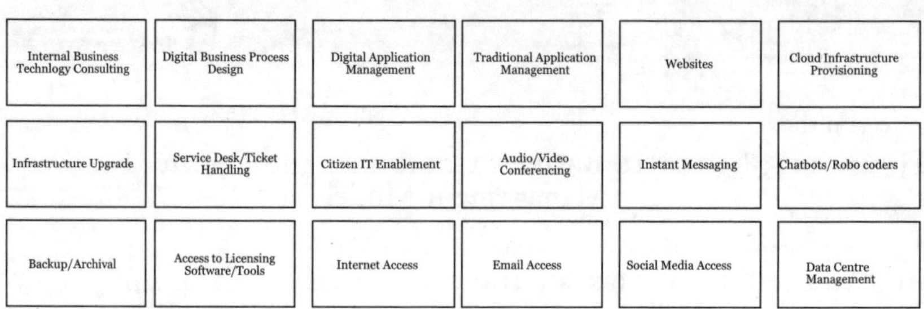

Source: Author

Figure 11.59: Technology Service Inventory

Auto Gear technology organization classified technology service as:

1. **Mission-critical:** Essential services for business survival
2. **Moderate:** Services which are critical but with localized impact on the business
3. **Low:** Services which are not critical and have a low impact on business

Technology Asset Inventory Management

Bob mentioned that to offer technology services without disruption, it was very important to safeguard the technology assets against vulnerabilities and threats. Bob shared the list of the current and future technology assets that will be maintained by Auto Gear technology units under the CIO. A few examples are given in Figure 11.60.

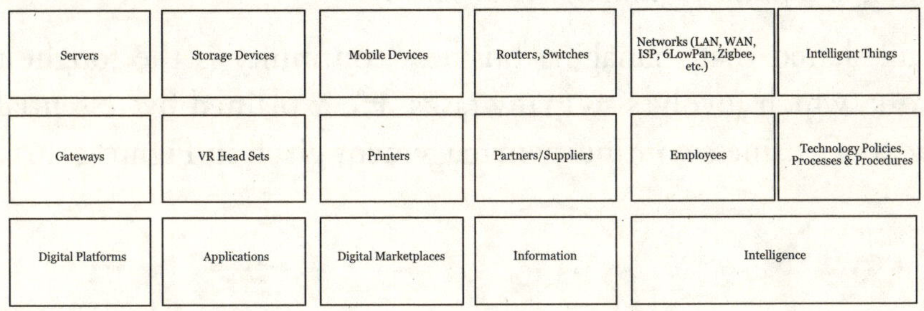

Source: Author

Figure 11.60: Technology Asset Inventory

Tom asked Bob, "Out of these assets, how do we know which is critical, which is not critical?" Bob responded that for each asset, they will rate the criticality in terms of continuing the technology services and corresponding business operations in the event of any incidents or disasters.

The list of assets needs to be updated continuously based on the changes in business and market trends. The assets (applications, data, servers, devices, etc.) are categorized into the following types:

1. **Mission-critical**: Requires immediate restoration
2. **Moderate:** Requires restoration in around 8+ hours
3. **Low:** Requires restoration in around 24+ hours

Paul added that capturing the interdependencies among different technology assets is very critical to understand the domino effect on business operations. Most of the assets are always interdependent, and they function only when the other assets are functional. For instance, application assets are dependent on servers. All dependencies must be analyzed likewise.

Tom expressed his concern regarding the many new technology services and assets on the list. They had been struggling even with the current assets and services and now to manage these new assets and services for digital business continuity would be a big challenge. He said, "We must work really hard and smart."

Theme 2: Business Continuity Management

Paul shared that managing business continuity is the toughest work, which involves so many tasks. He explained five essential areas of business continuity management given in Figure 11.61.

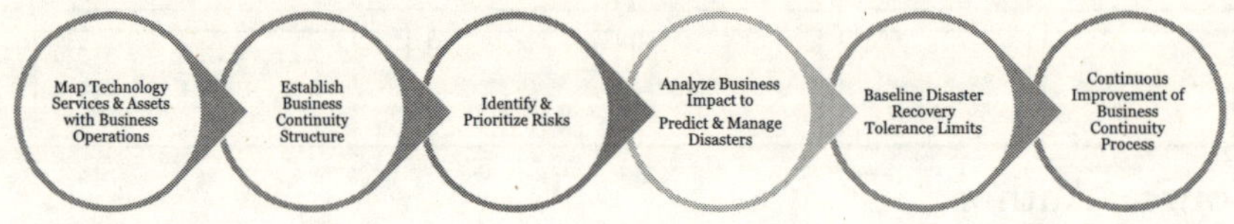

Source: Author

Figure 11.61: Pillars of Business Continuity Management

1. **Map technology services and assets with business operations:** Predicting possible business operations disaster paths based on potential disruptions of technology services and assets.
2. **Establish business continuity structure:** Establishing effective organization structure, roles and responsibilities.
3. **Identify and prioritize risks:** Identifying, preventing and controlling technology vulnerabilities, threats and risks, which may cause potential disasters in the future.

4. **Analyze business impact to predict and manage disasters:** Conducting the business impact analysis concerning vulnerabilities, threats and risks, and making sure that necessary and timely actions are taken.
5. **Baseline disaster recovery tolerance limits:** Finalizing the tolerance limits of various disasters which cause no/minimal business disruption.
6. **Continuous improvement of business continuity process:** Continuously improving the business continuity process based on the changing trends.

Map Technology Services and Assets with Business Operations

Greg raised the point that till now they had been talking about technology services and technology assets, but he still did not have any clarity on the mapping matrix of technology services, assets and business operations. Without this mapping, it was like jargon to him. Bob shared an illustrative mapping matrix for customer services and relationship management business function to make things clearer for Greg.

Business Function / Operation	Technology Services	Technology Assets (Digital Applications)	Technology Assets (Infrastructure)
Customer service and relationship management	• Develop digital applications • Manage digital applications • Manage digital technology infrastructure • Manage mobile devices • Manage networks	• Customer happiness mobile app • CRM software • Robocar adviser (chatbot) • Dealer AR/VR customer experience software • Digital car commerce portal	• AWS cloud infrastructure • Mobile devices, wearable devices • Bluetooth network • IoT at dealer showrooms (eg., beacons for messages and car information)

Source: Author

Table 11.39: Mapping Matrix Example: Technology Services, Assets and Business Operations of Auto Gear

A mapping matrix will be created for all critical business processes/operations. The example in Table 11.39 concerns only CRM function/process. Mapping matrix provides visibility on the disaster path. Business operation/function is disrupted when there is a disruption in the technology services and technology assets. For example, the customer order management process will be disrupted when there is a failure in CRM application management (service) due to inadequate cloud infrastructure (asset) to take the peak load.

Establish Business Continuity Structure

Paul stated that at this juncture, Auto Gear does not have an effective structure for business continuity. In a few cases, they found, there were overlaps and responsibility conflict. Hence, they needed to invest in leadership and specialists to effectively drive business continuity and disaster recovery activities at Auto Gear.

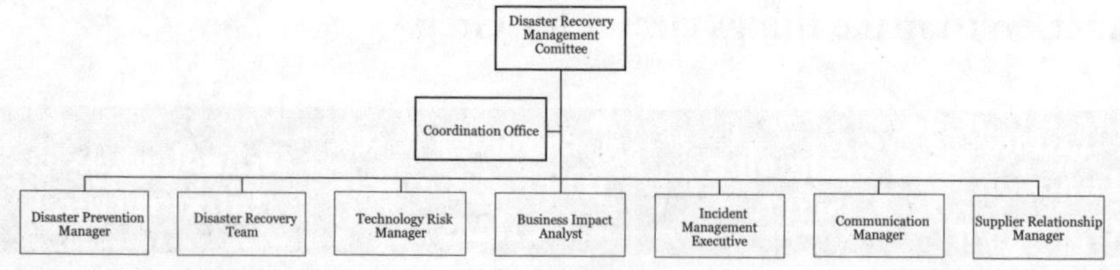

Source: Author

Figure 11.62: Business Continuity and Disaster Recovery Management Organization Structure

S. No.	Role	Responsibility
1	Disaster recovery management committee	Senior executives from different units of the organization
2	Disaster prevention manager	Identifies best practices, lessons learned, standards and policies to prevent disasters or reduce the impact of disasters

3	Disaster recovery team	Works 24x7 to recover business operations during disasters in collaboration with stakeholders by adhering to the protocols and processes defined
4	Technology risk manager	Identifies, analyzes and prioritizes the technology risks along with the risk response plan
5	Business impact analyst	Analyzes the impact of the various risks and identifies possible incidents that would cause a disaster
6	Incident management executive	Collects incidents during a disaster and manages the response and resolution; conducts root-cause analysis to define preventive practices for the future
7	Communication manager	Manages information and key messages, notifications and sends them to different stakeholders using various channels
8	Supplier relationship manager	Manages relationships with suppliers during disasters and monitors the contractual obligations to be adhered to by suppliers during disasters

Source: Author

Table 11.40: Business Continuity and Disaster Recovery Management Organization—Roles and Responsibilities

Identify and Prioritize Risks

Greg emphasized that proactively identifying the disaster paths due to disruptions in technology services and technology assets due to natural disasters or human-made disasters was of the utmost importance. Bob agreed with Greg that it was imperative to identify, prevent and control technology vulnerabilities, threats and risks, which may cause potential disasters in the future. He explained vulnerabilities, threats and risks in the business with a few examples:

1. **Vulnerability:** Inability or weakness in the system that can be exploited by a threat.
2. **Threat:** An intentional or unintentional action performed by exploiting the vulnerability, which results in unauthorized access to the confidential information, asset or causes damage, loss or destruction of the organization's assets.
3. **Risk:** The possibilities of damage or destruction to the organization's asset because of the threat that exploits the vulnerability.

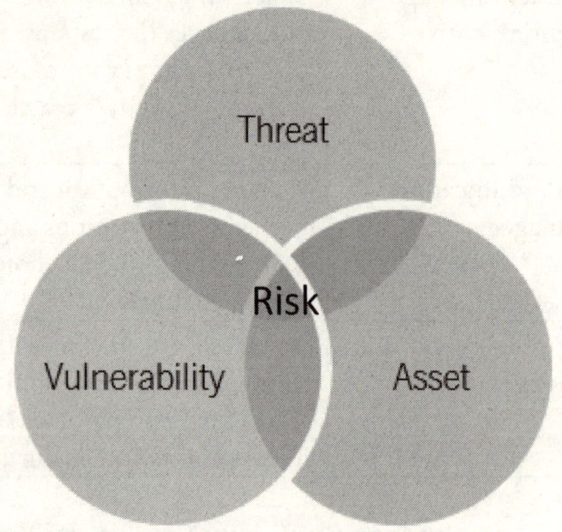

Source: Ralph P. Page

Figure 11.63: Threat, Vulnerability and Risk

S. No.	Vulnerability	Threat	Risk
1	Legacy software applications which do not have inbuilt cybersecurity features.	Cybercriminals may hack the supply chain system and change the delivery dates to delay the delivery of the product.	Cyberattack resulting in data privacy compromise and product delivery delays. Loss of enterprise's reputation in social media.

2	All servers hosted in one place, and there is no full-fledged disaster recovery site for any natural disasters.	Electric short circuit and fire in the server room may cause loss of data and major disruption to operations.	Loss of sensitive data in the data centre, customer complaints and high capital cost.
3	Weak contractual agreements with partners especially during crisis/disasters.	During the disaster, partners may take advantage of weak contracts and deny support.	Possible operational disruption due to supply failure and customer attrition.

Source: Author

Table 11.41: Examples of Technology Vulnerabilities, Threats and Risks

Bob and the core team identified a few indicative risks based on vulnerabilities and threats that had the potential to disrupt the business of Auto Gear. Bob emphasized that risk identification and analysis was a regular activity and they will be doing it every month.

The different categories of risks against the technology assets and services that needed to be identified were:

1. **External**: Related to government, regulations, environment or competition
2. **Internal**: Related to service, customer, cost, time or quality
3. **Technical**: Related to new technologies or obsolescence
4. **Unexpected**: Related to natural calamities or pandemic

Bob shared the technology risk impact scoring framework , which would be used by them to prioritize risks based on probability, severity and vulnerability, with the core team. The risk impact could be on cost, customer attrition, brand, service continuity, revenue and others.

S. No.	Risk	Probability (1=Low 3=High) (A)	Severity (1=Low 3=High) (B)	Vulnerability (1=Low 3=High) (C)	Risk Im[illegible] Score (A) × (B) × (C)
1	Possible operational disruption due to supply failure and customer attrition	3	3	3	27
2	Cyberattacks, data privacy compromise and product delivery delays	3	3	3	27
3	Data centre shut down, loss of sensitive data, customer complaints and high capital cost	3	3	3	27
4	Slow performance/bandwidth/computing during high demand periods	3	3	2	18
5	Negative impact on business due to latest technologies	3	3	2	18
6	Lack of support from external service providers	2	3	2	12
7	Non-availability of human resources	3	2	1	6
8	Inaccurate or incorrect AI algorithms	3	2	1	6
9	Error in coding blockchain smart contracts with respect to defined policies/rules	3	2	1	6
10	Internal technology services disruption	2	2	1	4
11	Fire	1	3	1	3
12	Natural calamities	1	3	1	3

Source: Author

Table 11.42: Technology Risk Impact Scores

Analyze Business Impact to Predict and Manage Disasters

Next, Paul stated that the most critical part of the technology service continuity was conducting the business impact analysis concerning vulnerabilities, threats and risks, and making sure that necessary and timely actions are taken to restore the business operations faster with minimal operational disruption.

Bob added that the business impact analysis was a very systematic process and they had identified a powerful framework to conduct the same every quarter. The business impact analysis framework, depicted in Figure 11.64, would help them in identifying the potential disaster paths after a detailed analysis of vulnerabilities, threats and risks.

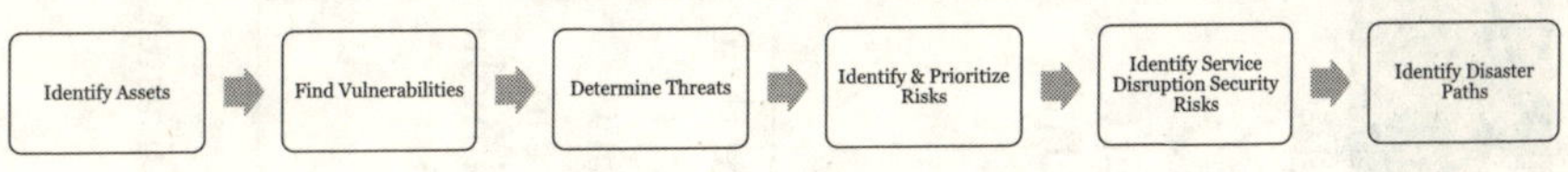

Source: Author

Figure 11.64: Business Impact Analysis Framework

Potential disaster paths identified using the business impact analysis framework are described in Table 11.43.

S. No.	Asset	Vulnerability	Threat	Risk	Potential Technology Service Disruption	Possible Business Impact (Potential Disaster)
1	Applications	Legacy software applications which do not have inbuilt cybersecurity features	Cybercriminals may hack the supply chain system and change the delivery dates to delay the delivery of the product	Cyberattacks, data privacy compromise and product delivery delays	Customers will not be able to track the order status due to the hacking of the supply chain application	Heavy load on the help desk and inability to take calls, too much waiting time, frustrated customers and brand dilution due to delivery failure
2	Servers	All servers are hosted in one place, and there is no full-fledged disaster recovery site for any natural disasters	Electric short circuit and fire in the server room may cause loss of data and major disruption to operations	Data centre shut down, loss of sensitive data, customer complaints and high capital cost	Infrastructure downtime and no access to software applications	Customers, employees and partners will fail to perform their work; business will be stopped for 48 hours due to infrastructure downtime

Source: Author

Table 11.43: Technology Disaster Path Samples

Thus, disaster path = [asset –> vulnerability –> threat –> risk –> potential technology service disruption –> potential business impact].

Bob also shared their plans to use AI and ML algorithms to predict the critical disaster paths that will emerge from the disruption of technology services and assets.

According to Bob, selecting the right disaster risk response strategy is essential for business continuity. Disaster paths identified proactively must be managed by choosing the right risk response strategy based on priority and its impact on the technology services, assets and business operations.

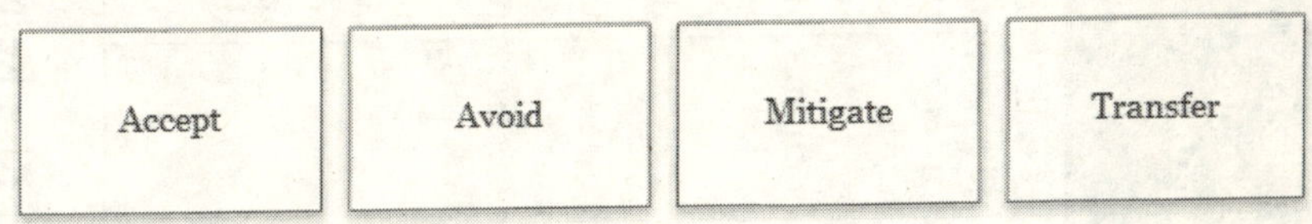

Source: Author

Figure 11.65: Risk Response Strategies

1. **Accept:** Accept the risk when it is realized, manage the same by relying on the budget (contingency reserve) and time. This includes a contingency plan and a fallback plan (usually for low priority risks).
2. **Avoid:** Eliminate the root cause of the risk and try to avoid the occurrence of the risk.
3. **Mitigate:** Put the effort and money to reduce the impact of the risk before its occurrence, which is difficult to eliminate.
4. **Transfer:** Transfer the risk to other suppliers or partners.

Bob also shared a few examples of the application of disaster risk response strategies against possible disasters (impact on business).

S. No	Asset	Vulnerability	Threat	Risk	Possible Technology Service Disruption	Possible Business Impact (Possible Disaster)	Disaster Risk Response Type	Disaster Risk Response
1	Applications	Legacy software applications which do not have inbuilt cybersecurity features	Cybercriminals may hack the supply chain system and change the delivery dates to delay the delivery of the product	Cyberattacks, data privacy compromise and product delivery delays	Customers will not be able to track the order status due to the hacking of the supply chain application	Heavy load on the help desk and inability to take calls, too much waiting time, frustrated customers and brand dilution due to delivery failure	**Transfer**	Transfer the cybersecurity risks to the insurance provider by taking security insurance
2	Servers	All servers are hosted in one place, and there is no full-fledged disaster recovery site for any natural disasters	Electric short circuit and fire in the server room may cause loss of data and major disruption to operations	Data centre shut down, loss of sensitive data, customer complaints and high capital cost	Infrastructure downtime and no access to software applications	Customers, employees and partners will fail to perform their work; business will be stopped for 48 hours due to infrastructure downtime	**Mitigate**	Proactively invest in an alternate data centre to get the technology infrastructure up and running for business continuity

Source: Author

Table 11.44: Examples of Disaster Risk Response

Tolerance/Threshold Management

Paul highlighted the significance of defining the tolerance limits or thresholds for **disaster recovery point** and **disaster recovery time**, based on which the disaster risk response strategies could be applied.

A disaster recovery point is nothing but **the level at which the business disruption/loss can be accommodated after a disaster** such that the negative impact and damage is minimal.

Disaster recovery time is **the time within which the business operations must be restored after a disaster** such that the negative impact and damage is minimal. Technology asset and service continuity plan must be implemented according to them.

1. **Define recovery point objective (RPO):** The threshold up to which the organization could bear the loss of business disruption (money, data, etc.). For example, losing data eight hours before the disruption is fine. If the disruption happened at 9:00 PM and if the backup of the data is available by 1:00 am, that's fine. This means, with this minimal data loss, the business may continue with minimal disruption.
2. **Define recovery time objective (RTO):** The duration of time within which an IT service/software application must be restored after a disaster. For example, within three hours of the disaster, the CRM app must be up and running.

Continuous Improvement of Business Continuity Management

Tom was concerned that Auto Gear's latest business continuity plan was designed three years ago and there had been no further updates. But in the last three years, there had been many changes in the technology and business landscape of Auto Gear. Hence, Tom suspected that the current business continuity management framework and plans needed to be completely redesigned.

Bob concurred with the suggestion and said that the business continuity management processes and practices must be modified and changed on a regular basis to align with the changes in the business and technology ecosystems. Then it would ensure the organization implements compelling and relevant disaster recovery strategies for business continuity.

The following changes to be factored in for improving business continuity management in areas like mapping between technology services, assets and business operations, risk management, disaster prediction, response strategies and tolerance limits.

1. Past disaster recovery experience (best practices, lessons learned)
2. Changes in the business processes/procedures
3. Changes in digital products (cars)
4. Changes in digital services (customer experience)
5. Changes in the organization structure or operating model
6. Changes in technologies
7. Changes in digital technology strategy
8. Changes in digital technology architecture
9. Changes in digital software applications
10. Changes in digital technology infrastructure
11. Changes in suppliers and partners

Theme 3: Disaster Impact Reduction Management

Paul stated that no organization can avoid disasters. The only thing one could do is minimize the impact of the disasters with proper planning and execution of business continuity and disaster recovery management processes.

Greg complained yet again about spending considerable money in the last couple of years, but failing miserably when it came to restoring business operations within the defined timelines successfully. This had impacted their brand, profit, revenue and quality of service.

Bob believed prevention is better than cure. Hence, he suggested that they must invest in disaster impact reduction management. Tom agreed with Bob and though Greg was still doubtful, he decided to listen to Bob. Bob began by sharing the four important pillars of disaster impact reduction management.

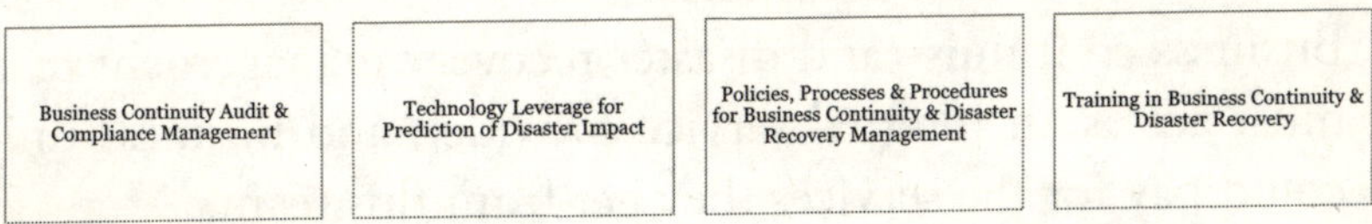

Source: Author

Figure 11.66: Pillars of Disaster Impact Reduction Management

Business Continuity Audit and Compliance Management

Greg interrupted Bob with the complaint that at Auto Gear, most of the inspections are done for the sake of compliance or as a ritual. He also raised the issue that these compliance checks have not helped Auto Gear improve business performance or operational excellence proactively. In most cases, business executives felt that these checks were overheads that took their precious time from day-to-day business.

While Bob agreed with Greg's view, he shared his candid opinion about why these compliance exercises had become a ritual. It was mainly due to a lack of senior management oversight and commitment. While revenue and profit are the direct measures of success, which senior management focuses on, ignoring these basic elements will cost them dearly during a crisis.

Tom intervened and asked Bob, "What do you think we must do to bring positive change here?" Bob offered the following suggestions:

1. Non-conformances must reflect the business unit performance.
2. The CEO and board of directors must review critical non-conformances that may cause a business discontinuity in future in their meetings.

3. Budget/cost incurred by the organization for business continuity and disaster recovery management must be apportioned among each business unit so that the senior executives get value for money.
4. Business continuity and disaster recovery management team must act as an internal service provider, and business units could pay for the services they get from this team.

Tom and Greg agreed with Paul's suggestions and asked Bob to go ahead and present his thoughts. Bob shared a comprehensive model for business continuity and disaster recovery management as depicted in Figure 11.67.

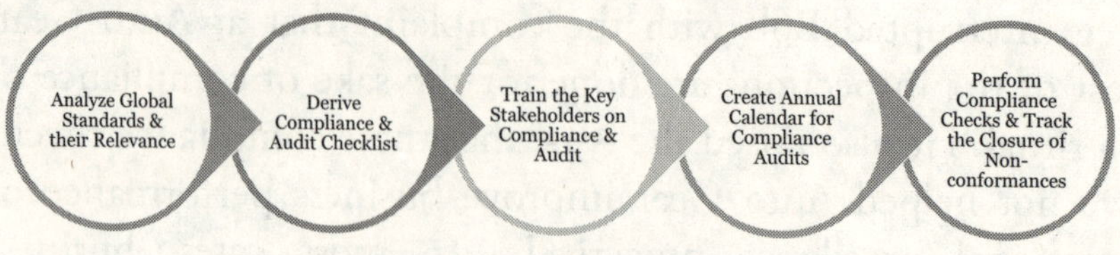

Source: Author

Figure 11.67: Business Continuity Audit and Compliance Management

Bob reiterated the fact that prevention is better than cure. In that sense, he shared that his team had spoken with experts and practitioners and identified a few standards using which they planned to derive checklists for performing the audit and compliance checks.

These checklists, given in Table 11.45, would help in identifying any non-conformances that help in taking corrective and preventive actions for effectively managing business continuity and disaster recovery management activities at Auto Gear.

S. No.	Standard	Description	Compliance and Audit Checklists
1	ISO 27001	This standard determines the requirements to protect information assets with adequate and proportionate security controls.	Information security checklist
2	ISO 27002	This advisory document recommends security control objectives to protect the confidentiality, integrity and availability of information.	Information confidentiality, integrity and availability checklist
3	NFPA 1600	This standard provides a framework for the protection of occupants, emergency response personnel, property and the environment by developing pre-incident plans to manage emergencies and destructive events.	Emergency preparedness checklist
4	HIPPA	This privacy standard helps protect patient's medical records and health information provided to doctors, hospitals and other healthcare providers.	Patient electronic health record privacy checklist
5	ISO 22301	This standard helps in business continuity in the following areas during disruptive events: • operations continuity, • delivery of products and services without disruption, • protection of the brand and • safeguarding the reputation of key stakeholders.	Business continuity management checklist

Source: Author

Table 11.45: Business Continuity Audit and Compliance Standards and Checklists

Timely closure of non-conformances would help in reducing the impact of the disasters and enable restoring the business operations faster, better and cost-effectively.

Technology Leverage for Prediction of Disaster Impact

The next important topic Paul and Bob wanted to discuss with Tom and Greg was how to leverage technology to bring proactive and preventive measures. Paul and Bob met all the senior executives of Auto Gear and asked them about their opinion on business continuity and disaster recovery activities being done at Auto Gear. They did not know whether they were ready to recover from disasters with minimal impact.

Most of them responded that:

1. Business continuity and disaster recovery management is a process overhead.
2. Business continuity activities are done for the sake of compliance.
3. Business continuity plans were never tested for their effectiveness.

Bob shared that in today's digital world, technologies helped in reducing the processing overheads with hyperautomation. Another critical trend he had observed was the simulation of disasters and predicting the impact of disasters using technology. Technology systems even gave recommendations to reduce the effects of disasters. AI and ML algorithms play a very vital role here.

Potential risk/disaster identification is made today mostly by experts who manually collate and analyze data points from various sources. Bob shared their plan to build a disaster predictor tool. This tool would collate data from ERP, CRM and SCM systems, machines, equipment, intelligent products and assets and external sources like satellite data, weather data, rainfall data or seismic data. Data is the gold mine and by using AI and ML, the disaster predictor tool would provide a list of potential disasters and response recommendations.

Bob shared their dream of digitalizing the whole business continuity and disaster recovery management process. A few examples are given in Table 11.46.

S. No.	Digitalization of Business Continuity and Disaster Recovery Management Process	Description
1	Business continuity compliance app	• Digitalize compliance and audit process and workflow by developing a software app to conduct regular compliance checks using the mobile app.
2	AI and ML algorithms to predict natural and human-made disasters	• Predict and simulate the flood scenarios based on rainfall data using ML algorithms and take corrective and preventive actions for business continuity. • Predict the location of potential earthquakes using the seismic data leveraging machine learning algorithms to predict the magnitude and pattern of earthquakes. • Leveraging the data logs received from various software applications, infrastructure and devices, predict and simulate different cybersecurity loopholes using ML algorithms, and take corrective and preventive actions for business continuity.

Source: Author

Table 11.46: Technology Leverage for Prediction of Disaster Impact

Bob emphasized that leveraging new age digital technologies was critical for optimizing human effort, predicting future disasters and reducing the impact of disasters. Tom asked an interesting question, "What does it mean for Auto Gear to leverage technology for business continuity and disaster recovery?" Bob replied that they must identify the changes required in existing applications and infrastructure to include business continuity features. This meant integrating new features in such a way that their systems automatically capture the data points to provide a list of potential disasters and recommendations. Also, they must mandate the leaders to include business continuity and disaster recovery management requirements while developing new software applications, products, things and infrastructure.

Policies, Processes and Procedures for Business Continuity and Disaster Recovery Management

Paul stressed the point that each stakeholder of Auto Gear must have a common purpose, shared understanding and common standards. This helped in implementing business continuity and disaster recovery processes.

Bob and the team firmly believed that without detailed policies, processes and procedures, there will be chaos, confusion and communication gaps. Bob shared a few representative policies and processes which Auto Gear would implement very soon. A few examples of service continuity and disaster recovery management policies are:

1. Legal policy during disasters
2. Asset and service criticality policy
3. Incident management policy
4. Regulatory policy
5. RTO and RPO policy (thresholds/tolerances)
6. Work from home (WFH) policy during disasters
7. Risks prioritization policy
8. Disaster recovery plan testing policy

9. Disaster management procurement approval policy
10. Risk response selection policy
11. Communication policy
12. Disaster management budget approval policy
13. Healthcare policy during disasters

Some examples of service continuity and disaster recovery management processes/procedures are:

1. Technology asset and service inventory management
2. Business impact analysis
3. Business continuity audit and compliance management
4. Business continuity audit training
5. Disaster recovery mock testing
6. Disaster recovery management
7. Disaster communication management

Bob reiterated the point that automation of policies, procedures and processes must be done to minimize process overhead, for visibility on potential disasters and to improve effectiveness.

Training in Service Continuity and Disaster Recovery

Paul admitted that the weakest link in Auto Gear was the poor learning strategy when it came to business continuity and disaster recovery. The learning strategy must help in preparing Auto Gear for better technology service continuity and business operations.

He believed in introducing on-demand and online collaborative learning, using their internal digital learning platform, which could accommodate audios, videos, group discussions, chatbots, Robo coders, etc. The plan was to conduct training and workshops for strategic, tactical and operational teams as mentioned here:

1. Business continuity and disaster recovery processes
2. Business continuity and disaster recovery policies
3. Business continuity and disaster recovery global standards
4. Disaster risk identification, prioritization and response

5. Business continuity and disaster recovery compliance checks
6. Business continuity and disaster recovery non-conformances management
7. Disaster incident management
8. Disaster impact reduction management
9. Disaster recovery mock tests and improvement
10. Emergency management and communication strategy
11. Software applications, AI and ML algorithms and disaster prediction simulation models

Theme 4: Disaster Recovery Management

Natural or human-made disasters are painful. These disasters put a lot of physical and mental strain on the organization's leadership, employees, partners, suppliers and other stakeholders.

CEO Tom shared a recent incident at Auto Gear when one of their data centres got flooded, and they failed to move to the alternate data centre site in the stipulated period. They lost data beyond their threshold. Tom was concerned about their preparedness for disasters.

Paul agreed with Tom and expressed his apologies. "This time," he confidently said, " we have worked with the world's best consultants while preparing for disaster recovery." He shared the four pillars of disaster recovery management he and his team had come up with.

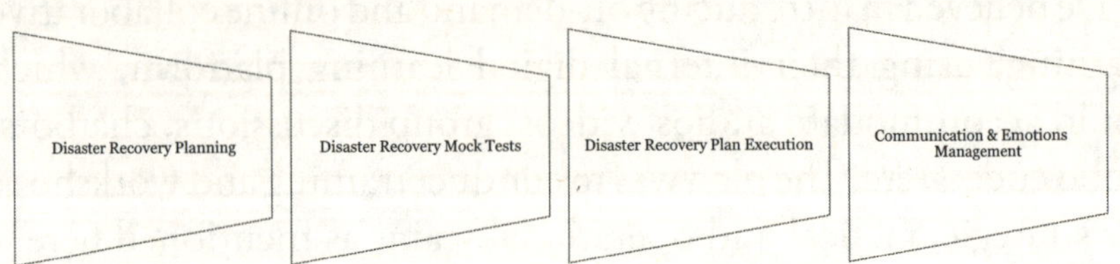

Source: Author

Figure 11.68: Pillars of Disaster Recovery Management

Disaster Recovery Planning

Bob commented that in today's uncertain world, the most critical activity of business continuity is how quickly organizations resume business operations and continue with the business after a disaster.

The ineffective disaster recovery planning, processes, policies and procedures lead to business disruption. Business continuity and disaster recovery preparatory activities cannot be done for the sake of complying with a few standards or frameworks. Disaster incidents may be expected or unexpected. The expected disaster incidents may emerge from identified risks. The unexpected disaster incidents may emerge during pandemics, wars, etc.

A disaster recovery plan may include the following components:

1. Technology services catalogue
2. Disaster recovery management responsible, accountable, consulted and informed (RACI) matrix
3. Technology asset inventory
4. Disaster recovery alternate site and infrastructure details (physical and digital)
5. Software and hardware inventory, license and warranty details
6. Business impact analysis and risk response strategies
7. Evacuation plan
8. Communication plan
9. Crisis management plan
10. CRM plan during a disaster
11. Employee emotion and behaviour management plan during disasters
12. Shareholders management plan during disasters
13. Public relationship management plan (if applicable) during disasters
14. Disaster recovery management team and contacts
15. Emergency management contact numbers
16. Employee contact numbers
17. Suppliers list, contracts catalogue and contact details

18. Help desk contact details and incident management process
19. Financial, legal and regulatory contact details
20. Insurance documents and contact details

Disaster Recovery Mock Tests

As emphasized by the senior business executives of Auto Gear, the organization did not have the necessary infrastructure, equipment and resources to test the effectiveness and efficiency of the disaster recovery plan.

Bob clarified, "Though conducting mock tests is an expensive and time-consuming process, it is mandatory to safeguard our business and customers. Mock tests will be conducted after communicating with the key stakeholders about the risks involved in conducting the mock tests and preparing them for any impact on business operations."

Business continuity and disaster recovery mock testing phases are described in Figure 11.69.

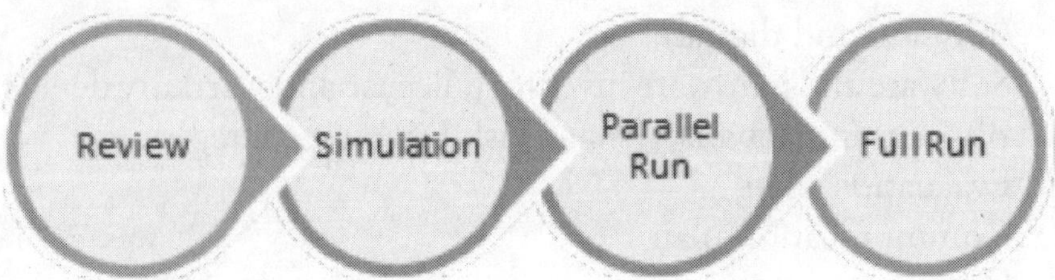

Source: Author

Figure 11.69: Phases of Disaster Recovery Mock Tests

1. **Review:** Review of business continuity and disaster recovery management processes, policies, plans and all other relevant documentation. Reviews are done for completeness, correctness, ambiguity and consistency.
2. **Simulation:** If the budget permits, create digital twins of the critical assets and business processes and simulate various disaster scenarios using AI algorithms and analytics. If digital twins are not available, physical simulation must be used as appropriate. Simulation is the most potent tool to virtually

test the effectiveness of business continuity and disaster recovery plans and documentation.

3. **Parallel run:** Testing the effectiveness of the disaster recovery plan is essential. For this, a few technology assets and services are shut down for a stipulated period of time. This is done to create a disaster situation. The disaster recovery plan is followed and the tasks are executed to restore operations in the alternate site to address the disaster situation. Thus, the effectiveness of the disaster recovery plan can be measured in terms of the effectiveness of communication, resources capability, approval delays and infrastructure provisioning at an alternate site and speed where the operations are up and running. This parallel run provides crucial information to identify improvement areas.
4. **Full run:** A full run means shutting down all the technology assets and services to create an artificial disaster situation and, as per the disaster recovery plan, restoring the entire operations to an alternate site. The disaster recovery plan is evaluated as per the defined RTO and RPO thresholds. A root-cause analysis for the failure points of the business continuity and disaster recovery plan is followed by the implementation of corrective actions.

Disaster Recovery Execution

Before Paul began to talk about disaster recovery execution, Tom said, "A bad plan is planning to fail, and a good plan without execution is only a dream." Nodding in agreement, Paul shared a three-step process for the effective execution of the Auto Gear's disaster recovery plan.

Source: Author

Figure 11.70: Steps of Disaster Recovery Execution

Steps	Details	Example
Assess	Assess the as-is situation of the disaster quickly in terms of cost, time and damage. Collect details, such as, when the disaster happened, which assets and services were impacted and must be restored, etc. Immediately send a disaster notification to the key stakeholders	**Disaster:** COVID-19 Pandemic **Technology services impacted:** Outsourcing IT provider in India stopped operations due to the spread of COVID-19 and hence, mission-critical applications and infrastructure management services are down. **Business impact:** • Auto component suppliers in China stopped production in China due to the pandemic. • Auto Gear wants to place orders with alternate suppliers in another country. • Auto Gear fails to place orders with alternate suppliers as the supply chain application is down as the outsourcing IT provider closes its operation. • Just-in-time inventory and no supply of automotive parts from China cause closure of production plants in the USA. **When:** Feb 2020 **Notification to:** Customers, employees, suppliers and logistics partners.

Implement	Conduct kick-off meeting and start the implementation of already defined disaster recovery plan, protocols, policies and procedures to restore mission-critical business operations. Manage the incidents effectively and make sure that the mission-critical technology assets and services are up and running as soon as possible.	**Kick-off meeting:** Conduct kick-off meetings with the disaster recovery management team, suppliers, partners and key stakeholders. **Disaster response strategy:** Auto Gear implements an already identified risk response strategy. Hence, they can configure a new and lightweight supply chain app from a cloud provider within 24 hours and be able to put order to alternate suppliers in another country. **Protocols:** Working from home for non-mission-critical activities, production plants and office hygiene and health protocols as well as government regulations and safety protocols. **Policies:** • Procure from alternate suppliers • Paid leaves for a few employees • Increased salary for critical employees, discounted prices to customers for delayed delivery of cars **Incidents:** One of the employees in the procurement department gets affected by the coronavirus. HR takes charge of the incident and the necessary procedure is followed to take the employee to the hospital to support him during the isolation period.
Restore	All original technology assets, services, systems are restored and are free from after-effects of the disaster and the whole business operations are brought to normalcy.	**Fully operational:** Once the impact of the pandemic comes down, IT providers from India can be outsourced and auto component suppliers from China can resume their operations. And Auto Gear production will gradually return to normalcy.

Source: Author

Table 11.47: Steps of Disaster Recovery Execution with Examples

Disaster Communication and Emotion Management

Bob shared a personal experience of a financial scam (a human-made disaster) that had happened in his previous organization. During that crisis, the management was forced to give pink slips to around 25,000 employees (50% of the organization) and 40% of the business was lost as a few customers left the organization claiming it was a tainted organization. Employee morale was too low and nobody knew whether they would get their salary the next month. Partners and industry analysts dissociated themselves from the organization abruptly. He explained how they effectively managed the crisis with effective communication and emotion management.

Disaster Communication Management

Bob shared a straightforward framework for communication management covering all the stakeholders during a disaster/crisis period. Effective communication provides a single source of truth, avoids rumours and fake news, and makes sure that all the stakeholders have a common understanding of the crisis and progress. The communication management is a systematic process that makes sure that all stakeholders have up-to-date information and news about the disaster.

Following are the key elements of communication management:

1. **What** information and key messages need to be shared?
2. **From whom** and **to whom** must the messages be sent?
3. **When** must the messages be sent (frequency)?
4. **How** must the information be sent (channels)?

What	From Whom	To Whom	When	How
Official communication about disaster	Disaster recovery management committee chairman	• Customers • Employees • Partners • Government agencies • Insurance agencies • Industry analysts • Public • Others	Immediately after the preliminary analysis of the disaster and its impact	• Mail • SMS • WhatsApp • Official letter through post • Facebook (if required)
Disaster recovery progress update	Disaster recovery management team leader	• Customers • Employees • Partners	Every 4 hours	• Mail • SMS • WhatsApp • Chatbot
Incident progress	Incident response manager	• Requester	• Priority 1: 4 hours • Priority 2: 8 hours • Priority 3: 24 hours	• Mail • SMS • WhatsApp • Chatbot
Motivational messages (appreciation, rewards, etc.)	CEO	• Employees	Weekly	• Mail • Post letters to family • Blog posts

Source: Author

Table 11.48: Disaster Communication Management

Emotion Management during Disasters

Emotion management refers to managing the emotions of the employees to reduce/eliminate stress, pain and psychological issues. Bob shared his experience of managing employee emotions when the senior leadership team was involved in a huge financial scam in his earlier organization. This scam had become the biggest disaster and employees were under tremendous pressure.

1. Ensure employees get their salaries on time each month
2. A six-month notice period for the employees who have been given the pink slip
3. Establish help desk to support the employees in getting jobs at other organizations
4. Parked ambulances at the organization premises to take care of emergencies that arise from stress and pressure
5. Closed entry to rooftops to avoid any potential suicides

Bimodal Technology Operations Management

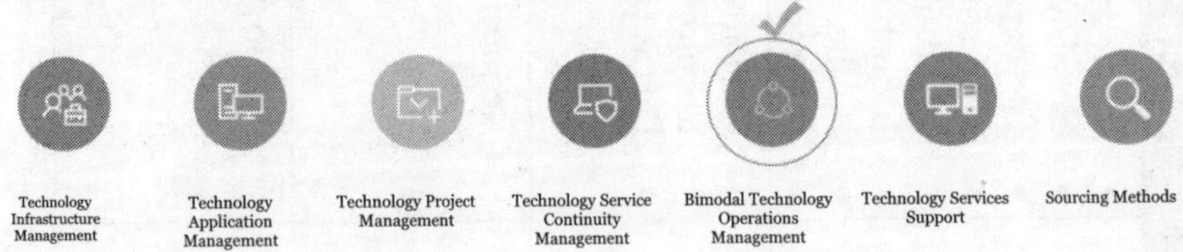

Source: Author

Figure 11.71: Technology Operations Management—Bimodal Technology Operations Management

A cynical Raj, one of the core members of Paul's team, mentioned that they were not doing a good job in managing the current technology operations to support the current line of businesses, and now they were talking about digital technology operations to support digital businesses, which is even more complex. They did not have the required resources, infrastructure, budget and leadership to support both current and digital line of businesses.

Hence, they may not be successful in adopting bimodal technology operations management.

Paul appreciated the concerns raised by Raj and agreed that they were going through a very tough time as there had been a decline in core auto business as well as a consistent decline in technology budget allocation by Auto Gear.

"In this current scenario, we can't defocus our existing technology operations which support Auto Gear's current line of businesses. But soon enough, our business executives will be pushed to transition to a digital business. Our technology organization must be ready to offer digital technology services and operations at that time."

Bimodal technology operations include the conventional IT operations and the digital technology operations aligning with Auto Gear's core businesses, as shown in Figure 11.72. The intersection in the figure highlights the importance of focusing on both conventional and digital technology operations at the same time.

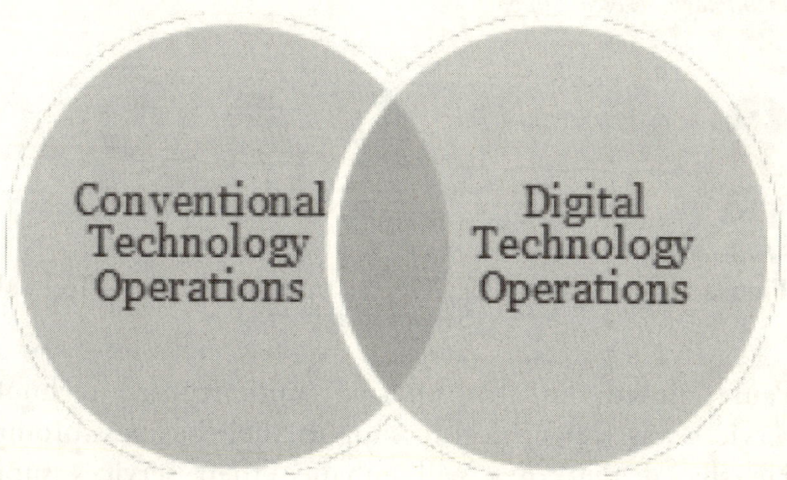

Source: Author

Figure 11.72: Bimodal Technology Operations

Raj asked Paul how they were going to drive bimodal technology services and operations to support both the current and the digital line of businesses. Paul shared his strategy and plan, which involved refining:

1. technology services,
2. technology assets/infrastructure,
3. policies,
4. processes,
5. organization structure,
6. leadership,
7. technology specializations and
8. service help desk.

Paul assured his team that Auto Gear's CEO was committed to the plan and had asked them to submit the strategy, detailed plan and budget required by the technology organization to offer both their current technology services and new technology services.

Technology Services Support

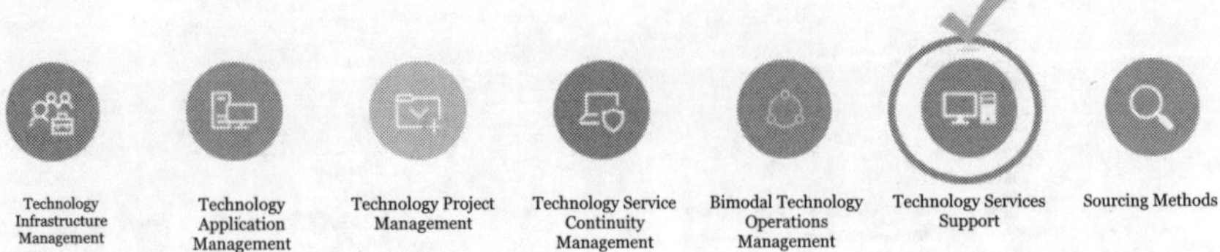

Source: Author

Figure 11.73: Technology Operations Management—Technology Services Support

Paul pointed out that offering uninterrupted technology services was very critical to support their business promptly. He shared that they will enhance their services support function to effectively manage bimodal technology services. Paul highlighted only the new features, which he planned to bring in, to support the current technology operations. See Table 11.49.

S. No.	Technology Services Support	Tools/Solutions
1	Service desk	• Robocall centre • Chatbot-based support
2	Incident management	• Incident management focuses on addressing the real incident in a faster and better way with very less impact on technology operations; an incident could be an unplanned interruption to a service or reduction in quality of service • AI and ML-based auto closure of incidents (no human interventions)
3	Problem management	• Problem management focuses on preventing incidents or reducing the impact of the incidents • AI and ML-based root-cause analysis of incidents and automatic recommendations help in reducing the incidents
4	Access management	• Cybersecurity policies and governance to effectively manage the access to data by employees, partners, suppliers and intelligent things
5	Service level management	• Define the service-level agreements separately for robot-based services and human-based services • Conduct root-cause analysis and implement the improvement action plan

Source: Author

Table 11.49: Technology Services Support

Sourcing Methods

Source: Author

Figure 11.74: Technology Operations Management—Sourcing Methods

Acquiring digital capabilities is one of the most significant steps of the process of preparing an organization for the digital era. Hence, organizations must move from traditional sourcing methods.

Paul shared his views on the need to change Auto Gear's sourcing decisions based on the new parameters. This would help in acquiring digital capabilities more effectively. Paul spoke with the Auto Gear procurement executives and planned to organize a detailed training session on the latest procurement decision-making process.

Paul also emphasized the point that they needed to acquire digital capabilities from new outsourcing partners, start-ups and other technology specialist firms. Table 11.50 describes the decision parameters considered for traditional sourcing and digital sourcing.

S. No.	Decision Parameters	Traditional Sourcing Decision Parameters	Digital Age Sourcing Decision Parameters
1	Cost savings	Yes	Yes
2	Only cost-based decisions	Yes	No
3	Compliance with regulatory and global standards	Yes	Yes
4	Risk management	Yes	Yes

5	Enhance competitiveness	Partial	Yes
6	Ease of integrating supplier systems with the overall digital business ecosystem	Partial	Yes
7	Improve business outcomes with supplier capabilities	Partial	Yes
8	Supplier's innovative business models that add value to the organization	No	Yes
9	Leverage technology service marketplaces like cloud providers AWS, Salesforce, Microsoft/IBM that provides IaaS, PaaS, SaaS and BPaaS	Partial	Yes
10	Outsourcing	Yes	Yes
11	Crowdsourcing/Gig economy	No	Yes
12	Offshoring	Yes	Yes
13	Insourcing	Yes	Yes
14	Cybersecurity	Yes	Yes
15	Biosecurity	No	Yes

Source: Author

Table 11.50: Sourcing Decision Parameters—Traditional vs Digital Age

Building Block 7: Technology Architecture

Stephen, the CMO of xRetail, was frustrated with the inability of xRetail's technology organization to respond to the dynamic needs of the business and customers. Stephen called up John, the CIO, and expressed his concerns about high customer attrition. He pointed out some of his observations as well.

1. Our e-commerce site was hacked and was down for two days, we have many vulnerabilities to be addressed.
2. Our competitors release new features every month, and we take about three months on an average.
3. We have serious skill gaps in the new technology areas like AI, AR, VR and CIoT and we are losing our business to the competition.
4. The technology team does not have the business acumen or an understanding of business criticality and is failing to add value to the business.
5. We have plenty of data from social media and wearable devices but we fail to get insights to offer a personalized experience.
6. Our CIO keeps reporting that our existing architecture is not agile enough to integrate new technologies.

John tried to explain the situation and assure Stephen that he was looking into this, but Stephen yelled at him. He decided to take this matter to the CEO for a review of the current situation of their technology organization. John knew how grave the situation was and decided to keep quiet.

On 3 March 2017, CEO Scott's office arranged a meeting with Stephen, John and all the business unit heads to discuss the readiness of digital xRetail. In the meeting, all business executives unanimously agreed that they probably had 18 months to exit from their business, and they all needed to find new jobs. xRetail had lost 25% of its business just in the last six months due to technology deficiencies. They had grossly failed to align their technology organization to meet the needs and wants of digital age consumers. Scott asked John about his game plan. It was very clear from that meeting that John had no clue about the way forward.

The CEO also took a lot of inputs from the business executives. After spending some time with John, Scott realized that he had to make some tough decisions.

He instructed his office to coordinate with the HR head to recruit a new CIO within six weeks and informed John to leave the organization after serving his notice period. Scott had realized that xRetail needed a new skilled CIO, someone who thinks beyond software applications and IT infrastructure and would become an integral part of the business strategy team.

It took 10 weeks for xRetail to recruit a new CIO who had extensive experience in managing a digital retail business. Scott set higher expectations for Philip, the new CIO, and requested him to work with business executives to turn xRetail purely digital.

Digital Technology Architecture Guiding Principles

Philip spent almost one month to understand the business **needs**, **wants** and **challenges** of xRetail and realized that there were many deficiencies in xRetail's digital technology strategy, digital technology governance and digital technology architecture. Philip realized that there were many skill gaps, a few of which had to be addressed immediately. He came up with a recruitment plan, extensively worked to induct new tech specialists and digital technology specialist partner firms.

As the first step, he formed a core team with a mix of business and technology executives and released a plan for strengthening digital technology strategy, digital technology governance and digital technology architecture.

On 10 July 2017, Philip conducted a kick-off meeting with the core team to work on digital technology architecture. Before this meeting, Philip and the team had worked closely with the business team and collated their needs, wants and problems.

Based on the business requirements, the team collectively agreed that there were two critical areas, which they had to baseline before finalizing xRetail's digital technology architecture. They were technology architecture guiding principles and building blocks.

Philip and the team spent a considerable time brainstorming and had come up with the following technology architecture guiding principles:

1. Flexibility to integrate with partner, supplier and third-party data platforms to build a unique business ecosystem.
2. Enhance business value in terms of customer experience, operational efficiency and business growth.
3. Build a fault-tolerant system for business continuity and faster disaster recovery.
4. Consider data as an asset and apply predictive and simulation models to enable manifold growth, new business opportunities, customer experience and profits.
5. Bring agility to embrace new technologies in a shorter period.
6. Build digital software applications at a faster pace with the convergence of operations, communication and collaboration and information technologies.
7. Establish technology infrastructure for the ease of integration across public cloud, private cloud, on-premise and edge computing.
8. Enable omnichannel access and immersive experience.
9. Support applications and infrastructure to work on short-range networks like 6LoWPAN, LoRAwan, 3G, 4G, 5G and fibre-optic networks.
10. Secure applications, information, intelligence, infrastructure, things and data privacy.
11. Componentize and leverage microservices for developing software applications in less time.

Applying Lean Digital Thinking to Identify Digital Technology Architecture Building Blocks

On 25 July 2017, Philip conducted a detailed session on **lean digital thinking** to develop xRetail's digital technology architecture holistically and derived 20 digital technology architecture elements from the 12 principles of **lean digital thinking**.

S. No.	Lean Digital Thinking Principles	Changing the Anatomy of Technology Architecture	Digital Technology Architecture Elements
1	Create a better today and better tomorrow	Design the technology architecture to best leverage the legacy systems while building new age digital systems	• Legacy digitalization architecture • Digital systems architecture
2	Embrace constant and instant change	Continuously collate and analyze the changing trends and shifts in technologies, customer needs, wants and continuously refine the digital technology architecture	• Digital technology architecture change management
3	Nurture digital business talent	Nurture digital technology architecture skills regularly to prepare the technology organization for digital business future	• Digital technology architecture talent management
4	Identify and eliminate non-value-added activities	Develop the technology architecture for faster, better and steadier development, implementation and management of software applications, things and infrastructure	• Lean architecture for applications • Lean architecture for infrastructure • Lean architecture for things

5	Predict, recommend and improve overall business performance	Predict the technology architecture performance by collating and mining the data received from various resources like software applications, infrastructure, devices and things, and implement actionable recommendations to continuously refine the architecture	• Digital technology architecture change recommendations
6	Optimize human interventions	Develop the technology architecture to help in building semi-autonomous and autonomous businesses and organizations	• Business ecosystem architecture • Blockchain smart contracts • Robot process automation • Chatbots • Virtual assistants • Cyber-physical systems (4IR)
7	Establish real-time and contextual communication and collaboration	Establish a real-time communication and collaboration system to design the digital technology architecture with the help of globally distributed internal and external specialists (collaboration tools, chatbots, virtual assistants, crowdsourcing tools, etc.)	• Digital technology architecture development with real-time communication and collaboration

8	Safeguard privacy, safety and security of business resources	Develop cybersecurity technology architecture to develop hacking-proof and secure software applications, infrastructure and things	• Cybersecurity architecture
9	Establish a technology start-up ecosystem	Build an entrepreneur mindset to develop the technology architecture in multiple sprints by continuously pivoting the same as per the business needs	• Digital technology architecture pivoting
10	Create socially responsible products, services and value chains	Design the technology architecture for optimal consumption of resources (people, electricity, computing power, etc.)	• Resource optimization architecture
11	Design and deliver unforgettable moments of experience to stakeholders	Develop the technology architecture to develop systems that offer an immersive experience to customers	• Experience management architecture
12	Build a human-centric digital culture	Establish a digital culture and set of values that enable the technology teams to consider business-first attitude while designing digital technology architecture	• Digital culture

Source: Author

Table 11.51: Applying Lean Digital Thinking Principles to Identify Digital Technology Architecture Elements

A few of the existing technology leaders of xRetail were very cynical about what Philip had shared with them. They said that they had already been working on all these elements and there was nothing new in what he had said. Philip asked a few questions in response and requested those cynical members to answer them.

1. How many algorithms and predictive models have been developed by us?
2. Do we have the skills to design unforgettable moments of experience for customers?
3. Are our business executives and CEO happy with our technology expertise and delivery capability?
4. How are we using the CIoT to offer better services?
5. Are our mobile apps, chatbots and virtual assistants developed in line with the business and customer needs?
6. Do you have any idea of how many customers we have lost in the last six months due to technology deficiencies?

Philip only saw blank faces in response to his questions. He requested them to join hands with him to build a digital xRetail in the coming months. Philip and his team brainstormed and grouped the 20 elements of digital technology architecture into 11 building blocks as mentioned in Figure 11.75.

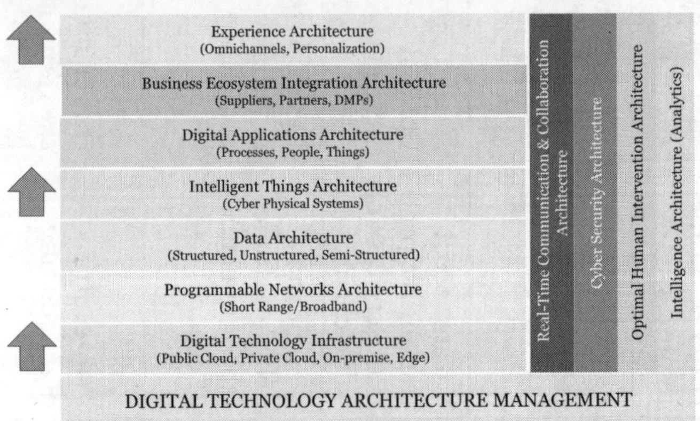

Source: Author

Figure 11.75: Digital Technology Architecture Building Blocks

S. No.	Digital Technology Architecture Building Blocks	Description
1	Experience architecture	Support delivering unforgettable moments of experience to stakeholders
2	Business ecosystem integration architecture	Integrate third-party systems and technology provider systems in plug and play model to build an extended business ecosystem
3	Digital applications architecture	Enable easy development of applications to digitalize processes considering transactional and emotional needs of customers, integrate business processes with intelligent things and providing insight, foresight using data collected from different sources
4	Intelligent things architecture	Support cyber-physical systems (IoT)
5	Data architecture	Support the storing, cleaning and retrieving high volume, high velocity and variety of data like structured data, semi-structured data and unstructured data
6	Programmable networks architecture	Establish short-range and broadband networks and enable real-time connectivity between people and things
7	Digital technology infrastructure	Support cloud, edge and quantum (in future)
8	Real-time communication and collaboration architecture	Enable real-time communication and collaboration while performing activities or work
9	Cybersecurity architecture	Secure all technology resources from cyberattacks/hacks and safeguard data privacy
10	Optimal human intervention architecture	Leverage digital technologies to optimize human interventions while performing activities/work

11	Intelligence architecture (analytics)	Leverage data from different sources and provide actionable insight/recommendations to take right decisions, corrective and preventive actions at the right time
12	Digital technology architecture management	Manage skills, culture and change to continuously refine the digital technology architecture

Source: Author

Table 11.52: Digital Technology Architecture Building Blocks

Philip conducted a meeting with Scott and the other senior business executives and shared the digital technology guiding principles and building blocks with them. He was confident that once the digital technology architecture building blocks are incorporated, xRetail will be ready to leapfrog the competition in the digital era. But he also emphasized the importance of testing the feasibility, scalability, agility and performance of the new digital technology architecture building blocks. Therefore, a few proofs of concept needed to be built for the identified use cases of xRetail using new digital technology architecture, which was a very complex task and required a considerable budget, skilled resources and infrastructure. A collaborative team had to be formed with the business and technology teams to complete the technology proofs of concept quickly.

Scott and the other executives were convinced that Philip had taken the right approach and was moving in the right direction. They felt that Philip has spent considerable time in the last couple of months to understand business needs, wants and problems and had been preparing the technology team to add value to xRetail businesses.

Philip returned to his office in a positive mood, thinking about his game plan for the next 90 days to accelerate the pace of making xRetail's digital technology architecture operational. The first step was to detail the 11 building blocks of digital technology architecture.

Austin, the head of infrastructure, and Harry, the head of software applications, who aspired to be the CIO of xRetail, were not able to work with Philip. They felt Philip had destroyed their plans. As they had been working for the last 10 years at xRetail, they had a great relationship with Scott and the other senior executives.

They started spreading propaganda against Philip by saying that he had been wasting too much time in conducting meetings and had not been able to deliver results as he had promised. They also went to the extent of saying that Philip's domain knowledge in retail was not too high. The news soon reached Scott.

But Philip was a brilliant executive; he kept Scott and the other senior executives updated on issues, risks, constraints and the progress made against the milestones he had committed to. There was no need for Scott or the others to take the inputs from Austin or Peter. In this way, Philip managed the expectations of senior management very well.

Scott thought that Austin and Harry were not doing the right thing, and if they continued to do so, Philip will fail and hence, the organization will fail too. Scott arranged a one-on-one meeting with Austin and Harry and discussed the situation with them.

Harry started the discussion in compliant mode and told Scott that Philip was not the right guy, and much time had been wasted due to Philip's incapability. Much time had been spent in meetings of no use.

Scott enquired about the meetings that Philip had conducted. Harry replied, "Philip held too many meetings with the business team to understand their needs, wants and problems, which we are already aware of. Internally, we spent many hours discussing technology architecture, which we had created a long time back. Philip micromanages and follows up every day to get the progress update, which is a real pain. We have an excellent technology team within xRetail, but Philip has been using external specialists from partner organizations. He might have some vested interests."

Scott smiled at Harry and asked, "Why did you think we lost 25% of our business to the competition?"

"I do not know."

"Why are we failing to introduce new features to the market?"

"It is because the business team was not clear and was confused."

During the conversation, Harry did not highlight the technology gaps within xRetail in the areas of AI, ML, AR/VR. But he shared that, they had built a few prototypes in AI, ML, MR areas. Scott stopped him and said, "After 18 months, you are still talking about prototypes, while our competition has already mastered in providing exceptional value to customers."

Scott asked Harry, "Did the meetings held by Philip help you get greater clarity on the business expectations and alignment with technology?" Harry said, "Yes!"

Scott raised his voice and asked Harry to support Philip and be a useful technology leader; else, he might have to look for a new job. Scott made it clear to Harry that he would not tolerate any negative propaganda. He arranged a similar meeting with Austin. Harry and Austin realized that there was no option except to work with Philip.

Digital Technology Architecture Building Blocks

Philip formed a cross-functional team to detail each building block and gave two weeks to them to come up with the first cut digital technology architecture. On 15 September 2017, there were some heated debates and discussions among the team. Finally, they baselined digital technology architecture. Philip made sure that importance was given to both Harry and Austin, and slowly they started warming up to Philip. He did not want to lose Austin and Harry as they had tremendous experience and business relationships that could help xRetail's business in the digital era.

Philip and team were quite happy that they had co-created the 11 building blocks of digital technology architecture that could unlock the potential of technology fusion and help xRetail become a digital leader in the coming years. Philip conducted a meeting with all the senior business executives and shared the detailed architecture diagram with all building blocks, given in Figure 11.76.

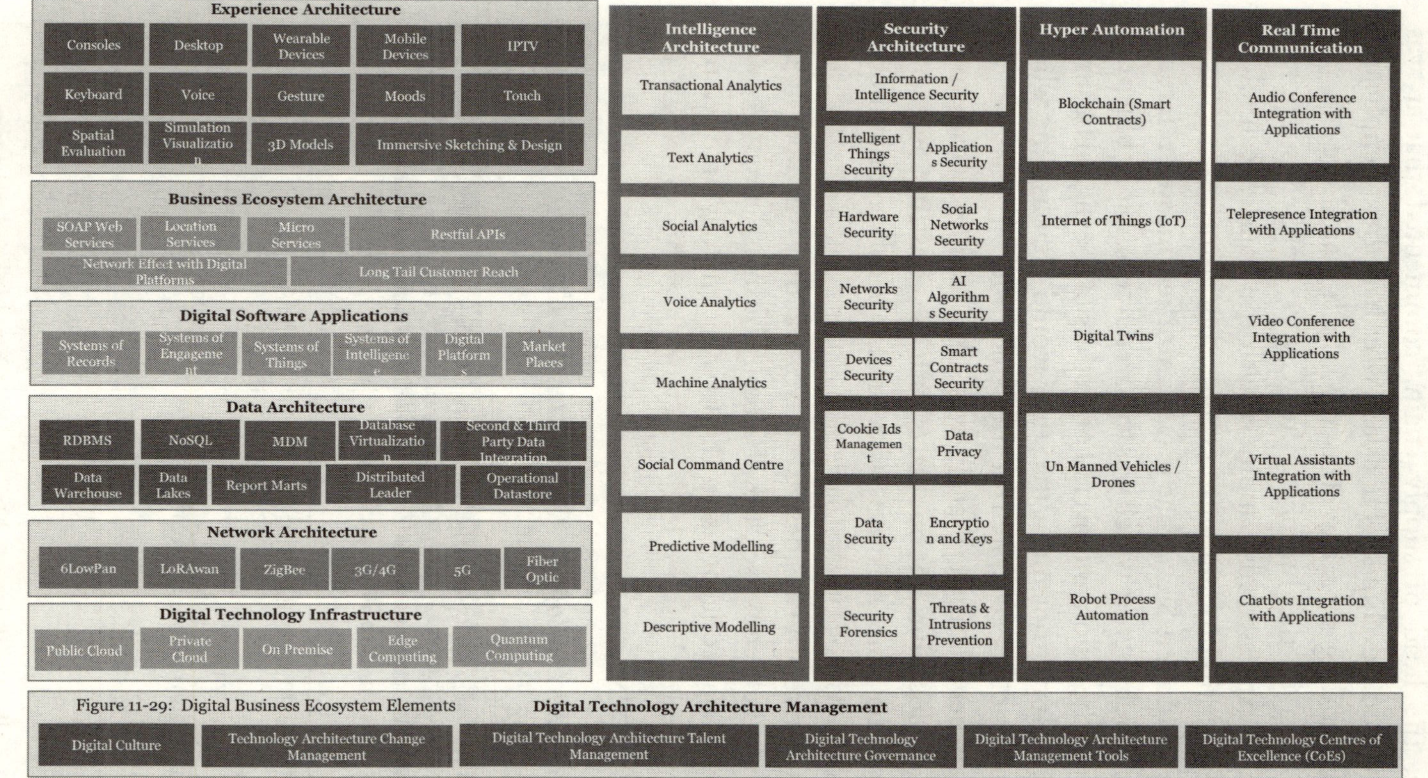

Source: Author

Figure 11.76: Unified Master Reference Blueprint for Digital Technology Architecture

Steve, the head of supply chain, commented that it was scary and looked clumsy. He asked why had they complicated things by adding so many elements which would create more complexities in the business. But CMO Stephen supported Philip and expressed his happiness with the holistic technology view that Philip brought within such a short time.

Philip informed Steve that it was a **unified master reference** for digital architecture which was the overall technology blueprint to support the vision and overall objectives of the digital business of xRetail. It did not mean that they would establish the whole digital technology ecosystem at the same time.

He further clarified that it was like having a five-storey building's blueprint, but one starts with the construction of only one floor at a time. As time passes and as per the requirements, other floors will be built.

Philip assured the business executives that they will focus on the top five digital retail use cases shared by business executives and build proofs of concept using the new technology architecture. Philip requested business leaders to work together to prepare for the digital future.

Digital Technology Architecture Building Blocks—Illustrative Tools

S. No.	Digital Architecture Building Block	Tools/Platforms
1	Digital technology architecture management	iServer, HOPEX, erwin EVOLVE, Abacus
2	Digital technology infrastructure	AWS, Azure, GCP, Nutanix HCI
3	Programmable networks architecture	Barefoot Networks, Junos Space from Juniper, and Cisco
4	Intelligent things architecture	Azure IoT Hub SDK, Cloud IoT Device SDK, Google cloud platform, ThingWorx, IBM Watson IOT, AWS

5.	Digital applications architecture	Microservices, SPA, PWA, Cloud Native
6	Business ecosystem integration architecture	TIBCO, Boomi, Workato, Zapier
7	Experience architecture	Vuforia, ARCore, Kudan, ARKit
8	Real-time communication and collaboration architecture	Microsoft Teams, Slack, Asana, HireVue, AI-driven chatbots
9	Cybersecurity architecture	Okta, SailPoint, CyberArk
10	Optimal human intervention architecture	Virtual assistants, chatbots, robots, Google cloud platform, ThingWorx, IBM Watson IOT, AWS
11	Intelligence architecture	IBM CloudPak, DataBricks, Unified Analytics, RapidMiner, Qlik Sense, IBM SPSS, TensorFlow, PyTorch, Keras

Source: Author

Note: the tools suggested in the above table are only illustrative and are not any recommendations to use.

Table 11.53: Digital Technology Architecture Building Blocks—Tools/Platforms

Digital Technology Centres of Excellence

Philip strongly believed that focused centres of excellence (CoEs) must be established to improve the digital technology skills internally and in collaboration with technology specialist firms.

Louis, the principal technology architect, was given the responsibility to define the structure, roles and responsibilities of the CoE. He worked with the technology and business executives of xRetail, technology specialists of the partner organizations and finalized the responsibilities of CoE. They were:

1. Conduct regular customer-focused group discussions.
2. Analyze market trends/competitors' strengths and weaknesses.
3. Regularly identify unique digital use cases.
4. Continuously acquire new digital technology skills.

5. Establish technology (horizontal) and business (vertical) CoEs.
6. Build digital retail solutions proofs of concept.
7. Pilot new digital retail solutions in collaboration with business executives and representative customers.
8. Build methods, accelerators and reusable components for faster and steadier delivery of apps/applications and features.
9. Train sales, marketing and retail operations team on new digital retail solutions and features.

Louis, after a detailed discussion and concurrence from various technology and business stakeholders, finalized 14 CoEs.

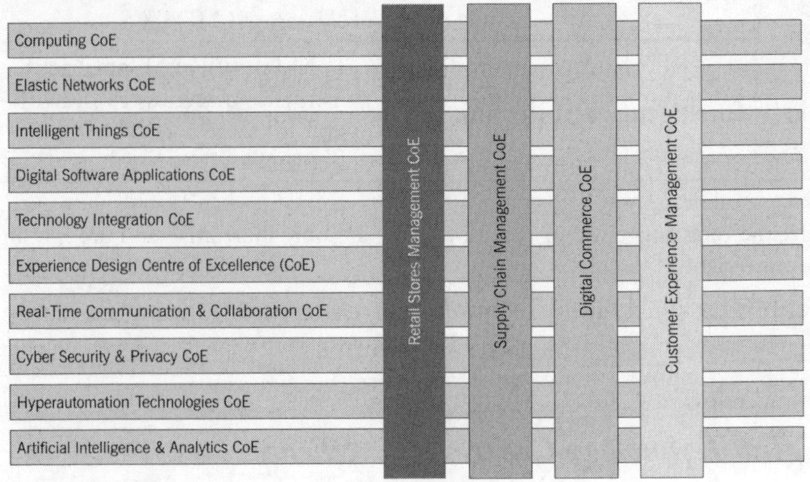

Source: Author

Figure 11.77: Digital Technology Centres of Excellence

Philip was quite pleased with the outcomes and released the digital retail solution proof of concept catalogue with a timeline to the business executives. They were quite happy with the progress made by Paul and the team.

The CoEs used the digital technology architecture foundation blocks. Business CoEs continuously focus on digitalizing products, services, customer experience, processes and operations to derive unique and relevant use cases. Each use case requires support from multiple horizontal CoEs to transfer the use case to reality. This horizontal and business CoEs matrix provided excellent results within xRetail.

In January 2018, Philip and his team celebrated the launch of two primary digital products/services, which became game changers for xRetail, with the business team. CEO Scott congratulated both technology and business executives for this turnaround in xRetail's story. He particularly thanked Philip for putting xRetail on the right track, during its digital journey. There was much work to be done in the future as the journey had just begun.

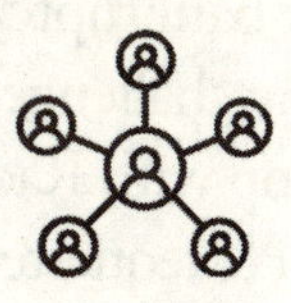

CHAPTER 12
Digitalization Cluster 3: Digitalize Offerings

Building Block 8: Products

Hayden, the CEO of Car Mac, a car manufacturing company shared with his business leadership council that one of the world's best mobile devices manufacturing company, Mobi Corp, had applied for bankruptcy recently. He asked his team, "Why do you think the company failed even though it had a great brand, financial strength and leadership?" Nick, the chief marketing officer (CMO), replied that probably they had failed to brand the product for the digital age customers. Hayden said that it was not branding, but a flaw in the product. The young generation saw better features being provided by other mobile device firms, which had started only a few years back.

According to Hayden, Mobi Corp had failed to understand the digital age customers whose needs, expectations and wants were different from traditional customers. These customers were hyperconnected. The digital age mobile devices have a large touch screen, millions of apps and an ecosystem that helps them generate revenue from complementary businesses. For example, the iOS ecosystem includes iPhone, thousands of developers, iOS platform and around 1.9 million apps. The ecosystem approach has helped Apple generate revenue not only from the iPhone but also from the apps which run on the iPhone. At the same time, users get better value.

Hayden warned his team that they were on the same path as Mobi Corp. "If we do not wake up and digitalize our existing products and build new digital products aligning with the needs and wants of digital age customers, we will soon need to apply for bankruptcy. Our undisputable leadership in the cars segment is getting challenged recently." He was planning to arrange a half-day workshop with Rao, a lean digital expert, on the anatomy of digital products. He requested the R&D head, Ram, to coordinate with Rao for the workshop.

On 25 November 2018, Rao met Hayden and the senior leadership of Car Mac and gathered a lot of information on their product, brands, competition, customer feedback, R&D initiatives and operational challenges. Three days later, Rao arranged focused group discussions with customers from representative segments like Gen Z, Gen Y, Gen X, women, etc., and gathered a lot of useful information. At the same time, Rao got a lot of inputs and suggestions from the customers directly using Car Mac's social media platform.

With this background, Rao conducted an offsite workshop with Hayden and Car Mac's senior leadership team on 30 November 2018 and started his session with a few critical comments.

Car Mac had excellent products, leadership strength and a good R&D establishment. Rao had observed that most of the customers expressed the view that their car brands had not been modernized like other new car manufacturers. In particular, the 20–30 years age group was not at all happy with the current models, which lacked many innovative digital features that other brands were offering. Rao also shared that there was no common understanding of connected cars or digital cars among various leaders and managers at Car Mac.

Hence, he set the agenda for the day to explain in detail about digital products and their features so that the Car Mac team could start innovating new features and introducing digital age cars. Rao used **lean digital thinking** and its 12 principles to explain the anatomy of a digital product.

Applying Lean Digital Thinking to Design a Digital Product

S. No.	Lean Digital Thinking Principles	Digitalizing Products	Features of Digital Product
1	Create a better today and better tomorrow	• Improve the existing product features and performance with digital technologies and build new digital products that meet the needs of digital age customers	• Retrofitting existing products for digitalization and building new digital products
2	Embrace constant and instant change	• Continuously collate and analyze the changing trends and shifts in the technologies, customer needs, wants and refine the products regularly	• Continuous newness
3	Nurture digital business talent	• Nurture skills to innovate new digital products (digital technologies, design thinking, computational thinking, etc.) and co-innovate products with customers and partners using crowdsourcing	• Continuous newness
4	Identify and eliminate non-value-added activities	• Design products in such a way that they can be accessed through an optimal number of channels (mobile, web, face-to-face, etc.), consume less material, use the optimal number of sensors and generate just required data	• Lean

5	Predict, recommend and improve overall business performance	• Design digital twins and establish simulation tools to design smart products • Implement automatic product diagnostic system to monitor the product performance in real time • Enable predictive maintenance by leveraging early warning system	• Predictable
6	Optimize human interventions	• Design the products with the following features: ○ Self-healing capability ○ Remote monitoring of performance ○ Remote access and troubleshooting ○ OTA updates of the software	• System of systems • Intelligent and autoupgradable • Usage-based (as a service)
7	Establish real-time and contextual communication and collaboration	• Establish a real-time communication and collaboration system to design digital products with the help of globally distributed internal and external specialists (collaboration tools, chatbots, virtual assistants, crowdsourcing tools, etc.) • Connect with other digital products in real time	• Hyperconnected

8	Safeguard privacy, safety and security of business resources	• Design the products with integrated cybersecurity features that will not harm the people and protect the data with the highest privacy standards	• Safe and secure
9	Establish a technology start-up ecosystem	• Design the products using the MVP model so that the right products can be released by pivoting multiple times (start small, fail fast and scale soon)	• Product development as a start-up
10	Create socially responsible products, services and value chains	• Optimize the usage of people, machines fuel while producing or utilizing the product • Reduce future maintenance costs of product • Reduce the consumption of power, water and other natural resources • Create no impact on health, air quality, water quality throughout the product life cycle	• Socially responsible

11	Design and deliver unforgettable moments of experience to stakeholders	• Design the products that offer unforgettable moments of experience to stakeholders (e.g., integrated music, rise auto trouble ticket and auto recommendations to repair minor problems)	• Emotionally engaging
12	Build a human-centric digital culture	• Establish a digital culture and set of values that enable the organization to adopt a customer-first attitude while designing digital products	• Customer-friendly product

Source: Author

Table 12.1: Applying Lean Digital Thinking Principles to Identify Digital Product Features

Digital Product Features

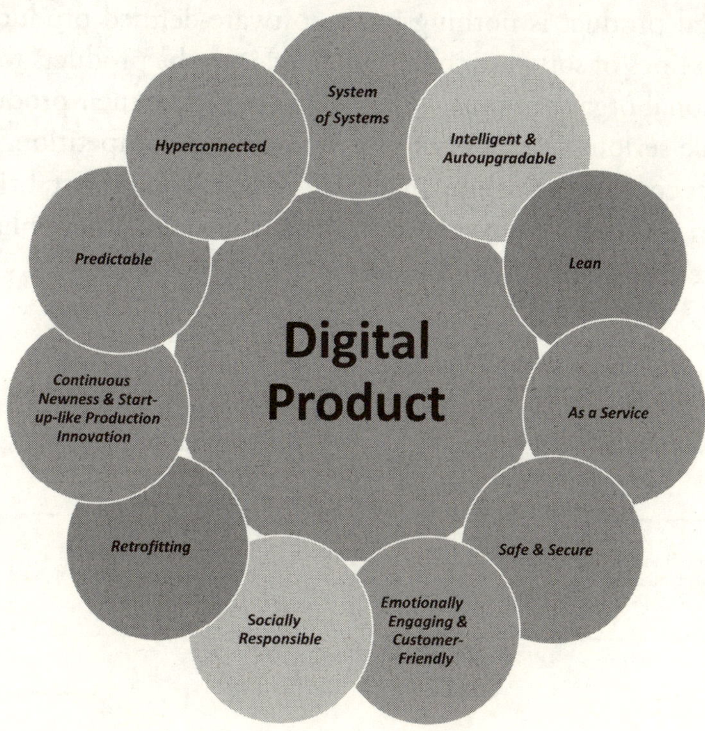

Source: Author

Figure 12.1: Digital Product Features

Digital Product Is a System of Systems

The product could be one system or a system of many systems. For example, a washing machine is a system that has an outer tub, cylindrical basket within the tub, motor, water pump, drainpipe and other components. But a car is a complex **system of systems** which includes braking system, battery system, emission system, exhaust system, fuel system, steering system, transmission system and others.

There is nothing new about this, but in the digital era, all these systems or system of systems are intelligent and are driven by a cyber system powered by technology fusion.

Digital Product Is Intelligent and Autoupgradable

A digital product is nothing but a software-defined product. A million lines of software code is injected into the products today. Traditional organizations, which fail to digitalize their products, will face serious challenges from the new age competition. The product could be a washing machine, vehicle, equipment, bridge, dustbin or others. Figure 12.2 shows the elements that enhance the intelligence of a digital product

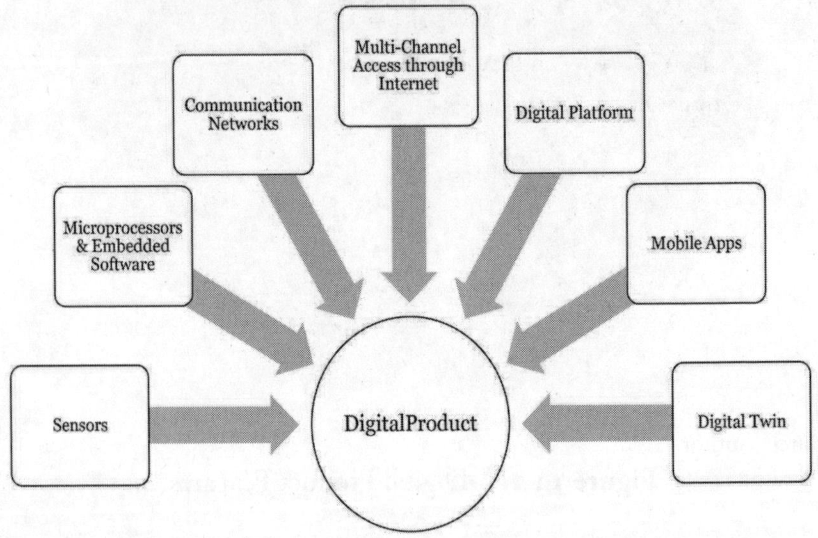

Source: Author

Figure 12.2: Elements of an Intelligent Digital Product

Intelligence Elements	Description
Sensors	Temperature sensors, pressure sensors, location sensors, gyroscopic sensors, impact sensors, flow sensors, proximity sensors, infrared sensors, touch sensors, smoke sensors and others. These sensors become an integral part of the product.

Microprocessors and embedded software	A minicomputer within the product, embedded with the software, which receives data from sensors and applies algorithms to process the data and perform the required actions. For example, one car may have 50 microprocessors. For instance, if speed sensor sends data to the accelerator system microprocessor and it finds that the vehicle is going beyond the speed limit, then it sends an instruction to the accelerator pedal to push back to reduce the speed.
Communication networks	Short-range networks like 6LoWPAN, ZigBee, LoRaWAN, and long-range networks like 3G, 4G and 5G enable the products to connect with other intelligent things/objects which are near or far from them through the internet. The communication networks establish hyperconnectivity between products and other smart things.
Multi-channel access through the internet	The product can be accessed from many channels using the internet, such as 1. mobile phone, 2. tablet, 3. web, 4. console and 5. others. Users can remotely monitor and control the product using any channel of their choice. For example, you could open the door lock of your home sitting at your office using a mobile app. You could start the AC in your car using an app in the mobile phone or mobile tab 15 minutes before you get into the car.

Digital platform	Any product without a digital platform today is not complete. The digital platform brings both consumers and producers together. The producers could be third-party service providers who enhance the value of the core product. There is a network effect between producers and consumers, that means when there are more producers, consumers get better value and vice versa. This means if there are many consumers, producers get more revenue; if there are many producers, consumers get a better choice of services. For example, in case of a product like a car, the producers could be gas stations, restaurants, insurance companies and telecom companies. The consumers here are drivers. When the driver is driving the car, he/she gets information about the nearest restaurants, malls, gas stations automatically. Telecom providers as a producer set up many hotspots within the car to connect to the internet. Digital platforms enhance the value of the core product and provide various services to consumers which had not been possible otherwise.

Mobile apps	The data available from the product is the real asset. Most of the product manufacturers or developers have been developing complementary mobile apps using their product data. These mobile apps provide further value to customers and other stakeholders. For example, a TV set-top box records the viewer's choices, attention span based on the use of remote control and additional information. On the basis of the set-top box data, the viewer gets TV recommendations on his/her mobile phone. The viewer can also use mobile apps to navigate through TV channels without a remote control.
Digital twin	Data, AI, ML algorithms and analytics are an integral part of each product today. These technologies help in creating a digital twin of the actual product. The digital twin or the virtual replica of physical product simulates different situations based on the actual data it receives from the real product. Then it predicts when the product may fail and thus introduces predictive and preventive maintenance. Digital twins are also useful for designing features and simulating the impact of the new features on the current product so that the right decisions are taken.

Source: Author

Table 12.2: Key Elements of an Intelligent Digital Product

In summary, today's products are highly intelligent. With no or less human interventions, they can perform a lot of tasks themselves.

Following are a few features:

1. The product can be operated remotely (on/off, lock/unlock and others).
2. The performance of the product can be monitored anywhere, anytime and using any device.
3. It automatically adjusts itself to consume less power, less water and other resources.
4. It enables over-the-air (OTA) software upgrades that enhance product features and capabilities.
5. It automatically raises trouble ticket and coordinates with customer and service representatives until it is closed.
6. For simple incidents, the product itself provides recommendations to the customer to solve the issues.

Digital Product Is Lean

Digital products must be lean with the following design considerations. They must

1. use an optimal number of sensors,
2. implement optimal embedded software to avoid performance degradation,
3. design the product to reduce rework and improve quality,
4. generate and utilize data optimally,
5. implement elastic storage, elastic computing and elastic network bandwidth to avoid high capital costs and
6. implement an adequate number of channels and mobile apps for the customers to access the product.

Digital Product as a Service

Due to technologically powered products, consumers can use limited features of products for a stipulated period. For example, medical device companies have been establishing rental centres at identified locations, which help hospitals avoid procuring costly medical equipment.

their operations and production accordingly. A start-up type work environment is expected to bring continuous newness to the products.

Digital Product Performance Is Predictable

Due to sensors, microprocessors, embedded software and internet, it is possible today to get data from all the subsystems of a system or components of a system in real time.

Now, product manufacturers must acquire new skills to mine the data and apply AI and ML algorithms that help in predicting future performance, issues and automatically recommends appropriate actions to the workers and managers.

Product diagnostics data is the next significant domain with which organizations get a lot of useful data from the product in real time. Organizations use this diagnostic data

1. to fix any problems proactively before the product fails and reduce the cost of maintenance, and
2. OTA software updates to fix issues or enhance capabilities.

Predictive maintenance of trains with IoT helped a train company, Tranquila, in Europe save millions of dollars.

Digital Product Is Hyperconnected

Digital products are intelligent and able to connect with other digital products or people over the internet or using short-range communication networks. For example, a car connects with another car and sends information about traffic congestion on the road, so that the other car driver can change the route.

Similarly, digital vehicles can connect with the digital parking meters at various parking places and inform the driver about open parking slots at a location, so that the driver can book the slot and directly go park the vehicle without waiting. A smart dustbin can connect with social media and display the tweets, Facebook posts, YouTube videos shared by the municipal corporation on the smart screen attached to the dustbin.

When Rao finished explaining, Hayden conveyed his thanks to Rao for the beautiful session on digital products. He directed his team to work on their digital car product design and asked the R&D head to work with all the stakeholders to come out with the next big digital product designs.

Building Block 9: Services

During a meeting with the senior leaders of Digi Global Bank, the CEO, Jonathan, read out the quote painted on a wall of the conference room:

"A customer is the most important visitor on our premises. He is not dependent on us. We are dependent on him. He is not an interruption of our work. He is the purpose of it. He is not an outsider of our business. He is part of it. We are not doing him a favour by serving him. He is doing us a favour by giving us the opportunity to do so."

Jonathan asked his senior leadership team a question. "How many of us practice this philosophy?" Not many leaders responded positively. Jonathan alerted his team that there is an unknown competition in the digital era. They had been struggling to understand the changing needs of digital age customers, which were very dynamic. While the customers had been looking for continuous newness, they were focused on profit, quality and productivity. They needed to change their attitude to outside-in, which meant making the customer the focal point for business decisions.

About three months ago, Nancy, a lean digital specialist, had conducted a workshop on lean digital thinking for the leaders of Digi Global Bank. Jonathan asked his team whether they had been practising **lean digital thinking** philosophy shared by Nancy. Again, the response from the team was not very encouraging.

Jonathan asked Krish, the CMO, to manage the customer services and customer engagement initiative like a programme and produce results in the next 4–6 months. He asked Krish to select his team and share with him to get necessary approvals.

Krish was a young and passionate CMO who was excited by the challenge given by Jonathan and finalized the core team with support from the various business units of Digi Global Bank. He wanted to finalize the critical attributes of digital age customer service by leveraging **lean digital thinking** which would be a standard reference for the leadership team.

Applying Lean Digital Thinking to Design a Digital Service

S. No.	Lean Digital Thinking Principles	Digitalizing Services	Elements of Digital Service
1	Create a better today and better tomorrow	Improve the existing service features and performance with digital technologies and build new digital services that meet the needs of digital age customers.	Retrofitting existing services for digitalization and building new digital services
2	Embrace constant and instant change	Continuously collate and analyze the changing trends and shifts in the technologies, customer needs and wants and refine the customer services regularly.	Continuous newness

3	Nurture digital business talent	Nurture skills to innovate new digital services (digital technologies, design thinking, computational thinking, etc.). Co-innovate services with customers, partners using crowdsourcing.	Continuous newness
4	Identify and eliminate non-value-added activities	Design digital services using digital technologies that help the customers to avail the services with: • less or no waiting time, • low process overheads, • no or low movement of customers and • high quality of service and no rework.	Hyperautomated
5	Predict, recommend and improve overall business performance	Predict the service quality in various dimensions: customer complaints, customer retention rate, conversion rate, average response time and resolution time, social media sentiment, and get automatic recommendations to improve customer delight.	Predictable performance
6	Optimize human interventions	Design customer services in such a way that customers get their choice of services without depending on service representatives/workers of the organization.	Hyperautomated

7	Establish real-time and contextual communication and collaboration	Establish a real-time communication and collaboration with customers using various channels like chatbots, mobile apps, virtual help desk, face-to-face interaction using video calls.	Real-time response and resolution
8	Safeguard privacy, safety and security of business resources	Design customer services with integrated cybersecurity features that will safeguard the safety and privacy of customers.	Safe and secure
9	Establish a technology start-up ecosystem	Design the new services using the MVP model so that the right services are released by pivoting multiple times (start small, fail fast, scale soon).	Services development as a start-up for continuous newness
10	Create socially responsible products, services and value chains	Design customer services which do not impact health, environment and biodiversity. For example, high radiation from cell towers may impact the health of people and birds. Telecom companies must be socially responsible in this respect.	Socially responsible
11	Design and deliver unforgettable moments of experience to stakeholders	Design customer services in such a way that customers get unforgettable moments of experience at the interaction points.	Unforgettable moments of experience to customers

12	Build a human-centric digital culture	Establish a digital culture and set of values that enable the organization to acquire a customer-first attitude when designing digital services.	Unforgettable moments of experience to customers

Source: Author

Table 12.3: Applying Lean Digital Thinking to Identify Digital Service Features

Digital Service Features

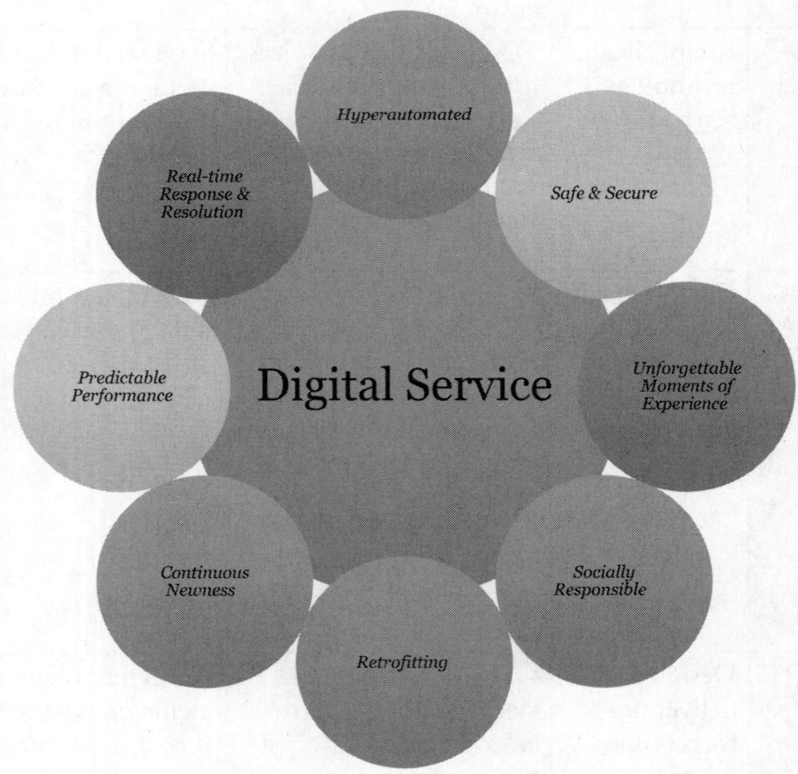

Source: Author

Figure 12.3: Digital Service Features

Digital Service Is Hyperautomated

Hyperautomated services mean optimizing human interventions while delivering services to customers. Following are a few examples of hyperautomated services:

1. **Humanoids** could be placed in banks to welcome customers and use face recognition software to authenticate the identity of the customer. They could help customers with their questions, opening accounts, transferring money and making investments.
2. **Video teller machines (VTMs)** in place of ATMs to help banks reduce the number of branches which are expensive to manage. Banks could establish VTM kiosks at different locations. The customer gets into a video conversation with bank employees using a VTM and performs banking transactions.
3. **Near-field communication (NFC)** wearable devices to help customers make payments for various transactions. This would encourage contactless payments.
4. **Blockchain smart contracts** automatically implement business rules, and with multi-node consensus automatically verify the bank account, credit rating to approve the loan amount.
5. With **peer-to-peer lending,** powered by blockchain, individuals can borrow and lend money without banks.
6. **Open banking** allows banks to partner with account information service providers (AISPs) and payment service providers (PSPs) by using open banking and PDS2. It allows fintech start-ups, social media networks, telecom providers and other technology providers to analyze customer spending, create a personal budget, pay customers' bills and transfer money.

Digital Service Is Safe and Secure

While offering digital services, the two major risks are concerning cybersecurity and data safety. While customers get great value from digital services, hackers may hack and steal the data, perform illegal financial transactions and may infect the system with ransomware.

Hence, it is important that banks design their services factoring cybersecurity and data privacy concerns throughout the service life cycle management. This means software applications and platforms must be strengthened with cybersecurity and data privacy devices, software and policies.

Digital Service Provides Unforgettable Moments of Experience

Banks have been working and focusing their efforts to provide unforgettable moments of experience to customers. Following are a few examples:

1. **Personalized services:** Banks offer exclusive brand-focused campaigns based on customer spending patterns on those brands. If a customer spends money on a perfume every month, he/she gets information about that perfume in a personalized manner.
2. **Location-based discounts:** Banks integrate social media and GPS to identify the customer's location and inform them about any offers in the nearby shops.
3. **Personal finance:** Banks help customers using mobile apps to estimate monthly budgets based on prior spending patterns.
4. **Non-banking services:** Banks enrol car dealers and real estate providers on their digital loan platforms and provide access to customers with matchmaking algorithms. Matchmaking algorithms help customers find their preferred choice of car or house based on the parameters given by them. Once the customer finalizes his preferred car or house, the bank approves the loan. In this way, banks help their existing customers interact with third-party providers using their loan platform and increase their loan business.

Digital Service Is Socially Responsible

Corporate social responsibility (CSR) does not mean that banks offer socially responsible banking services. With CSR, banks may invest in many society-friendly activities. Most of the CSR activities may be non-banking activities (child education, women health, green plantation and others). Therefore, banks must design their services, infrastructure and other assets which will not negatively impact society and the environment. CSR is an offline activity. A few examples are:

1. Digital banking platforms to take banking services to rural locations and offer better services to farmers and contribute to economic growth.
2. Smart buildings for bank offices and branches, using IoT and technology-based solutions, which reduce the consumption of water and electricity.
3. Online services which contribute to pollution control as consumers do not need to travel to banks in person. For example, cheques could be deposited using a mobile app.
4. Millions of trees are cut every year to produce paper. Banks have the greatest potential to reduce paper usage by digitalizing various banking processes.
5. Discontinuing use of plastic cards and promotion of digital wallets and cashless transactions.

Digital Service Is Also Retrofitting a Conventional Service

Conventional services without digital features will sunset as they fail to meet the needs of digital age customers and also may be expensive to operate in the digital environment. Hence, conventional services must be assessed for compatibility with digital features. This is called retrofitting. For example, today, social media is used as an extension of help desks to sort customer issues and complaints.

Digital Service Brings Continuous Newness

In the digital age, the needs and wants of customers change within a shorter span of time. Continuous service innovations are expected to leapfrog the competition and to enhance customer satisfaction. A start-up type work environment is expected to bring continuous newness in the services offered at a faster pace.

Digital Service Is Predictable

Measuring the current performance and predicting the future performance of customer services is important. Figure 12.4 depicts the key performance indicators (KPIs) that measure the effectiveness of customer services.

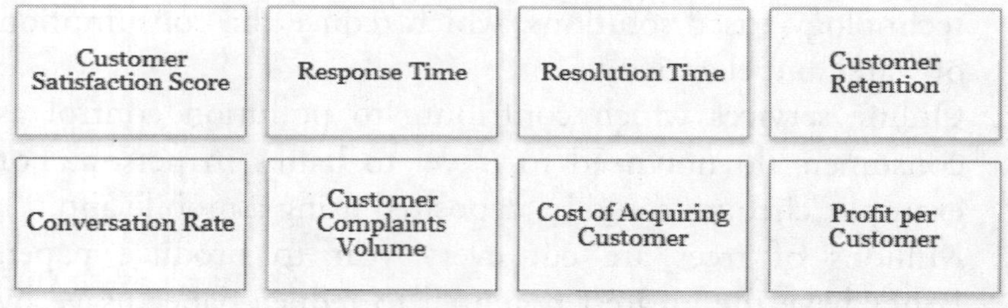

Source: Author

Figure 12.4: Service Effectiveness KPIs

Data collated from various sources like customer relationship management (CRM) systems, customer feedback, social media channels is mined with the help of AI and ML algorithms to predict the performance against the identified KPIs. These systems are intelligent enough to provide recommendations to organizations to take corrective and preventive actions at the appropriate time.

Digital Service Supports Real-Time Response and Resolution

Digital age customers, in general, have a lower tolerance limit, and they expect response and resolution to their issues immediately. It is humanly impossible to handle thousands of customer requests in real time. This led to the emergence of digital workers, as mentioned below, that enable real-time conversation, issue resolution collaboration and communication.

1. **Robo compliance officers** make sure that there are no non-conformances with regulatory policies while delivering services to customers.
2. **Robo insurance policy managers** advise the customers to choose the right policies and provide resolutions to their issues and clarifications to their queries.
3. **Robo investment advisers** advise the customers to choose the right investment plans and engage with them on a regular basis to effectively manage their customer's portfolio.

All these Robo coders are nothing but software programs that use mathematics and algorithms on client data for decision-making without human interventions. Amelia from IPsoft is one of the digital workers popular today, offering many services to banking customers. Amelia offers support to non-English speaking users as well.

At the end of the session, Jonathan conveyed his thanks to Krish and core team and directed the CIO to establish a digital service centre of excellence to re-engineer and design the digital services offered by Digi Global Bank.

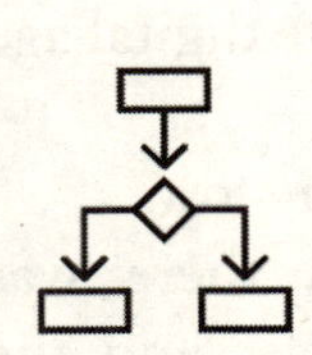

CHAPTER 13 Digitalization Cluster 4: Digitalize Business Operations

Building Block 10: Workplace

Angelina, the chief operating officer (COO) of ThinkDigital, an electronics design and manufacturing company, was under tremendous pressure as they had lost the best employer of choice ranking for two successive years. She attended a digital workplace conference at which, James, the keynote speaker and a well-known lean digital specialist, had presented his views on the challenges and opportunities in modern digital workplaces. She made detailed notes of his keynote speech and wanted to share the same with her operational team to start working on establishing a digital workplace at ThinkDigital.

On 5 January 2020, at 7:00 AM, Angelina conducted a breakfast meeting with HR, IT and a few senior business leaders to discuss the various challenges at ThinkDigital's workplace. Angelina asked each of them about the precautions being taken by them to safeguard themselves and their families from the COVID-19 pandemic. HR Head Catherine shared a few sensitive issues being raised by employees about COVID-19. Angelina proposed to discuss all the issues during the breakfast meeting. Some of the striking issues on overall employee engagement at ThinkDigital were:

1. Annualized attrition had gone up to 18% from 5% in the last two years.
2. They had failed to understand Gen Z behaviour, attitude and expectations.
3. They had not modified their workplace to suit the digital age employees.
4. Employee engagement quotient was at an all-time low.
5. Employee productivity had gone down, and there was no emotional connection between the employees and the organization.
6. Many complaints were received from Gen Z and a few Gen Y employees on the leadership style of Gen X leaders. The command and control style of leadership had been widening the gap.
7. Most of the employees complained that they had access to excellent digital tools and technologies outside the office but they still had to use outdated technologies and software at ThinkDigital.
8. Employees could not see new roles and career progression paths as per the changing needs of the organization while other organizations had been continuously refining their employee skills and roles according to the changing needs in the digital era.
9. There were a few safety concerns both in their production plants and the site. In the recent past, a few employees had met with accidents on the shop floor and on site while addressing service requests.
10. Employees and their families did not yet feel confident about ThinkDigital's plan and implementation schedule to safeguard the employees at the workplace against the COVID-19 pandemic.

Angelina requested business leaders to provide the required support to bring back the glory of ThinkDigital. Angelina shared the CEO's mission of improving the workplace of ThinkDigital in the next 12 months and the aim of becoming one of the top 10 best employers of choice in the next 24 months.

In the meeting, Angelina proposed a new initiative called "Intelligent and Immersive (I&I) Workplace".

She requested the business leaders, instead of the HR, to drive this I&I programme so that employees gain confidence in them and become more involved. She asked Jackson to champion the programme and assured that he will get all the support in terms of budget and resource. Jackson was known as one of the employee-friendly business leaders and had championed many internal programmes at ThinkDigital in the past.

Defining the Workplace

On 12 January 2020, Jackson conducted a meeting with the core team representing business units, technology units, HR, corporate administration, legal and finance departments. He wanted to bring a common understanding of the workplace among the core team members.

According to Jackson, the workplace is where people perform their day-to-day activities as per their role, domain and industry. The workplace is highly contextual and domain-centric. In general, the workplace must be safe, secure, healthy and provide an environment that builds harmony among co-workers to deliver high-quality services, results, benefits, products, assets and others.

Usually, employees spend 70–80% of their time in a day at the workplace. Hence, designing and building a better workplace is essential for their physical and psychological wellness. Figure 13.1 depicts typical workplaces covering all industries. In the digital era, technologies have the potential to build intelligent workplaces that are better than conventional workplaces which have been there for many decades. There are a few workplaces which are not safe and are hazardous too. It is important to understand each workplace and its implications on the employees.

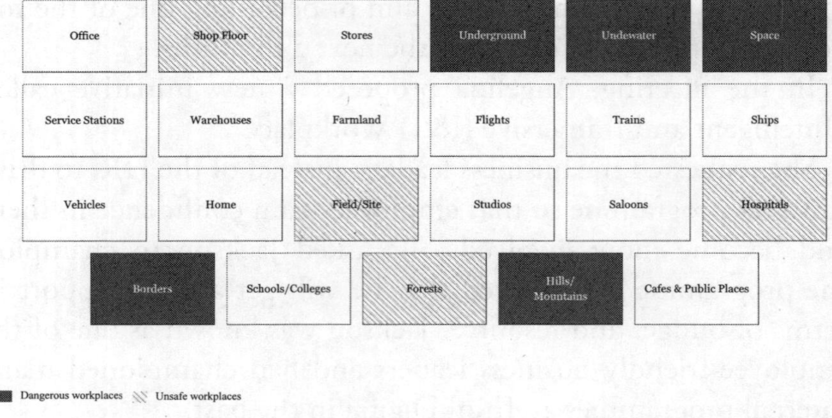

Source: Author

Figure 13.1: Types of Workplace

Designing Digital Workplace Powered by Lean Digital Thinking

Jackson had received training on **lean digital thinking** and its 12 principles. He was a lifelong learner and had many contacts in the business world. He had a good idea about how **lean digital thinking** had helped other organizations to be ready for the digital era.

He wanted to use **lean digital thinking** to design an I&I workplace at ThinkDigital. As the first step, he conducted a workshop with the core team and explained each principle in detail. After the team arrived at a common understanding of the principles, he moved on to a brainstorming session to apply **lean digital thinking** to workplaces.

The brainstorming session helped the core team to get an idea of what changes/improvements they needed to bring in the workplace of ThinkDigital leveraging **lean digital thinking** principles that would help them bring ThinkDigital back to its deserved place.

Jackson and the core team summarized the workplace anatomy of ThinkDigital powered by **lean digital thinking** principles, as given in Table 13.1.

S. No.	Lean Digital Thinking Principles	Changing the Anatomy of Workplace
1	Create a better today and better tomorrow	Build a workplace that suits different age groups, genders, cultures, promotes harmony and emotional connection by using various digital tools and technologies
2	Embrace constant and instant change	Continuously collate and analyze the changing trends and shifts in the business and bring changes in the workplace by leveraging digital technologies
3	Nurture digital business talent	Nurture digital dexterity skills for the leaders and employees to perform well in the digital era
4	Identify and eliminate non-value-added activities	Design human capital management processes using lean methods, tools and techniques for eliminating non-value-added activities within the processes
5	Predict, recommend and improve overall business performance	Predict employee performance by collating and mining the data received about employee performance from various systems and implement actionable recommendations to improve employee productivity, performance and social responsibility
6	Optimize human interventions	Digitalize physical workplace and integrate with digital tools; digitalize human capital management processes

7	Establish real-time and contextual communication and collaboration	Establish a real-time communication and collaboration system to improve employee performance and morale (collaboration tools, chatbots, virtual assistants, crowdsourcing tools, etc.)
8	Safeguard privacy, safety and security of business resources	Design and build workplaces using digital technologies that are secure and safe and safeguard the privacy of various stakeholders
9	Establish a technology start-up ecosystem	Build workplaces that are agile and flexible like technology start-ups to embrace any change that emerges from business as well as market changes; ensure a flatter organization structure to move away from command and control model to collaborate and coach model
10	Create socially responsible products, services and value chains	Design the workplace with no or low negative impact on the environment and society by leveraging digital technologies
11	Design and deliver unforgettable moments of experience to stakeholders	Create employee journey maps; design and implement unforgettable moments of experience at each interaction point using digital technologies
12	Build a human-centric digital culture	Promote digital etiquette and digital professionalism at the workplace

Source: Author

Table 13.1: Applying Lean Digital Thinking Principles to Design a Digital Workplace

Digital Workplace Guiding Principles

Based on the workplace anatomy designed using **lean digital thinking** principles, Jackson and the team wanted to finalize a template for designing their digital workplace. After a detailed analysis, they finalized a few guiding principles for the digital workplace:

1. Build human digital capital to outperform in the digital era.
2. Safeguard health, safety, security and privacy of employees.
3. Augment the intelligence of employees to perform better.
4. Bring agility and flexibility to the workplace.
5. Promote new etiquette and values aligning with the digital era.
6. Provide guided learning paths to employees to prepare for the digital future.
7. Build a flatter organization structure with real-time collaboration and communication.

Jackson and his team clearly articulated the difference between a traditional workplace and a digital workplace, as given in Table 13.2.

Workplace Elements	Traditional Workplace	Digital Workplace
Human capital	Human capital	Human digital capital
Learning	One medicine fits everyone learning model—means no personalization of learning as per the learning pace of the employee	Personalized on-demand learning with guided learning paths
Physical assets	Static physical infrastructure	Intelligent physical infrastructure powered by technology
Automation	Less automation and considerable manual work	Hyperautomation and less human interventions

Work location	Fixed work location	Any time, any place work location powered by technology
Code of conduct	Traditional workplace code of conduct	Digital workplace code of conduct
Safety and security	Manual dependent and less technology leverage	Highly automated, preventive and proactive management of health, safety, security and privacy
Performance management	Assessment of employee performance using traditional methods, which has an element of the subjectivity of assessors	Assessment of employee performance using gamification techniques, which improve objectivity and transparency

Source: Author

Table 13.2: Traditional Workplace vs Digital Workplace

Digital Workplace Components

Figure 13.2 depicts the various components of a digital workplace.

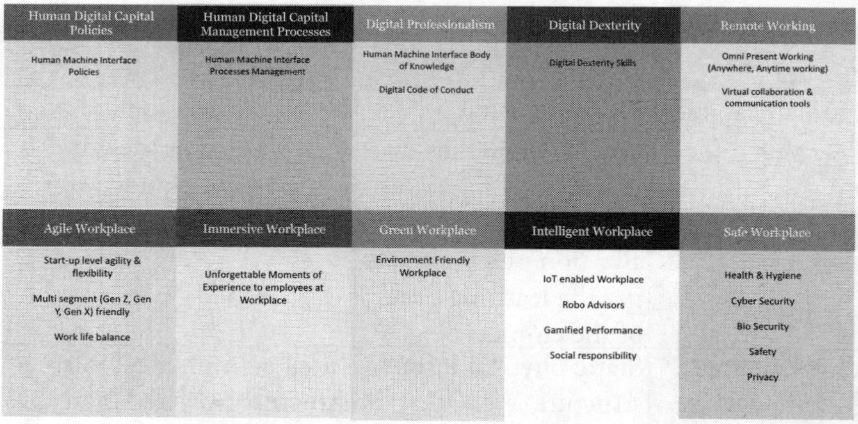

Source: Author

Figure 13.2: Digital Workplace Components

Jackson and his team wanted to create a process handbook explaining the 10 components of the digital workplace. Angelina and other business leaders were quite happy with the template proposed by Jackson as it covered a 360-degree view of digital age workplaces.

Sebastian, head of supply chain management and one of the core team members, asked Jackson, why was human capital changed into human digital capital in the digital workplace. He explained that in the digital workplace, the workforce was not simply constituted by human beings anymore.

Human capital is the knowledge, skills and experience possessed by an individual or organization. Mature organizations continually improve their knowledge, skills and provide opportunities to employees to acquire rich experience in their domain areas. Since, the digital workforce has extended beyond human beings, human capital needed to be redefined.

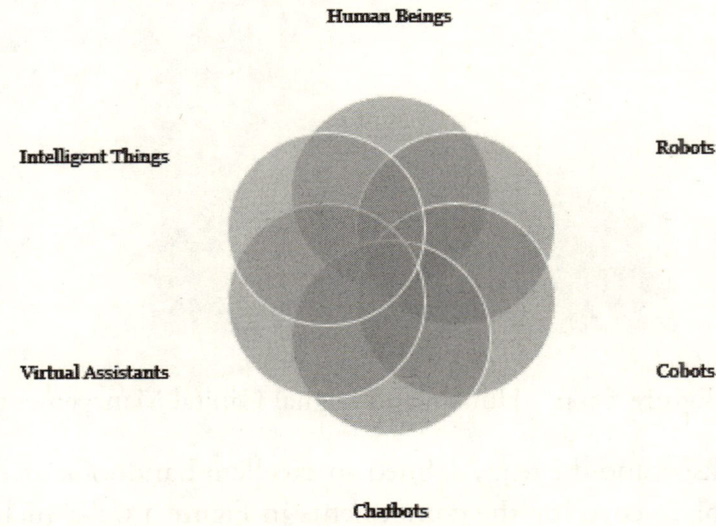

Source: Author

Figure 13.3: Workforce in a Digital Workplace

Human digital capital means the knowledge, skills and experience possessed by human beings, robots, cobots, chatbots, virtual assistants and intelligent things (cars, machines, etc.). With this context, the strength of the digital workforce is very different from the traditional workforce.

For example, if an organization has 10,000 employees and 2,000 digital workers, then the total workforce strength is 12,000. This implies both human workers and digital workers will work together in an entirely new work environment.

Jackson stated that unless ThinkDigital improved its human digital capital, their survival itself will be uncertain. Managing human digital capital was new for them and hence, they needed to introduce new policies, processes to best leverage the same.

The elements of human capital and human digital capital are shown in Figure 13.4.

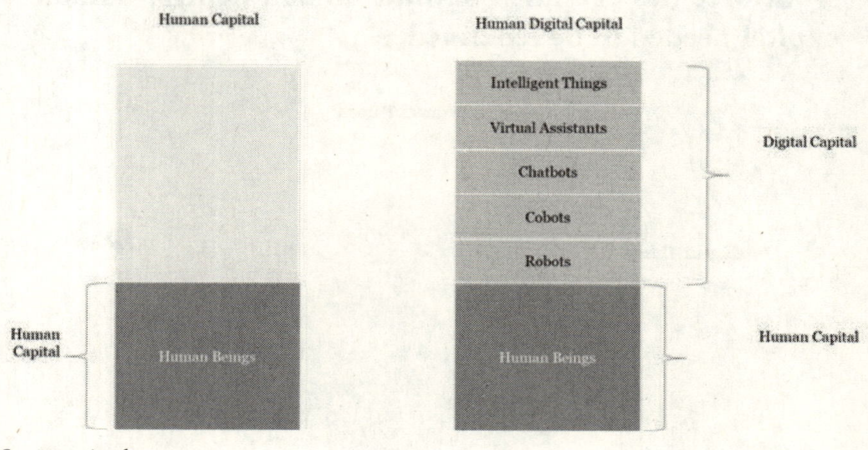

Source: Author

Figure 13.4: Human and Digital Capital Management

Jackson and the team defined an excellent handbook on digital workplace covering the components in Figure 13.2, which were derived based on **lean digital thinking** principles.

Human Digital Capital Policies

Human digital capital management is quite different and new to many organizations and that requires new policies. A few policies are:

1. Digital worker acquisition policy
2. Digital worker induction policy
3. Digital worker performance measurement policy
4. Human and digital worker interaction policy
5. Remote working policy
6. Green workplace policy
7. Safe workplace policy
8. Intelligent workplace policy
9. Agile workplace policy
10. Digital detoxification policy
11. Freelancer induction policy
12. Crowd innovation policy

Human Digital Capital Management Processes

Human digital capital management processes have a few specific process items, which are different from human capital management. A few of the processes, mentioned in Figure 13.5 are described in detail here. While the generic processes like payroll, compensation and benefits are also part of human digital capital management, they have not been covered in this section.

HUMAN & DIGITAL WORKERS ACQUISITION
HUMAN & DIGITAL WORKERS SKILL MANAGEMENT
HUMAN & DIGITAL WORKERS INTERACTION MANAGEMENT
HUMAN & DIGITAL WORKERS PERFORMANCE MANAGEMENT
EMPLOYEE SAFETY, SECURITY AND PRIVACY MANAGEMENT
VIRTUAL LEADERSHIP DEVELOPMENT

Source: Author

Figure 13.5: Human Digital Capital Management Processes

Human and Digital Worker Acquisition

This process focuses on determining human and digital worker demands, acquiring human workers and digital workers as well as defining the roles and responsibilities of both human and digital workers.

1. **Workforce:** In the digital era, workforce includes human workforce and the digital workforce. Digital worker gets an identification number just like a human worker.
2. **Roles and responsibilities:** Define roles and responsibilities for both human and digital workforce. For example, if a chatbot plays the role of a customer service representative, the job description must be clearly articulated. Similarly, define the role and responsibility of each digital worker in addition to the human workforce.
3. **Workforce planning**: Plan the resource requirement as per business demands. The workforce here includes both human and digital workforce.
4. **Workforce acquisition:** The workforce acquisition process is different in the digital era. There are three sources:
   a. Regular recruitment of human workforce
   b. Recruitment of freelancers or consultants on need basis (gig economy)
   c. Recruitment of the digital workforce

   AI/ML algorithms are used in the recruitment of human workers now. Hilton International uses a chatbot AI-backed interview platform for its recruitment process.

   Recruitment of the digital workforce is quite different, and since HR has never handled this process before, it must be trained to work with internal technology organizations and external technology vendors to induct the digital workforce as per the requirements.

Human and Digital Worker Skill Management

Learning strategy defined by most organizations is ineffective. The human and digital worker skill management processes focus on creating a list of skills to be nurtured every year as per the business demand. Also, it focuses on necessary skill enhancement programmes to be introduced.

Figure 13.6 highlights the skills required by human and digital workers in different functions.

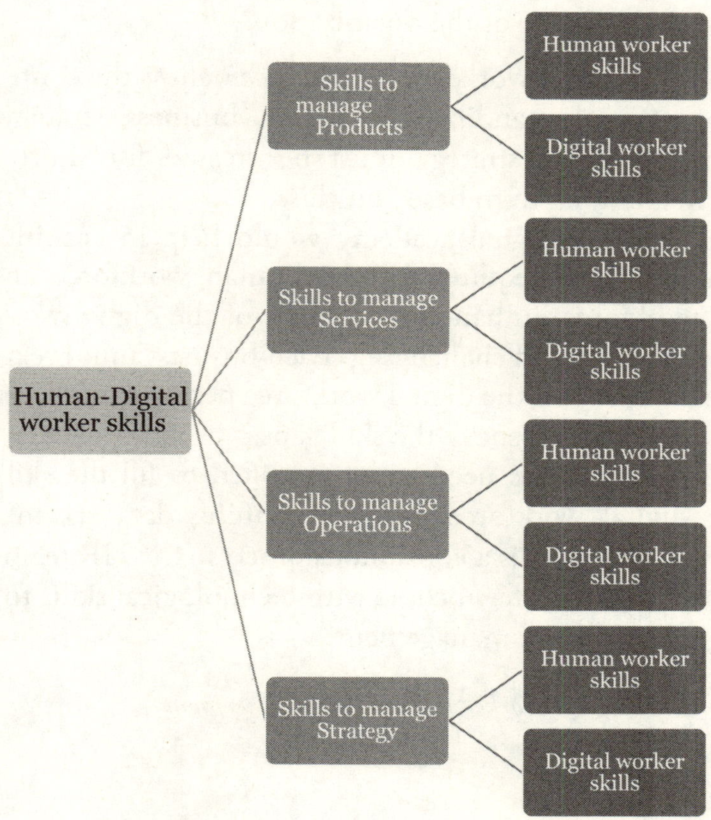

Source: Author

Figure 13.6: Human and Digital Worker Skills

Human and digital workers skill enhancement focus areas are:

1. **Product management:** What roles and skills are required to deliver the products faster, better, cost-effectively and consistently.

2. **Service management:** What roles and skills are required to deliver the services faster, better, cost-effectively and consistently with a superior experience?
3. **Operations management:** What roles and skills are required to manage business operations effectively and efficiently?
4. **Strategy management:** What roles and skills are required to design the future of the organization?

Human and digital workers skill enhancement is the continuous process of understanding the digital business strategy and demand. Learning strategy must be created for short term, midterm and long term based on this.

Establishing a digital academy would help in continuously honing the skills required by the human workforce and the digital workforce. Enhancing the skills of the digital workforce is quite different and challenging. Each business unit is expected to give feedback on the digital workforce performance as per the defined goals and identify the skill gaps.

AI/ML algorithms need to be modified to fill the skill gaps of the digital workforce and the training data set must be improved as well. It is a continuous process. The HR team must be enhanced with a few people with technological skills to work on digital workforce management.

Human and Digital Worker Interaction Management

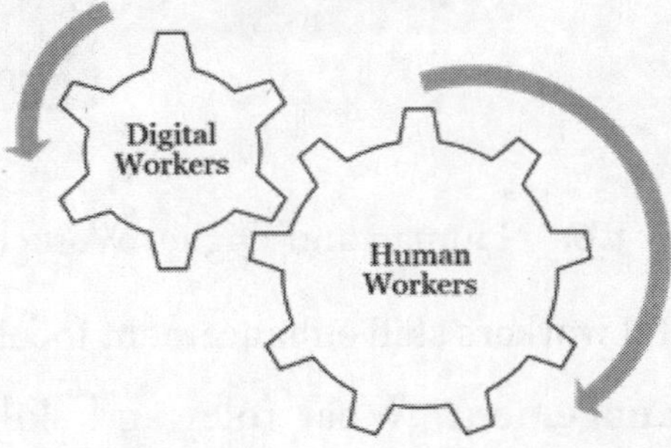

Source: Author

Figure 13.7: Human and Digital Worker Interaction

While driving digital business, it is imperative that human and digital workers work together to perform tasks. This is entirely a new work environment and organizations have not understood the new risks that will emerge along with many other benefits. A few of the risks and issues are:

1. A human worker may not work at the pace of a digital worker. Humans need rest and digital workers may work round the clock with self-charging capabilities.
2. Digital workers may go on strike if hackers take the controls of digital workers.
3. Digital workers may cause harm to human workers if they malfunction due to unaudited and non-certified algorithms.

For example, if a chatbot fails to respond to a customer on a particular query, then the human customer service representative immediately interferes and responds to the customer without delay. In some cases, the human worker may finish his/her part of the job and hand over the rest of the work to a digital worker. Human to digital worker interaction diagrams for each task can help identify exceptions and risks for each interaction and manage them effectively.

Human and Digital Worker Performance Management

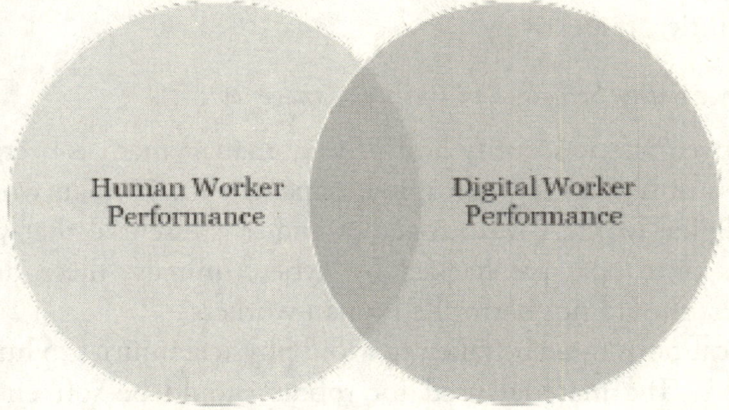

Source: Author

Figure 13.8: Human and Digital Worker Performance

Performance management is a sensitive and complicated subject in traditional organizations. With the digital workforce, there are many unknowns and measuring the performance of a digital worker may further complicate the work.

Subjectivity, lack of transparency and bias are the three significant villains, which cause disappointment to employees when their performance is assessed.

Following are a few techniques that could be used to assess the performance of human and digital workers:

1. **Performance management of human workers:** Integrate gamification techniques with the software systems within the organization, which automatically tracks the performance of a human worker without human interventions. The system automatically provides the statistics of worker performance, which is very objective. In addition to this, leaders can assess the soft/behavioural aspects of the human worker.
2. **Performance management of digital workers:** AI-powered software systems can automatically record how many times a digital worker failed to deliver the expected result/outcomes and then rate their performance accordingly.

Performance improvement plan for both human workers and digital workers must be released to enhance their skills and capabilities, respectively.

Employee Safety, Security and Privacy Management

Employee safety, security and privacy management is a critical process in human digital capital management. Human workers and digital workers work together and it is essential that when robots or cobots are hacked by cybercriminals, these digital workers should not harm the human workers.

Precautions must be taken to avoid physical injuries to human workers. The material used for robots should be soft enough such that even if they hit the human workers, the impact must be minimal.

Organizations must also safeguard employee personal data against cyberattacks by putting stringent measures and avoid the misuse of employee data by any individual or department.

Virtual Leadership Development

Virtual leadership has become the norm in the digital era. Leaders are expected to manage employees and freelancers (gig economy) distributed across multiple countries, with different cultures, accent and time zones. For example, a leader sitting in India must be able to discuss, debate, motivate and provide direction to distributed employees in different countries using video conferencing. This is called virtual leadership. Leaders are also expected to monitor the effectiveness of human and digital worker interaction while working on a job/task. Organizations must focus on building virtual leadership skills.

Digital Professionalism

The two components of professionalism are

1. body of knowledge (BOK) and
2. code of conduct.

BOK is the collection of knowledge areas of a profession. For example, project management BOK has 10 knowledge areas and supporting tools and techniques. They are

1. integration management,
2. risk management,
3. scope management,
4. quality management,
5. resources management,
6. time management,
7. cost management,
8. stakeholder management,
9. communication management and
10. procurement management.

If you want to become a project management professional, you are expected to be proficient in these 10 knowledge areas and apply them in your day-to-day work. Another critical component of any profession is to adhere to the code of conduct relevant to that profession. For example:

1. Doctors work as per healthcare BOK and code of conduct
2. Lawyers work as per legal BOK and code of conduct
3. Engineers work as per the engineering BOK and code of conduct

In today's digital era, each professional is a digital professional. This means, in addition to their core domain body of knowledge and code of conduct, each professional is expected to follow the digital domain body of knowledge and code of conduct.

In Figure 13.9, digital professionalism is depicted as the convergence of the BOK and code of conduct of both digital domain and core domain.

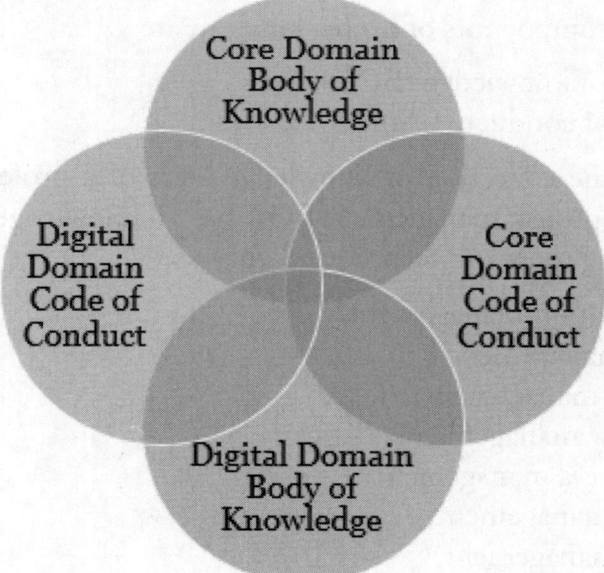

Source: Author

Figure 13.9: Digital Professionalism

For example, a project manager must be proficient in project management BOK, project management code of conduct, the digital BOK and digital code of conduct to become a **digital age project manager.**

The digital BOK consists of 12 **lean digital thinking** principles, 12 digital business building blocks and relevant templates and tools, which is explained in this book in detail. A digital code of conduct would look something like this:

As a digital professional,

1. I shall not compromise the data privacy of others;
2. I shall not use technology for the destruction of humans and nature;
3. I shall obey the digital policies and laws of the country in which work is being performed;
4. I shall continuously hone my digital skills to add value to business and society;
5. I shall use technologies responsibly to safeguard my physical and mental health as well as others'.

Digital Dexterity

67% of business leaders say their company will no longer be competitive if it can't be significantly more digital by 2020.[5]

Dexterity means using the hands skilfully in doing something. In the era of the Second Industrial Revolution, dexterity played a very vital role as there was a high need for manual intervention. In the age of the Fourth Industrial Revolution, we need digital dexterity to use our brains skilfully by augmenting human intelligence with digital technologies.

Without acquiring digital dexterity skills, it would be difficult for organizations to successfully navigate through their digital journey. Digital technologies have been evolving at extremely

5 Wiles, Jackie. 2018. 'Mobilize Every Function in the Organization for Digitalization'. Available at: https://www.gartner.com/smarterwithgartner/mobilize-every-function-in-the-organization-for-digitalization/ (accessed on 3 December 2018).

high speed, and it is becoming difficult for organizations to acquire new technology skills and innovate new products, services and operating models.

Digital dexterity means:

1. Not just acquiring technology acumen or skills, it is the ability to best leverage digital technologies to
    a. create new business opportunities,
    b. new products,
    c. new services,
    d. new revenue streams,
    e. improve customer retention and
    f. create new operating models.
2. Building a digital mindset of business leaders and creating an entirely new culture that is required to drive the digital initiatives effectively and efficiently.
3. Taking data-based decisions using AI and analytics.

The organization leadership must focus heavily on improving the digital dexterity of the organization, without which the digital vision becomes only a myth. The digital dexterity charter could be implemented by each business unit and supported by the HR and technology teams for successful implementation.

Remote Working

In the UK, 86% of the British businesses have at least one staff member working from home.[6]

The remote working trend is picking up due to various reasons.

1. Reduced commuting
2. Lower environmental impact
3. Work-life flexibility
4. Reduced cost of operations (real estate, meals, transportation, etc.)

6 BCC and BT Business. 2015. 'Making Technology Work for Your Business'. Available at: https://business.bt.com/content/dam/bt/business/Business-Solutions/Downloads-PDF/BCC%20-%20Booklet.pdf

Gartner CFO Survey reveals 74% intend to shift some employees to remote work permanently.[7]

Remote working is not always from home, it could be carried out in a cafeteria, hotel room, vehicle or any other public place. Various components of remote working are given in Figure 13.10.

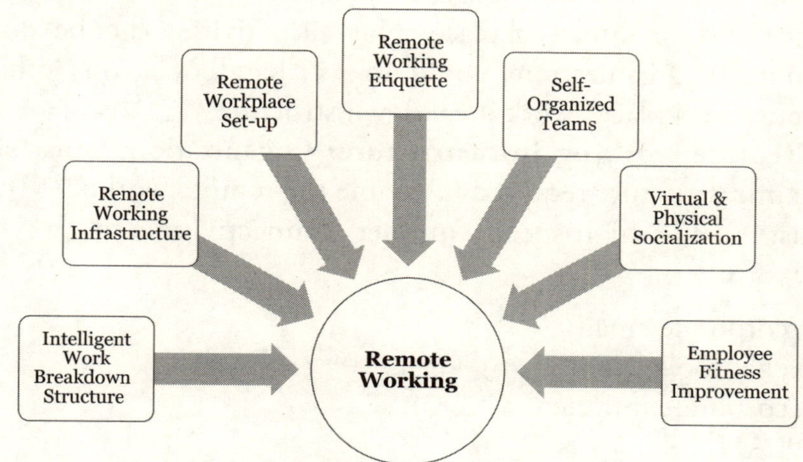

Source: Author

Figure 13.10: Remote Working Framework

Intelligent work breakdown structure: A few decades back, when the outsourcing trend first started, organizations achieved many benefits. Now, in the digital and pandemic era, remote

---

7 Arlington, VA. 2020. 'Gartner CFO Survey Reveals 74% Intend to Shift Some Employees to Remote Work Permanently'. Available at: https://www.gartner.com/en/newsroom/press-releases/2020-04-03-gartner-cfo-surey-reveals-74-percent-of-organizations-to-shift-some-employees-to-remote-work-permanently2

working has become a norm due to the availability of digital infrastructure and other constraints (like the pandemic).

In a few cases, remote working requires necessary approvals from the customers. Many change management activities need to be performed by the organizations to prepare their employees to work remotely. Where required, organizations must discuss the impact and implications of remote work on suppliers and partners and refine the contractual terms as appropriate.

The first step is to intelligently breakdown the work between office and remote workplace. Not all activities can be done remotely. Organizations could release handbooks on "office-remote workplace" work breakdown structure.

Remote working infrastructure: Organizations must size the infrastructure required to enable the remote working. They must enable high-speed internet connectivity and provide access to

1. corporate email,
2. chat tools
3. corporate intranet,
4. LAN,
5. VPN access,
6. collaboration tools,
7. communication tools and
8. technical support.

Employees must get an option to work from a remote place or at an office that gives a lot of flexibility. Based on the capacity at the office, the organization must provide approval to employees to work at the office.

Remote workplace set-up: Employees and other stakeholders must take the necessary steps to make sure that the remote workplace will not distract them and others working remotely. Following components help in setting up a good remote workplace.

1. Background of the remote workplace (light colours may be preferable)
2. Lighting at the remote workplace
3. Camera positioning
4. Person sitting position
5. Fewer distraction places

Remote working etiquette: Remote working using virtual collaboration and communication tools requires a few etiquettes to avoid distractions and disturbance. A few examples are:

1. When in a virtual meeting using video conferencing/audio conferencing, put your mic on mute.
2. Use the options like raise hand in video calls to speak anything rather than interfering abruptly. The moderator will unmute you to speak.
3. Maintain remote workplace (especially home) like your office space and prevent family members and maids from getting into the room abruptly.
4. If taking bio breaks or coffee breaks, publish the status explicitly in the internal messengers/communicators.
5. Maintain a good distance between the camera and your seating position.
6. Use a good microphone to avoid low voice quality.
7. Do not spam the internal communicators/messengers with too much information.

Self-organized teams: It is difficult to monitor the work being done by the employees working remotely. Hence, organizations must build a system to promote self-organized teams. Self-organization system improves the accountability of the employees. This means that the employee does not need to wait for the manager to assign the work. The teams find their work and manage the same promptly. Managers divide the work into manageable tasks and publish the same.

When taking up any task, the employee must inform all the concerned team members that he/she has taken up a task. The employee updates the progress in a collaborative platform to make sure that everyone concerned is informed. Employees use discussion forums to get any clarifications or guidance while working on a task.

Nurturing self-organized teams is not easy. It depends on the country and its culture. In a few countries, the culture is that younger people expect guidance and instructions from the elders. They do not take decisions themselves. It is essential to understand cultural differences among the employees of various countries and design a system and train the employees on self-management.

Virtual and physical socialization: As most of the employees work from remote locations, they miss the social networking and coffee break gossip, lunch break personal discussion and smoking breaks. Such employees miss this whole informal environment. Hence, it is essential to schedule virtual happy hours and social networking hours using video conferencing. Also, every 60 days, organizations could plan high-tea networking sessions at the office for face-to-face interaction. This helps in rejuvenating the energy of the employees.

Employee fitness improvement: Organizations must advocate fitness improvement programmes. They must request each employee to use a wearable device and share the daily physical exercise details. Organizations may design a new rewarding system for the employees, who have been regularly doing daily exercise.

If two employees are performing equally well in the business, there will be a dilemma on whom to give promotion? Organizations may consider an employee for promotion who is health-conscious. That means an employee who has been doing meditation, physical exercise and yoga may differentiate him or her from other employees.

Agile Workplace

Build workplaces that are agile and flexible like technology start-ups to embrace any change that emerges from business and market changes. Ensure a flatter organization structure to move away from the command and control model to the collaborate and coach model.

Flatter structure allows cross-functional collaboration and communication system to get the required guidance, mentoring and support from anyone in the organization, instead of being limited to the reporting manager and a few identified specialists.

Organizations could implement internal job postings. Business units publish their internal projects or initiatives and ask employees to bid for the project. The bidding here is that each employee submits a brief proposal covering timelines, quality, the scope of the work that they will deliver. Business units may choose an employee on the basis of the best proposal.

The organization must define policies to ensure that there are no conflicts that emerge while sharing the employees among the business units. In order to establish a cohesive work environment, the different generation of employees like Gen Z, Gen Y and Gen X must be trained to work cohesively by understanding the differences in their working habits and leadership styles.

The workplace must be designed to have the required flexibility to accommodate standing desks, sitting desks, bright lights, dim lights and other features as per the employee's preference.

Immersive Workplace

In the digital era, organizations must try to provide unforgettable moments of experience to their employees throughout their journey in the organization. It is like customer experience management. Employee journey maps must be created for different types of jobs and at each interaction point, emotional needs (beyond task needs) of the employees must be identified. AI, augmented reality (AR), virtual reality (VR) and other digital technologies can be used to design the experience at each interaction point.

Green Workplace

Organizations must design and build both physical and virtual workplaces that consume less energy, water and other natural resources. Organizations must install sensors to measure the air quality, carbon emissions at workplaces, and take preventive steps to minimize the impact on the environment. Organizations must have a special focus on effectively managing liquid, solid and electronic wastes using digital technologies.

Intelligent Workplace

More and more organizations have been building intelligent workplaces, which need fewer human interventions. For example, Amazon uses more than 200,000 robots in its warehouses. Russia's state-owned television channel, Russian-24, introduced an AI news anchor to deliver the news. In some hotels and restaurants in Japan, most of the tasks are operated by robots.

With the help of blockchain smart contracts, it is possible to build a distributed autonomous organization in which manual approvals are not required, and the system is intelligent enough to validate the policies and business rules to provide necessary approvals or recommendations

Workplaces today are digitalized with the internet of things (IoT). Organizations implement digital chairs, digital conference rooms, digital helmets and Fitbit-like wearable devices. These digital things send alerts to employees to take a break, to visit doctors and avoid working at insecure places.

Within organizations, mobile software applications with analytics and AI provide automatic recommendations to employees and augment their intelligence so that employees can perform their job effectively and efficiently. Human advisers are being replaced with Robo coders in financial institutions to make the right investment decisions and to take loans.

Intelligent workplaces are supported by systems of record, systems of engagement, systems of things and systems of intelligence (digital software applications). Technology organizations must have a clear strategy to build intelligent workplaces. Robots/digital workers play a vital role in insecure places like country borders, mines and other areas.

Safe Workplace

We are living in quite a different world. Due to industrialization and digitalization, new challenges have emerged which are causing many security and safety issues. With the COVID-19 pandemic, it is quite clear that a pandemic may emerge again and again, and workplaces must be safeguarded in such a way that employees will not be impacted by viruses. Biosecurity measures leveraging technologies must be taken up by organizations.

There will be an increase in cyberattacks with the increase in digitalization and hence, workplaces need to be designed in such a way that the required policies, processes and governance are established for strong cybersecurity.

A few workplaces like mines, country borders and construction sites are unsafe and use digital technologies like IoT and analytics to minimize accidents by sending early warning alerts before the worker reaches an unsafe point.

Surveillance with IP-enabled cameras and video analytics helps in identifying any abnormalities and alert the security to take preventive steps. Organizations have also been promoting digital detoxification programmes to safeguard employees from physical and mental illness due to too much digitalization of their lifestyle.

Jackson and the team presented a detailed handbook on digital workplaces to Angelina and other senior business executives. They were quite impressed with the digital workplace template based on **lean digital thinking** principles, which gave a comprehensive view of the digital workplace. The digital workplace handbook would become a guiding document to nurture a positive work environment by leveraging digital technologies with exceptional behaviour and leadership skills. Next, Angelina requested the HR and technology teams to implement the digital workplace handbook in an evolutionary manner with the support of the business executives.

Building Block 11: Business Processes

Srini, the COO of Green Engineering Corp, a small-scale auto spare parts manufacturing company, stopped George, the head of business processes, to ask him when he could start the internal project to digitalize their business processes. Srini expressed his disappointment with the slow progress and emphasized the importance of reimagining, redesigning and re-engineering business processes by leveraging various digital technologies to align them with the digital business model.

Srini cautioned George about developing any new business processes or changing existing business processes without aligning with the latest digital business models created by Green Engineering Corp. He gave George two weeks to prepare a detailed plan and the standard definition of digital business process within their organization.

George spent some quality time with his business process management team and other stakeholders. He made an appointment with the COO and other senior business leaders to share the framework for digitalizing Green Engineering Corp's business processes.

He was well-versed in **lean digital thinking** principles as he had been a part of the team that had created the digital business models and digital business strategy map of Green Engineering Corp. George began the meeting by welcoming Srini and the other leaders and assured them that he will try to provide clarity on the following questions, which had been unanswered for a long time:

1. What is the digital business process pyramid?
2. How digital business value chain can be defined based on the digital business model?

3. How to define the digital features of the digital business process tasks?
4. How to define the standard operating procedure for a digitalized task?

Digital Business Process Pyramid

George explained step by step, the process of defining the digital business process pyramid that aligned with the digital business model. Without establishing the business process pyramid, organizations fail to develop a reliable business ecosystem.

The digital business process pyramid is one of the most important elements of digital business operations, which has a five-layered structure as mentioned in Figure 13.11.

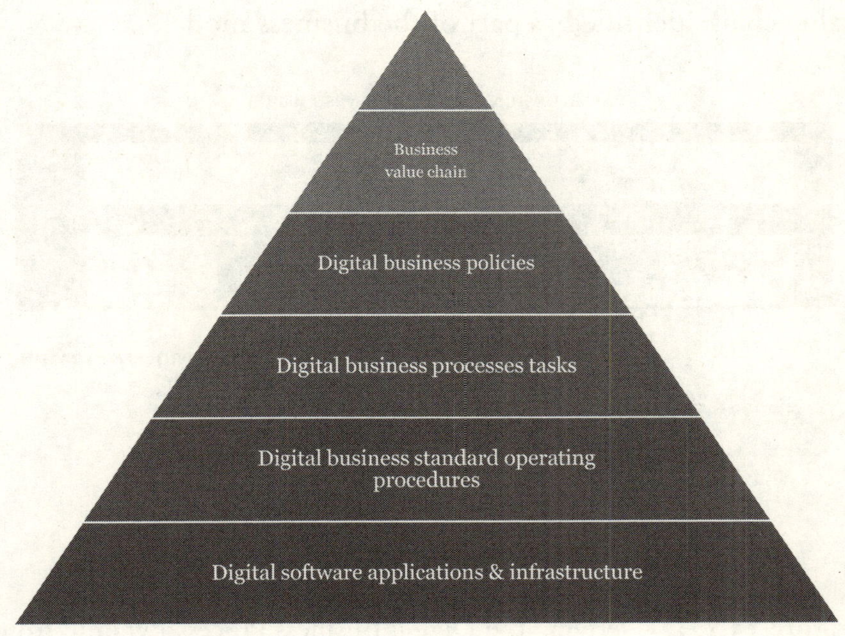

Source: Author

Figure 13.11: Digital Business Process Pyramid

Business models help in identifying the customer segments, their problems, unique products and services to address the problems and the value proposition to be delivered to the customers. They also help in identifying key resources, **business value chain,** key partners to deliver the products, services and value proposition. Business value chain, as identified in the business model, is nothing but the business processes and the tasks that are required to manage the business operations.

Hence, defining business processes and tasks without aligning them with the business model does not yield expected operational results. The relationship between the business model and the business process pyramid is depicted in Figure 13.12. The business process pyramid must be derived from business value chain identified as part of the business model.

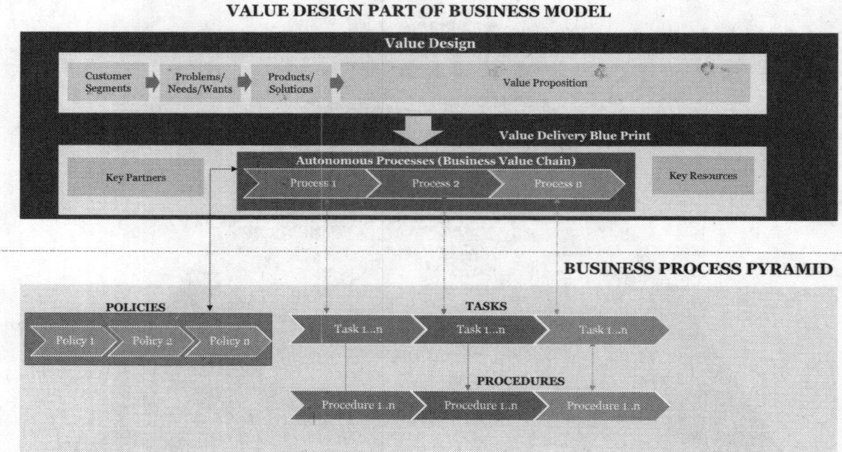

Source: Author

Figure 13.12: Deriving the Digital Business Process Pyramid from the Business Model

George also shared the essential elements of business process as defined at Green Engineering Corp. Figure 13.13 depicts the anatomy of business processes described by George.

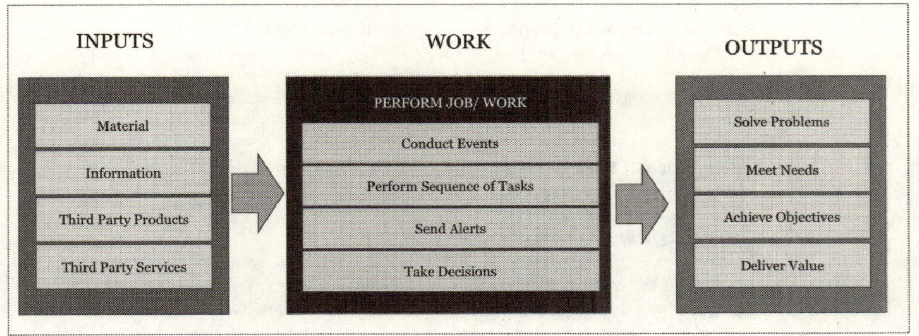

Source: Author

Figure 13.13: Business Process Anatomy

Business process solves the problems, meets the needs, delivers the results and value to stakeholders by

1. taking inputs from different sources,
2. executing a sequence of tasks and events,
3. leveraging the organization's assets and
4. taking the right decisions at the right time.

Steps to Digitalize Business Processes

George and his team put in extraordinary efforts to come up with a five-step framework to design effective and efficient digital business processes for Green Engineering Corp.

In the interest of time, George used the example of the **research and development (R&D) process** in the manufacturing value chain to explain the process of digitalization. The same framework was to be applied to all other processes. Steps to digitalizing business process are given in Figure 13.14.

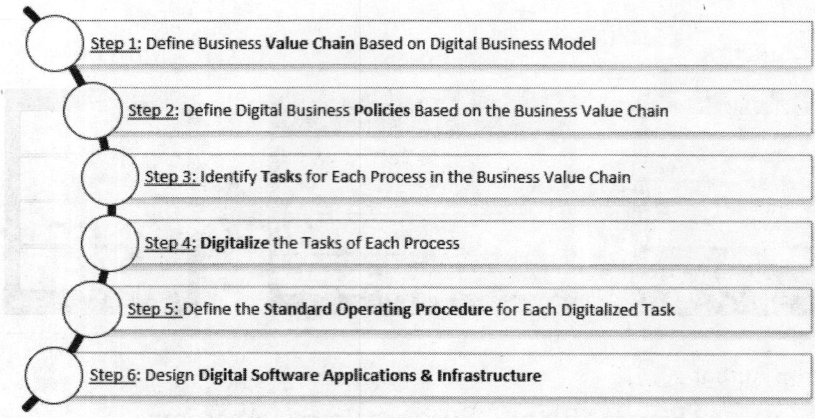

Source: Author

Figure 13.14: Steps to Digitalize Business Processes

Step 1: Define Business Value Chain

Figure 13.15 illustrates the seven business processes of the manufacturing value chain at Green Engineering Corp.

Source: Author

Figure 13.15: Example—Manufacturing Value Chain

Step 2: Define Digital Business Policies

S. No.	Digital Manufacturing Value Chain	Digital Manufacturing Policies
1	R&D	• Crowd innovation policy • Experience design policy • 3D printing files IP policy

2	Sales and marketing	• Personalized branding policy • Digital commerce policy • Contactless sales policy (online)
3	Demand forecasting	• Demand sensing policy
4	Supply chain	• 3D printing material selection policy • Supplier selection policy • 3D printing kiosks policy (near customer)
5	Manufacturing	• Cybersecurity policy • Personalized products policy • Digital infrastructure policy • Digital manufacturing worker policy (robots, intelligent machines, intelligent equipment, etc.)
6	Customer experience	• Data privacy policy • Customer engagement policy
7	Services	• Enterprise reputation policy in social media • Digital services worker policy (chatbots, Robo coders) • Digital payments policy • Remote monitoring policy • Predictive maintenance policy

Source: Author

Table 13.3: Example—Digital Business Policies

Step 3: Identify Tasks

Once the digital business processes in an organization have been defined, the next step is to identify the tasks for each of these processes. The process can be categorized into three types.

1. **Physical processes:** Most of the manufacturing processes are physical processes. Intelligent machines, equipment, tools, forklifts and other physical objects communicate and coordinate with each other and perform the following tasks:
   a. Production (traditional or additive 3D printing manufacturing)
   b. Internal movement of material/products on the shop floor using robots/intelligent forklifts
   c. Movement of finished goods to the warehouse by internal drones
2. **Core business processes**: The core business processes vary according to industry. Manufacturing processes include customer relationship management (CRM), demand sensing, field service management, supply chain management (SCM) and others. Retail industry processes include merchandising, in-store management, optimal inventory placement, point of sales (POS) and others.
3. **Shared processes:** These include HR, finance, legal, regulatory and compliance, innovation management, learning management and others.

Once tasks have been identified for each of these processes, they are optimized and then the organization kick-starts the digitalization with which the tasks will be refined and optimized further.

Along with the seven processes in the Green Engineering Corp manufacturing value chain, George shared a list of tasks for the R&D process, given in Figure 13.16. The team wanted to digitalize the other six processes in the manufacturing value chain similarly.

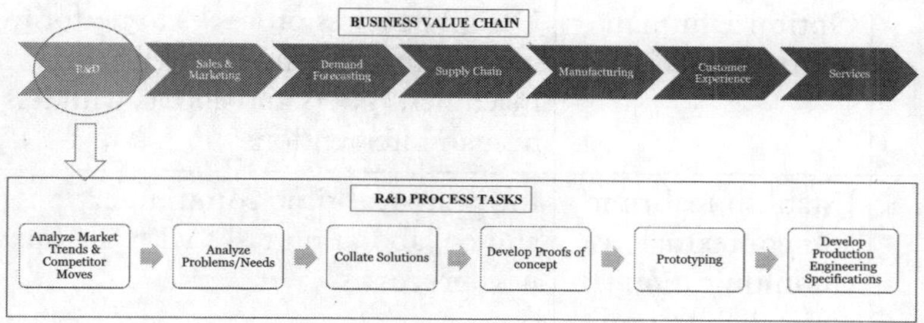

Source: Author

Figure 13.16: Example—R&D Process Tasks

Step 4: Digitalize Tasks

George gave a beautiful explanation on digitalizing business process tasks using 12 **lean digital thinking** principles as described in Table 13.4.

S. No.	Lean Digital Thinking Principles	Digitalize Business Process Elements
1	Create a better today and better tomorrow	Refine the existing business process tasks and design new business process tasks to support both traditional and digital businesses at the same time
2	Embrace constant and instant change	Continuously collate and analyze the changing trends and shifts in the business and bring changes in the business process tasks by leveraging digital technologies
3	Nurture digital business talent	Nurture digital talent that is required to design and execute the digital business process tasks
4	Identify and eliminate non-value-added activities	Design the process tasks using digital technologies to eliminate the wastes while performing the tasks
5	Predict, recommend and improve overall business performance	Predict the digital business process tasks performance by collating and mining the data received from various systems and implementing actionable recommendations to improve the business processes

6	Optimize human interventions	Design business processes to perform most of the tasks using intelligent machines, robots and cobots with less human interventions
7	Establish real-time and contextual communication and collaboration	Integrate real-time communication and collaboration tasks with the business processes
8	Safeguard privacy, safety and security of business resources	Design the digital business process tasks that are secure and safe and safeguard the privacy of various stakeholders
9	Establish a technology start-up ecosystem	Integrate technology start-up practices, tools and techniques while designing business process tasks to make them more agile and flexible in absorbing business changes
10	Create socially responsible products, services and value chains	Design the business process tasks by leveraging digital technologies that would have a positive impact or very less negative impact on society and the environment
11	Design and deliver unforgettable moments of experience to stakeholders	Design business process tasks to engage stakeholders while performing different tasks
12	Build a human-centric digital culture	Integrate digital etiquette and digital professionalism to embrace and perform digital business process tasks effectively

Source: Author

Table 13.4: Applying Lean Digital Thinking Principles to Identify Digital Business Process Tasks

Template for Digitalizing Business Process Tasks

Based on the **lean digital thinking** principles, the important elements to digitalize business process tasks are mentioned in Table 13.5

Business Process	Lean	Hyperautomation	Secure	Real-Time Engagement	Social Impact	Real-Time Insight	Real-Time Communication and Collaboration	Agility
Task 1	Eliminate waste while performing a task	Leverage IoT, blockchain smart contracts, AI, ML, chatbots and robots to perform the task with no or low human intervention	Ensure cybersecurity and privacy while performing a task	Emotionally engage stakeholders while performing a task	Ensure low consumption of natural resources and less impact on air, water, land and nature while performing a task	Leverage data, analytics, AI and ML to take the right decisions at the right time while performing a task	Enable real-time communication and collaboration while performing a task	Build highly flexible processes to embrace changes
Task 2								
Task 3								
Task n								

Source: Author

Table 13.5: Template for Digitalizing Business Process Tasks

Sample Digitalization Matrix for R&D Process Tasks

R&D Process Tasks	Lean	Hy-perau-toma-tion	Se-cure	Re-al-Time Engage-ment	Social Im-pact	Re-al-Time Insight	Real-Time Communica-tion and Col-laboration	Agility
Analyze market trends and competitor moves						√	√	
Analyze problems/needs				√	√		√	
Collate solutions						√	√	√
Develop proofs of concept	√	√				√		
Prototyping		√				√	√	
Develop production engi-neering specifications		√						

Source: Author

Table 13.6: Sample Digitalization Matrix for R&D Process Tasks

Example: Digitalizing R&D Process Tasks

R&D Process Tasks	Lean	Hyperautomation	Secure	Real-Time Engagement	Social Impact	Real-Time Insight	Real-Time Communication and Collaboration	Agility
Analyze market trends and competitor moves	—	—	—	—	—	Use AI to crawl through the web and internal sources automatically and create a market analysis report	Use real-time communication and collaboration tools to interact with the customers diligently	—
Analyze problems/ needs	—	—	—	Leverage virtual tools to empathize while discussing the customer's needs and problems	Select the problems which generate profits and revenue while creating a positive impact on society	—	Use real-time communication and collaboration using digital tools to work with the customers closely	—

Collate solutions	—	—	—	—	—	Use software to analyze each solution against organizational objectives and remove irrelevant ideas	Get as many solutions from globally distributed solution specialists using a digital platform; use real-time communication and collaboration digital tools among various stakeholders	Enable problem category-wise solution selection parameters using software to support various problem categories
Develop proofs of concept	Use digital technologies to simulate the products and avoid wasting materials and tools	Develop a digital twin of the proposed product/solution and optimize the same	—	—	—	Generate automatic recommendations to improve the proof of concept (recommendation engine)	—	—

Prototyping	—	Use 3D printing to develop the sample product	—	—	—	Analyze the testing logs and provide actionable insights to improve the product	Use real-time communication and collaboration tools while beta testing the product	—
Develop production engineering specifications	—	Based on the final digital twin, generate product engineering specifications automatically	—	—	—	—	—	—

Source: Author

Table 13.7: Example—Digitalizing R&D Process Tasks

Example: Consolidated Actions Taken to Digitalize R&D Process Tasks

S. No.	R&D Process Tasks	Digitalization of R&D Process Based on the Template		
1	Analyze market trends and competitor moves	Use AI to crawl through the web and internal sources automatically and create a market analysis report	Use real-time communication and collaboration tools to interact with the customers diligently	
2	Analyze problems/needs	Leverage virtual tools to empathize while discussing the customer's needs and problems	Select the problems which generate profits and revenue while creating a positive impact on society	Use real-time communication and collaboration digital tools to empathize and work with the customers closely
3	Collate solutions	Use software to analyze each solution against organizational objectives and remove irrelevant solutions	Get as many solutions from globally distributed solution specialists using a digital platform; use real-time communication and collaboration tools among solution seekers and solution providers	Enable problem category wise solution selection parameters using software to support various problem categories

4	Develop proofs of concept	Use digital technologies to simulate the products and avoid wasting materials and tools	Develop a digital twin of the proposed product/solution and optimize the same	Generate automatic recommendations to improve the proof of concept (recommendation engine)
5	Prototyping	Use 3D printing to develop the sample product	Automatically analyze the testing logs and provide actionable insight to improve the product	Use real-time communication and collaboration tools among stakeholders while beta testing the product
6	Develop production engineering specifications	Based on the final digital twin, generate product engineering specifications automatically		

Source: Author

Table 13.8: Example—Actions Taken to Digitalize R&D Process Tasks

Step 5: Define Standard Operating Procedure

The fifth step is to define the standard operating procedure for each task. George chose to explain the standard operating procedure for the collating solutions among the R&D process tasks.

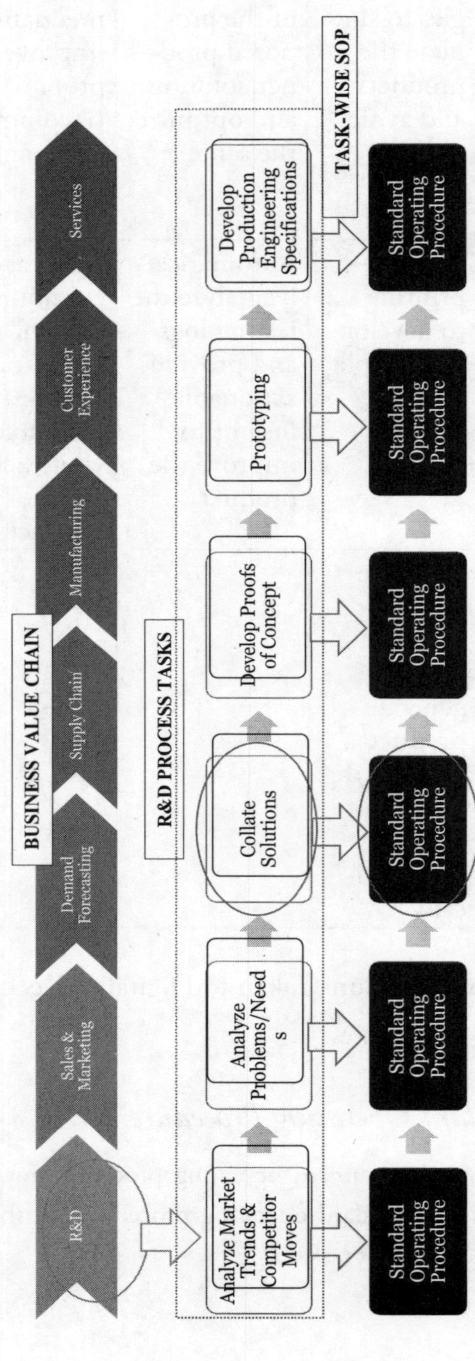

Figure 13.17: Example—Defining Standard Operating Procedure for Each R&D Process Task

Source: Author

The standard operating procedure for collating solutions R&D task is explained in Table 13.9.

S. No.	Step	Templates	Standards/ Policies	Duration	Outputs Expected	Supporting Documentation
1	Appoint solution manager/ solution head	—	Solution manager selection criteria	Start date + 1 Week	Moderation	Solution manager/head responsibilities
2	Provide access to the cloud infrastructure to implement open innovation digital platform	—	Cybersecurity	Start date + 1 Week	Infrastructure access	—
3	Configure solution seekers within the digital innovation platform	Solution seeker information configuration template	—	Start date + 2 Weeks	—	Solution seeker responsibilities
4	Identify problem statements		Problem identification criteria	Start date + 2 Weeks	—	—
5	Elaborate problem statements in the prescribed formats	Problem creation template	—	Start date + 3 Weeks	Complete and unambiguous problem statements	Videos on problem details and other supporting documents

6	Upload problem statements and supporting documentation into the digital innovation platform	—	—	Start date + 3 Weeks	—	—
7	Create legal terms, conditions to be uploaded into digital innovation platform (generic, between solution seekers and solution providers)	Legal agreement templates	Organization's legal, regulatory and IP protection policies	Start date + 4 Weeks	Compliance and IP protection	Leverage existing legal agreement templates
8	Finalize the reward system and cash prizes for solution providers and solution assessors	—	Reward criteria and fee structure for solution providers and solution assessors	Start date + 4 Weeks	Motivation, commercial benefits	—
9	Send invitations to solution providers	Solution provider invitation template	—	Start date + 5 Weeks	—	—
10	Send invitations to solution assessors	Solution Assessor invitation template	—	Start date + 5 Weeks	—	—

11	Configure the solution providers within the digital innovation platform	Solution provider information capture template	—	Start date + 6 Weeks	—	—
12	Configure the solution assessors within the digital innovation platform	Solution assessor information capture template	—	Start date + 6 Weeks	—	—
13	Conduct virtual workshops to provide orientation to solution providers and solution assessors on problems	Virtual workshops agenda template	Virtual workshops meeting protocols	Start date + 7 Weeks	Clarity on requirements and engagement model	—
14	Arrange face-to-face meetings between all solution providers and solution seekers as per the requests and agreed calendar	Face-to-face meeting agenda template	—	Start date + 7 Weeks	Clarity on requirements and engagement model	—

15	Receive solutions	Solution design template	Solution design standards and guidelines	Start date + 10 Weeks	Many alternate solutions for one problem	—
16	Solution review, assessment and prioritization	—	Solution assessment parameters weightage; solution rating and prioritization policies	Start date + 12 Weeks	Prioritized solutions	—
17	Solution approval	—	Solution approval guidelines	Start date + 12 Weeks	Go/no go decision	—

Source: Author

Table 13.9: Sample Standard Operating Procedure for R&D Process Task—Collate Solutions

Step 6: Design Digital Software Applications and Infrastructure

George explained that identification of the portfolio of digital software applications and required infrastructure must be based on the digital business process pyramid. All the physical processes, core business processes and shared processes will be automated using digital technologies. Figure 13.18 depicts the different type of digital software applications which form the digital software infrastructure.

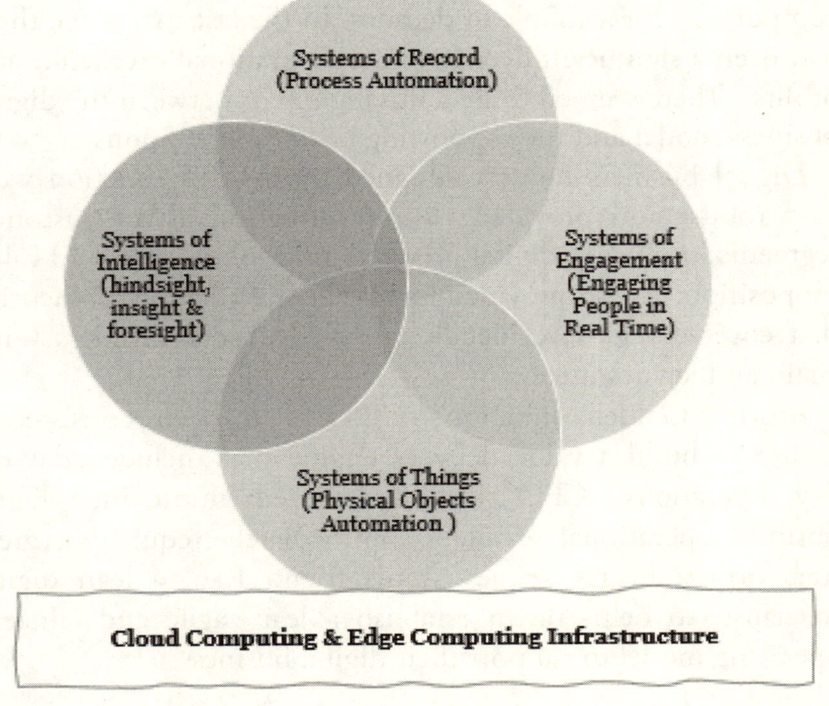

Source: Author

Figure 13.18: Digital Software Applications and Infrastructure

Srini and other senior leaders of Green Engineering Corp had gained greater confidence in their abilities to digitalize the business processes after the meeting with George. Srini approved the completion of the digitalization blueprint for the whole manufacturing value chain.

George cautioned Srini that considerable investments in their technology organization would be required to build various digital software applications and infrastructure based on the digital business process pyramid.

Building Block 12: Business Operating Model

Golden Junction had been manufacturing automotive components for a couple of decades. In the last two years, there had been a significant decrease in its operational excellence and profits. There seemed to be a misalignment between the digital business model and the supporting business operations.

Digital business models designed by Golden Junction were very robust and provided clear information about customer segments, problems, digital products/solution features and value proposition. It also provided an excellent operational blueprint that covered high-level details on key partners, business value chain and key resources.

Austin, Golden Junction's COO, had been consistently failing to build a value delivery engine that included day-to-day operations. CEO Russell believed improving digital business operational excellence immediately required a great transformation. He requested his friend Rao, a lean digital specialist, to help Austin establish a lean, agile and efficient operating model to support their digital business.

Applying Lean Digital Thinking to Design Digital Business Operating Model

Rao believed that the 12 principles of **lean digital thinking** would help in designing and establishing an effective digital business operating model. As a team, Rao, Austin and other senior executives brainstormed and identified the basic elements of digital business operating model.

S. No.	Lean Digital Thinking Principles	Changing the Anatomy of Business Operating Model	Digital Business Operating Model Elements
1	Create a better today and better tomorrow	Design the organization structure to drive both traditional business and digital business (bimodal business) at the same time; streamline suppliers and partners to the extent possible to support bimodal businesses	• Flatter and technology-driven structure
2	Embrace constant and instant change	Continuously collate and analyze the changing trends and shifts in the business and bring changes in the structure, assets, locations, things, digital software applications, suppliers and partners	• Flatter and technology-driven structure • Intelligent assets/resources • Digital software applications and infrastructure
3	Nurture digital business talent	Nurture digital talent that helps human resources (employees, suppliers, partners) to collaboratively work with intelligent assets, machines, things and bots	• Flatter and technology-driven structure • Digital software applications and infrastructure
4	Identify and eliminate non-value-added activities	Optimize data, sensors, devices and other things being used for driving business operations to minimize cost escalations and effort overrun; streamline the existing suppliers using digital technologies (AI/ML/analytics) to avoid overhead costs of managing contracts with many suppliers	• Intelligent sourcing of suppliers • Intelligent assets/resources • Digital business value chain

5	Predict, recommend and improve overall business performance	Predict the operating model performance by collating and mining the data received from various systems and implement actionable recommendations to improve operational efficiency	• Business insight • Digital software applications and infrastructure • Smart monitoring and control
6	Optimize human interventions	Digitalize the operating model in such a way that digitalized assets, machines and other things optimize the human interventions in day-to-day work; design flatter organization structure that optimizes human efforts by leveraging digital technologies	• Intelligent sourcing of suppliers • Flatter and technology-driven structure • Digital business value chain • Digital software applications and infrastructure
7	Establish real-time and contextual communication and collaboration	Integrate real-time communication and collaboration tools with the organization structure to promote cross-functional collaborative work	• Flatter and technology-driven structure • Digital software applications and infrastructure
8	Safeguard privacy, safety and security of business resources	Secure the digitalized machines, tools, equipment of the organization and partner ecosystem	• Intelligent assets/resources • Digital software applications and infrastructure

9	Establish a technology start-up eco-system	Integrate technology start-up practices, tools and techniques with the organization structure to become more agile and flexible in absorbing business changes	• Flatter and technology-driven structure
10	Create socially responsible products, services and value chains	Design and implement digital assets, machines and other things that consume fewer natural resources and create minimal impact on the environment	• Intelligent assets/resources • Digital software applications and infrastructure
11	Design and deliver unforgettable moments of experience to stakeholders	Design and implement digital assets, machines and other things, which are intelligent enough to emotionally engage the stakeholders (employees, customers, partners and suppliers)	• Intelligent assets/resources • Digital software applications and infrastructure
12	Build a human-centric digital culture	Integrate digital etiquette, digital professionalism while working with customers, suppliers and partners	• Flatter and technology-driven structure

Source: Author

Table 13.10: Applying Lean Digital Thinking Principles to Design Digital Business Operating Model

Rao, Austin and team consolidated the business operating model elements by applying **lean digital thinking** principles as shown in Figure 13.19.

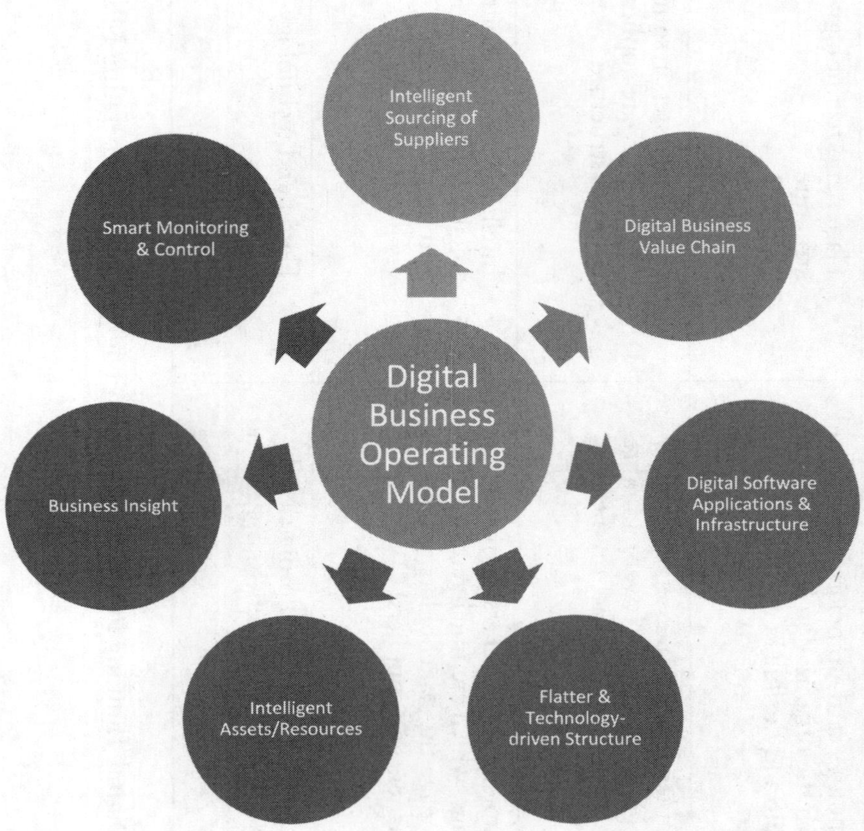

Source: Author

Figure 13.19: Digital Business Operating Model Elements

Digital Business Operating Model

Identifying digital business operating model elements is not enough, these elements must be aligned with the digital business model. Rao shared a very closely-knit business operating model, as given in Figure 13.20.

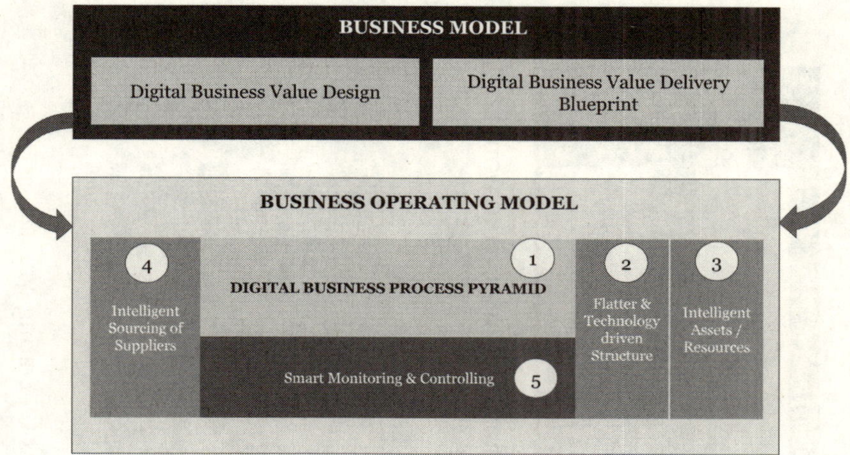

Source: Author

Figure 13.20: Digital Business Operating Model

Rao concluded that

1. the **business model** is like a building blueprint that provides the **value design and value delivery blueprint** and
2. the **business operating model** is like an actual building built according to the blueprint created as part of the business model. It consists of organization structure, business processes, policies, procedures, suppliers, partners, assets, resources, infrastructure and software applications, which are used effectively to perform day-to-day tasks in the organization.

Expanded View of Digital Business Operating Model

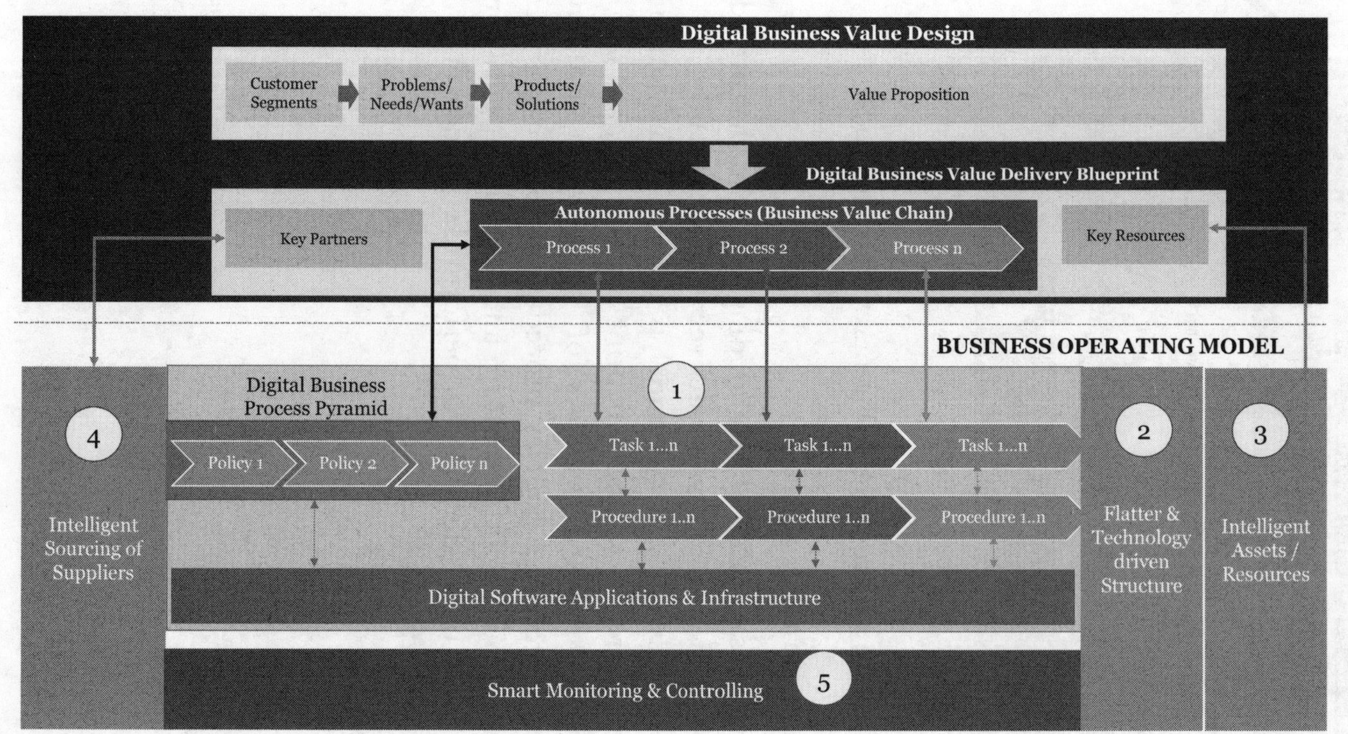

Source: Author

Figure 13.21: Expanded View of Digital Business Operating Model

Digital Business Process Pyramid

Refer to Figure 13.11 for more details on the digital business process pyramid, which is an essential element of the digital business operating model.

George explained that the detailed execution of the tasks will be performed to drive day-to-day operations based on the digital business process pyramid. Digital software applications like systems of record, systems of engagement, systems of things and systems of insight enable nearly autonomous operations (hyperautomation).

Flatter and Technology-Driven Structure

It is a myth that a flatter organization means there is no hierarchy. It would be challenging to drive an enterprise-scale organization without some level of hierarchy. A few layers of structure are required to distribute the humongous amount of workload so that operations can be scaled up. Figure 13.22 shows the schematic structure at multiple levels. The organization designs the levels as per their organization structure guiding principles. For example, Level 2 could be geography (North America, Europe, ASEAN, MEA, etc.) or a product line-based (cars, tractors, bikes, etc.). But there will be shared functions like HR, SCM, finance, learning and others.

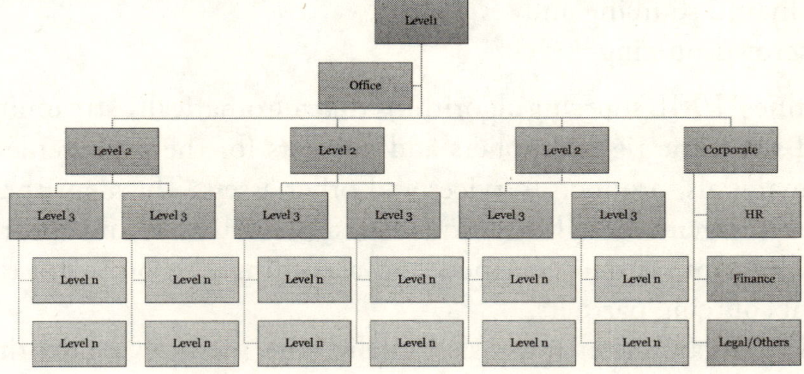

Source: Author

Figure 13.22: Business Operating Model—Organization Structure Template

Sunita, a business analyst at Golden Junction, asked what did Rao mean by a flatter organization; she said that her understanding was that there will not be any hierarchy in a flatter organization. Rao corrected her and shared the following details:

1. Flatter means there is a free flow of two-way information from top to bottom and bottom to top.
2. There are no barriers for employees, irrespective of their position; they can communicate freely without bureaucratic interference.
3. Technologies will be leveraged for real-time communication and collaboration among business stakeholders using virtual tools.
4. There will be the implementation of start-up practices that improve agility in the structure to absorb the changes.
5. Data-based decisions using analytics will help in taking corrective and preventive actions promptly.

Intelligent Sourcing of Suppliers

Sourcing options must be identified to select/streamline the right suppliers and partners as per the business model, products and services. They include

1. outsourcing,
2. insourcing,
3. hybrid sourcing and
4. crowdsourcing.

Define AI/ML sourcing algorithms that automatically streamline and select the right suppliers and partners for the procurement of materials, products, services and other assets. The algorithms are intelligent enough to find the past performance and future-readiness of sourcing partners and accordingly recommend the right sourcing partners.

The organization must also finalize the method to identify, assess, acquire and integrate digital technology start-ups that help in improving the operational capabilities of an organization.

Intelligent Assets/Resources

In the digital era, human interventions are optimized more and more. With technology fusion, many of the human tasks are now being performed by intelligent things like machines, equipment, tools, forklifts and other digitalized infrastructure.

Another category of key resources is digital workers, like chatbots, virtual assistants and cobots, which have become part of the organization's workforce. The digital business operating model leverages both humans and intelligent things to perform day-to-day operations. The design of the digital business process pyramid factors the execution of the tasks considering both human and intelligent things. Digital software applications provide another valuable resource, information and intelligence to improve the operational efficiency.

Rao proposed a template, depicted in Table 13.11, to identify the assets/resources, which could be digitalized to reduce the human interventions as well as improve the quality, speed and productivity of operations.

Asset	Cost of Digitalization	Time Taken to Digitalize	Business Value
Machines	High/Medium/ Low	High/Medium/ Low	High/Medium/ Low
Equipment	High/Medium/ Low	High/Medium/ Low	High/Medium/ Low
Forklifts	High/Medium/ Low	High/Medium/ Low	High/Medium/ Low
Tools	High/Medium/ Low	High/Medium/ Low	High/Medium/ Low
Devices	High/Medium/ Low	High/Medium/ Low	High/Medium/ Low
Others	High/Medium/ Low	High/Medium/ Low	High/Medium/ Low

Source: Author

Table 13.11: Template to Identify Intelligent Assets/Resources

Digitalization of assets = sensors, microprocessor, embedded software, short-range/long-range networks to connect with the internet;
Digital workers = AI-driven chatbots and virtual assistants that perform some of the tasks performed by humans like customer interaction, employee engagement, etc.

Smart Monitoring and Control

One of the most critical elements of the business operating model is business insight. Digital software applications powered by AI, ML and analytics provide hindsight, insight and foresight on overall business operations. Without business insight, organizations fail to improve operational performance.

Data received from processes, intelligent assets/things, people and digital workers will be mined and analyzed to provide greater insight to managers and leaders to take the right decisions at the right time.

The management will get early warning alerts, auto-generated risk register, suitable mitigation plan and auto-generated dashboard as per their role. Most importantly, the system will provide appropriate recommendations (actionable insights) to the management to augment their intelligence.

Smart monitoring and control is done with the power of digital technologies. It helps organizations to proactively manage risks and address issues with the most suitable solutions so that operational performance improves manifold.

Execution Method of Digital Business Operating Model

Rao, Austin and the team concluded the session by sharing the steps of the execution method of the digital business operating model, as listed in Figure 13.23.

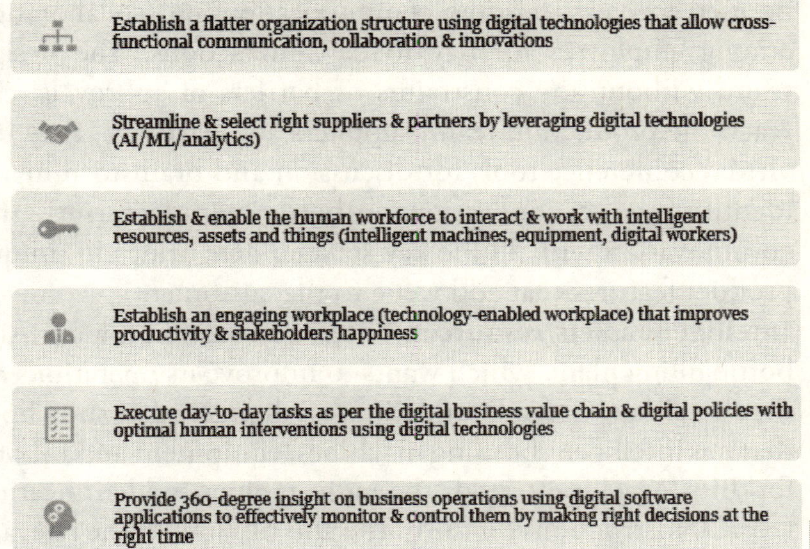

Source: Author

Figure 13.23: Execution Method of Digital Business Operating Model

Austin updated the CEO that he and his team had gained an exceptional understanding of the digital business operating model and implementation method. Austin assured him that they would make sure that, in the next 6–12 months, they will bring a considerable change in the operating model to help realize Golden Junction's digital vision and objectives.

Example: Digital Business Operating Model for a Bottling Plant

Digital technologies play a very vital role in re-engineering the current business operating models. A few examples of how business operating model works with the power of digital technologies are given below:

1. **Flatter and technology-driven structure**: Waiting time due to bureaucratic overheads causes many delays in the organization. Without a hierarchical structure, employees could collaborate and communicate to perform their job well. Technologies could

be used to have real-time communication and collaboration among employees for day to day interactions. The design team, without any constraints or barriers of approvals, can reach the production team, suppliers and partners. They use virtual conference tools for discussion and brainstorming to identify product innovations in bottle manufacturing. This co-innovation with all the key stakeholders brings in unique product features that add value to the customer.

2. **Intelligent assets/resources**: Green Foundation is a mid-size bottle filling plant, which wants to improve its operations on the shop floor. Green Foundation has digitalized its shop floor that has intelligent bottling machines, equipment and robotic forklifts. Machine 1 reads the radio-frequency identification (RFID) instructions on the bottle and fills it with the relevant fluids as per the pre-defined mix of chemicals and sends it to the next machine. Machine 2 automatically closes the bottle with a cap and puts it on the conveyor. The robotic forklift grabs the bottles and keeps them at the pick-up point. Indoor drones take the bottles and put them in the warehouse.
3. **Digital business process pyramid:** AI-, blockchain- and analytics-based digital software applications automatically identify the best suitable suppliers based on past performance and current delivery ability. For a few selected items (not all), procurement approvals can be automatically generated by blockchain smart contracts based on the bill. Customers use the chatbot to order the products of their choice and can have audio or text conversations with the bot for clarifications and recommendations that help in selecting the right products for them.
4. **Smart monitoring and control:** The management, as per their role, gets complete visibility on business operations performance. The digital software applications are intelligent enough to provide the performance of KPI's, list of potential risks and mitigation plan.

The system sends early warning alerts on sales, costs, profits, employee attrition, machine capacity utilization and others. For example, for sales/revenue KPI, the following information will be provided by the system automatically.

Sales Operations	System Insight and Recommendations
What happened?	Overall revenue declined by 15%
Why did it happen?	Sales in India and Africa has declined
What will happen?	Sales in the ASEAN region may decline by 10% due to changes in the government policies, and high sales growth of new competition like Gen Mac
If it happens, what will we do?	The system provides a few actionable insights automatically using AI/ML/analytics technologies. • Introduce design changes to align with the digital age customer requirements • Competition SWOT analysis report and strategies report is attached here, take necessary action • The digital age customer needs report is attached here; connect with customers and identify their actual needs, wants and pains to bring changes in our product mix

Source: Author

Table 13.12: Example—Digital Sales Operating Model

PART 4

How to Digitalize?

CHAPTER 14
Digitalization Roadmap: Five-Phase Lean Digital Journey

- In the Time of Crisis
- Identifying Digital Champions
- Five Phases of Lean Digital Journey

CHAPTER 14

Digitalization Roadmap: Five-Phase Lean Digital Journey

In the Time of Crisis

Green Stores was one of the best retail stores in the US. Stephenson the CEO of Green Stores often found himself stuck in an embarrassing situation at the quarterly board meetings. For the last three quarters, the share price of Green Stores had been going down consistently due to bad business performance. "The crisis has just begun, I will step down as CEO of Green Stores," said Stephenson to his close friend Paul at the dinner table.

Paul observed that Stephenson was not in his usual mood. When Paul asked what was wrong, Stephenson sadly replied that he had failed to turn around Green Stores' situation. Aware of Stephenson's "can-do" attitude, Paul felt that Stephenson should not take a hasty decision.

Stephenson probably felt that a new generation technology-savvy CEO was required to reshape Green Stores. The digital tech start-ups and modern retail stores were giving Stephenson sleepless nights. He was puzzled about the reasons why Green Stores had not been performing well even after he had invested heavily in new age digital technologies.

Paul advised Stephenson to meet Rao, a well-known lean digital specialist, and confidently shared that Rao was known for many turnaround stories. He was an execution champion and a practitioner, unlike other consultants. Stephenson literally shouted and said, "No more consultants. Enough damage has been done by those so-called consultants. We spend a lot of money, they come, collate data from us, package those data points well and present great transformation blueprints. However, they miserably fail to translate the grandeur of blueprints during execution."

Paul convinced Stephenson to meet Rao once. Stephenson reluctantly agreed to meet Rao the next day at the Peacock Innovation Centre in the city, in the late evening.

After obtaining detailed background information about Green Stores, Rao asked Stephenson to provide a few more details.

Rao	**:**	**Does the organization define a common purpose of digitalization?**
Stephenson	**:**	**We have not defined any separate core purpose for our organization for the digital journey.**
Rao	**:**	**Do all employees have a common understanding of digitalization?**
Stephenson	**:**	**I do not think so. There are a few leaders who were trained on digital business strategy and leadership, and out of which a few leaders have already left. At this juncture, I am not confident enough to say that our leaders and employees have a common understanding of digitalization.**
Rao	**:**	**Does the organization baseline the current digital state (digital strengths, digital weaknesses and business outcomes)?**

Stephenson : We do not have clarity on the digital readiness of our business building blocks like processes, products, services, workplaces, business models, etc. We do not have a comprehensive list of digital strengths and weaknesses within our organization to baseline the current digital state.

Rao : Does the organization benchmark the future digital state? (What is to be strengthened digitally? What are the business outcomes you want to improve?)

Stephenson : Yes, we partially did the benchmarking exercise for our outcomes like revenue, profit and scale. But we have not benchmarked our business building blocks (business model, business processes, products, services, etc.) for their future digital state. Hence, we do not have complete clarity on the changes that we need to bring in our business building blocks.

Rao : What is the process followed to identify digital initiatives/solutions?

Stephenson : We identified a few digital strategies and digital solutions based on our sales, marketing team inputs, industry analyst's information and our consultant's guidance. As such, there is no method used to identify the digital initiatives, probably that could be the reason we are not getting the return on our investments.

Rao : Are the current digital initiatives unique/innovative and high-value-added?

Stephenson	**:**	**Probably a few. But we are not getting the expected results from our digitalization initiatives, hence, my frustration. That means our digital initiatives are not effective.**
Rao	**:**	**How are the digital solutions implemented, in an incremental manner or the big bang way?**
Stephenson	**:**	**We do not have any portfolio management system to prioritize and implement digital solutions. We, as a team, collectively discuss in the meetings and work on various digital initiatives. Most of the decisions are made collectively, but we do not have any formal method to identify, prioritize and implement digital initiatives.**
Rao	**:**	**What is the organization's structure to drive digital initiatives?**
Stephenson	**:**	**We do not have any separate structure to drive digitalization initiatives. The existing leaders spend some amount of time on digital initiatives. We have established a monitoring office to oversee the progress of these digital initiatives.**

Stephenson was happy that, unlike most of the external consultants who just came and sold their solutions and products, Rao had asked the right questions to understand Green Stores' current situation.

Rao explained that digital transformation is an overly complex programme because most of the business leaders must acquire the technical acumen to understand the possibilities and opportunities that emerge from digital technologies. The business leaders and domain specialists are the real champions, not the consultants.

He offered Stephenson an interesting proposal. He and his team wanted to spend 40 days to first baseline the current digital status of Green Stores using **lean digital thinking**. This would include digital strengths and weaknesses of the 12 business building blocks as well.

Once they obtain a detailed list of digital strengths and weaknesses of the 12 business building blocks, they will work on the right digital strategies and solutions. Rao committed that for the first 40 days, they will not charge any consulting fee.

Rao believed that aiming at reaching the digital benchmark without an understanding of the digital baseline information does not make any sense. After baselining the current digital state of Green Stores, he would charge a premium fee to prepare Green Stores for the digital journey.

It would take three years to complete the digital transformation programme, but there will be digital deliverables every 90 days that would incrementally add value to Green Stores, its customers and help it leapfrog the competition.

By the end of the meeting, Stephenson was convinced about the plan and requested Rao to take up the challenge. Rao smiled and said, "We are getting into an exciting and challenging journey. Be ready to experience surprises, cultural shocks and finally, great benefits."

Stephenson cautioned Rao, "If we fail to show incremental value from our digital programme, the board will be upset, and we all will be proven as inefficient and incapable." Rao assured Stephenson that the board will be happy with their progress.

Identifying Digital Champions

After two weeks of extensive work, Stephenson, Rao and the other business leaders released the qualification criteria for the team to work for the digital transformation programme. Rao emphasized the importance of committed and focused effort.

The team didn't need people who take additional work as per their convenience with a volunteer mindset. They needed a focused effort. Stephenson finalized the digital transformation programme structure in consultation with Rao and his business leaders.

They conducted a series of open house meetings and webinars. After a series of internal interviews and discussions, Stephenson and his senior leaders identified the digital champions, who would work full time on the Green Stores digital programme. They also finalized the recruitment plan to induct a few digital specialists to enhance digital capabilities.

Five Phases of Lean Digital Journey

Rao conducted an executive workshop with all the digital champions on the lean digital journey. There are five phases of lean digital journey, as illustrated in Figure 14.1. The five phases of lean digital journey are executed in the sequence depicted in the figure. The reverse arrows mark the feedback to the previous phases and the necessary improvement actions to progress further is communicated throughout the process.

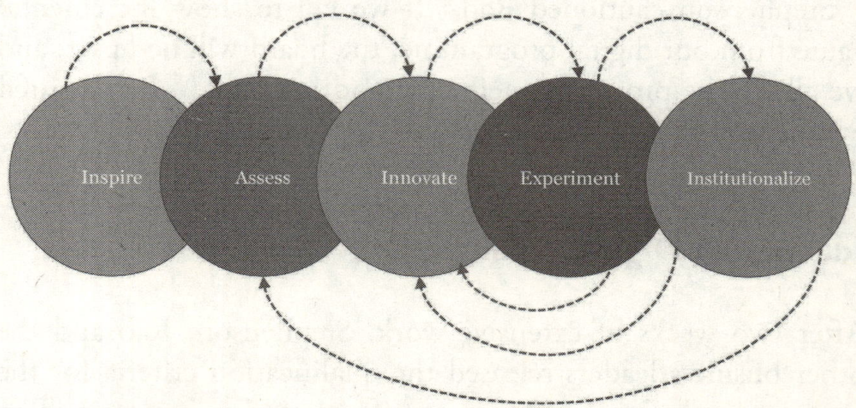

Source: Author

Figure 14.1: Five Phases of Lean Digital Journey

This journey is an **iterative** process. Rao briefly explained each phase of the lean digital journey, as mentioned in Table 14.1.

Phase	Details
Inspire	• Establish a common digital purpose • Bring a common understanding of digital • Establish a digital language • Ignite the spark for a digital movement in the organization
Assess	• Assess the digital strengths and weaknesses of the organization and baseline the current digital state • Identify problems and possibilities based on digital weaknesses and digital strengths and prioritize them
Innovate	• Innovate unique digital technology solutions to resolve the problems and leverage the possibilities • Prioritize digital technology solutions • Get approvals to experiment with the prioritized digital technology solutions
Experiment	• Pilot the identified digital technology solutions • Analyze the results • Pivot and improve the outcomes of digital technology solutions
Institutionalize	• Operationalize the outputs of piloted digital technology solutions • Manage the change effectively • Identify the lessons learned • Identify the best practices • Identify the next practices • Continuously improve the digital business building blocks

Source: Author

Table 14.1: Five Phases of Lean Digital Journey

Digitalization Roadmap Phase 1: Inspire

Rao emphasized that **inspire** phase was extremely critical to create a momentum for digital transformation in an organization. It includes the following:

1. Establish a common digital purpose
2. Bring a common understanding of digital
3. Establish a digital language
4. Ignite the spark for a digital movement in the organization

Rao explained the **Why, How and What Framework** to inspire the core team of Green Stores about digital transformation.

WHY	To survive and excel in the digital era by meeting the needs of digital age customers. To enhance employee skills for them to become relevant in the digital age.
HOW	By leveraging **lean digital execution**, organizations will be able to complete their digital journey effectively. Lean digital execution is a digital business excellence framework that ensures the effective execution of the right digital initiatives.
WHAT	Create new business models, business opportunities, new revenue streams, new services, new products, new operating models and workplaces by leveraging digital technologies.

Introduction to Key Elements of Digital Business Transformation

Some of the core team members of Green Stores were a bit cynical and frustrated as they had a different understanding of digital and different expectations from the senior leaders of the organization. This was mainly due to the lack of common understanding among the leaders about digital businesses. Hence,

Rao wanted to make sure that all the core team members had a common understanding of the digital using the key elements given in Figure 14.2.

- Business Building Block
- Instance of Business Building Block
- Lean Digital Principle
- Lean Digital Practice
- Digital Strength & Possibilities
- Digital Weakness & Problems
- Lean Digital Quotient

Source: Author

Figure 14.2: Key Elements of Digital Business Transformation

Seven Key Elements of Digital Businesses Transformation

1. Twelve business building blocks
    a. Business vision
    b. Business model
    c. Business strategy
    d. Business outcomes
    e. Technology strategy
    f. Technology governance
    g. Technology architecture
    h. Products
    i. Services
    j. Workplace
    k. Business processes
    l. Business operating model
2. Instances of business building block
   An organization may have multiple building models, multiple products, multiple services, multiple processes, which means

each business building block may have multiple instances. A few examples are given below:

a. **Instances of business process building block:** If the business process is one of the building blocks of an organization, then it will have instances like customer relationship management (CRM), supply chain management (SCM), finance, human capital management (HCM) and others. All these instances (processes) are required to drive the business operations of an organization.
b. **Instances of product building block:** If the product is one of the building blocks of an organization, then it will have instances like a car, washing machine, two-wheeler and others (an organization will have multiple product lines).

To assess digital readiness, all instances or critical instances of business building block must be assessed as per business requirements.

3. Twelve **lean digital thinking** principles
   a. Create a better today and better tomorrow
   b. Embrace constant and instant change
   c. Nurture digital business talent
   d. Identify and eliminate non-value-added activities
   e. Predict, recommend and improve overall business performance
   f. Optimize human interventions
   g. Establish real-time and contextual communication and collaboration
   h. Safeguard privacy, safety and security of business resources
   i. Establish a technology start-up ecosystem
   j. Create socially responsible products, services and value chains
   k. Design and deliver unforgettable moments of experience to stakeholders
   l. Build a human-centric digital culture
4. Lean digital practices

Each **lean digital thinking** principle is implemented with a set of lean digital practices. Each **lean digital thinking** principle may have 3–5 corresponding lean digital practices. For example:

Business building block: Product
Lean digital thinking principle: Design and deliver unforgettable moments of experience to stakeholders
Lean digital practices:

1. Design the product in such a way that customer safety, security and privacy is not compromised.
2. Design the product in such a way that the customer gets an immersive experience while using the product (AR, VR, etc.).

There are around 500 lean digital practices covering 12 business building blocks and 12 **lean digital thinking** principles, which help in assessing the digital readiness of an organization holistically.

However, it is not necessary to assess the digital readiness of all the12 business building blocks at a time. One may assess one or more business building blocks for their digital readiness as per their business priorities. The number of lean digital practices depends on the number of business building blocks one plans to assess.

5. Digital strength and possibilities
   If a business building block is functioning as per the defined **lean digital thinking** principles and lean digital practices, then the digital strengths will be identified to create new possibilities.
6. Digital weakness and problems
   If a business building block is not functioning as per the defined **lean digital thinking** principles and lean digital practices, then the digital weaknesses will be identified to define new problems.
7. Lean Digital Quotient (LDQ)
   There are four maturity levels of LDQ in an organization, which have been derived on the basis of digital strengths and

weaknesses of business building blocks as shown in the figure 14.3:

Source: Author

Figure 14.3: Lean Digital Business Maturity Levels

Lean Digital Quotient - Maturity Levels

LDQ	Description
Digital Beginner	The organization lacks an understanding of digital, most of the time avoids discussion on digital. A few sections of leaders know the importance of digital, but there will not be any organization-level direction or blueprint. A few leaders implement digital initiatives due to their passion.
Digital Opportunist	The organization introduces a few digital initiatives on an ad hoc basis, without a proper digital business blueprint. The organizations in this category imitate the competition. If the competition introduces any digital products/services, they too implement the same to be a part of the race.

Digital Emergent	There is an organization-wide digital business blueprint. Focused initiatives will be in place too. A few business building blocks reach their digital mature state, but it may take some more time for other business building blocks to become purely digital. Such organizations start tasting the benefits of digitalization, and there is a return on digital investments.
Digital Cultivator	The organization knows its current digital state and an aspirational future digital state. There is a digital rhythm and it unlocks the fullest potential of digitalization and leapfrogs the competition and will be in the trajectory of exponential growth.

Source: Author

Table 14.2: Lean Digital Business Maturity Levels

Training on Digital Assessment Framework

After clarifying the critical elements of digital business transformation, Rao explained the relationship between the critical elements of digital businesses by training the employees on the digital assessment framework. Each business building block was to be assessed against the 12 **lean digital thinking** principles, with around 40 lean digital practices for each block. Hence, to find the digital readiness of the whole organization, the framework covers 12 building blocks, 12 **lean digital thinking** principles and about 500 lean digital practices. This is illustrated in Figure 14.4.

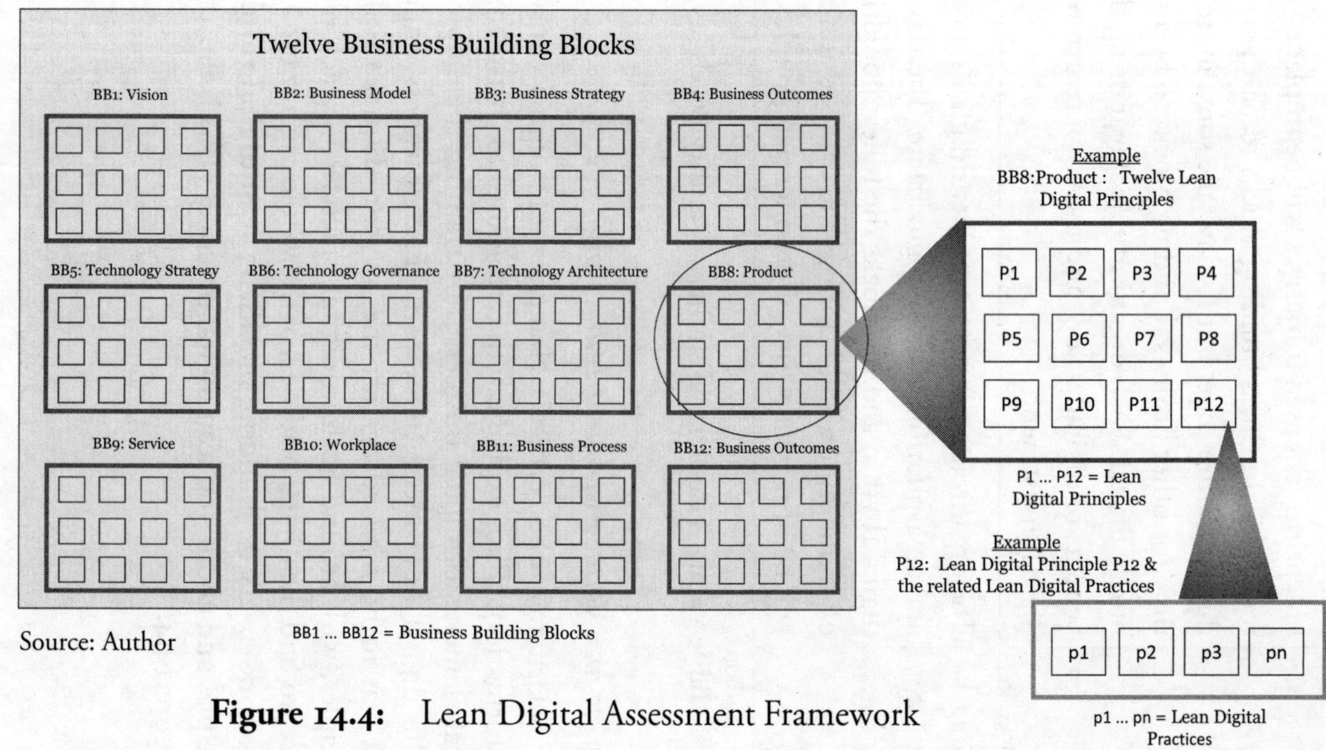

Source: Author

Figure 14.4: Lean Digital Assessment Framework

Business Building Block	Instance	Lean Digital Thinking Principle	Weightage	Lean Digital Practice	Weightage	LDQ of an Instance	LDQ of a Business Building Block
Business Building Block 1	Instance 1	Lean Digital Principle 1	%	Lean Digital Practice 1	%	Digital Beginner	Digital Opportunist
				Lean Digital Practice 2	%		
				Lean Digital Practice 3	%		
				Lean Digital Practice 4	%		
		Lean Digital Principle...	%	Lean Digital Practice 1	%		
				Lean Digital Practice 2	%		
				Lean Digital Practice 3	%		
				Lean Digital Practice 4	%		
		Lean Digital Principle 12	%	Lean Digital Practice 1	%		
				Lean Digital Practice 2	%		
				Lean Digital Practice 3	%		
				Lean Digital Practice 4	%		
	Instance n	Lean Digital Principle 1	%	Lean Digital Practice 1	%	Digital Opportunist	
				Lean Digital Practice 2	%		
				Lean Digital Practice 3	%		
				Lean Digital Practice 4	%		
		Lean Digital Principle...	%	Lean Digital Practice 1	%		
				Lean Digital Practice 2	%		
				Lean Digital Practice 3	%		
				Lean Digital Practice 4	%		
		Lean Digital Principle 12	%	Lean Digital Practice 1	%		
				Lean Digital Practice 2	%		
				Lean Digital Practice 3	%		
				Lean Digital Practice 4	%		

Business Building Block …	Instance 1	Lean Digital Principle 1	%	Lean Digital Practice 1	%	Digital Emergent	Digital Emergent
				Lean Digital Practice 2	%		
				Lean Digital Practice 3	%		
				Lean Digital Practice 4	%		
		Lean Digital Principle…	%	Lean Digital Practice 1	%		
				Lean Digital Practice 2	%		
				Lean Digital Practice 3	%		
				Lean Digital Practice 4	%		
		Lean Digital Principle 12	%	Lean Digital Practice 1	%		
				Lean Digital Practice 2	%		
				Lean Digital Practice 3	%		
				Lean Digital Practice 4	%		
	Instance n	Lean Digital Principle 1	%	Lean Digital Practice 1	%	Digital Opportunist	
				Lean Digital Practice 2	%		
				Lean Digital Practice 3	%		
				Lean Digital Practice 4	%		
		Lean Digital Principle…	%	Lean Digital Practice 1	%		
				Lean Digital Practice 2	%		
				Lean Digital Practice 3	%		
				Lean Digital Practice 4	%		
		Lean Digital Principle 12	%	Lean Digital Practice 1	%		
				Lean Digital Practice 2	%		
				Lean Digital Practice 3	%		
				Lean Digital Practice 4	%		

<table>
<tr><td rowspan="24">Business Building Block 12</td><td rowspan="12">Instance 1</td><td rowspan="4">Lean Digital Principle 1</td><td rowspan="4">%</td><td>Lean Digital Practice 1</td><td>%</td><td rowspan="12">Digital Emergent</td><td rowspan="24">Digital Cultivator</td></tr>
<tr><td>Lean Digital Practice 2</td><td>%</td></tr>
<tr><td>Lean Digital Practice 3</td><td>%</td></tr>
<tr><td>Lean Digital Practice 4</td><td>%</td></tr>
<tr><td rowspan="4">Lean Digital Principle...</td><td rowspan="4">%</td><td>Lean Digital Practice 1</td><td>%</td></tr>
<tr><td>Lean Digital Practice 2</td><td>%</td></tr>
<tr><td>Lean Digital Practice 3</td><td>%</td></tr>
<tr><td>Lean Digital Practice 4</td><td>%</td></tr>
<tr><td rowspan="4">Lean Digital Principle 12</td><td rowspan="4">%</td><td>Lean Digital Practice 1</td><td>%</td></tr>
<tr><td>Lean Digital Practice 2</td><td>%</td></tr>
<tr><td>Lean Digital Practice 3</td><td>%</td></tr>
<tr><td>Lean Digital Practice 4</td><td>%</td></tr>
<tr><td rowspan="12">Instance n</td><td rowspan="4">Lean Digital Principle 1</td><td rowspan="4">%</td><td>Lean Digital Practice 1</td><td>%</td><td rowspan="12">Digital Cultivator</td></tr>
<tr><td>Lean Digital Practice 2</td><td>%</td></tr>
<tr><td>Lean Digital Practice 3</td><td>%</td></tr>
<tr><td>Lean Digital Practice 4</td><td>%</td></tr>
<tr><td rowspan="4">Lean Digital Principle...</td><td rowspan="4">%</td><td>Lean Digital Practice 1</td><td>%</td></tr>
<tr><td>Lean Digital Practice 2</td><td>%</td></tr>
<tr><td>Lean Digital Practice 3</td><td>%</td></tr>
<tr><td>Lean Digital Practice 4</td><td>%</td></tr>
<tr><td rowspan="4">Lean Digital Principle 12</td><td rowspan="4">%</td><td>Lean Digital Practice 1</td><td>%</td></tr>
<tr><td>Lean Digital Practice 2</td><td>%</td></tr>
<tr><td>Lean Digital Practice 3</td><td>%</td></tr>
<tr><td>Lean Digital Practice 4</td><td>%</td></tr>
<tr><td colspan="6">The organization's LDQ (against around 500 lean digital practices)</td><td colspan="2">Digital Emergent</td></tr>
</table>

Source: Author

Table 14.3: Lean Digital Assessment Framework

Each business building block is to be assessed against a set of lean digital practices to find its digital strengths and weaknesses. As given in Table 14.1, the 12 business blocks will be assessed against **lean digital thinking** principles from which a comprehensive list of digital strengths and weaknesses of the whole organization will be derived.

Overview of Lean Digital Quotient Finder

Weightage of Lean Digital Thinking Principles: Based on the industry segment, the weightage of each **lean digital thinking** principle varies. For example, for an oil and gas production industry, the weightage of "design and deliver unforgettable moments of experience to stakeholders" principle might be zero or extremely low. But for the same principle, the weightage will be extremely high in the retail or hospitality industry.

Weightage of Lean Digital Practices: Each **lean digital thinking** principle will have 40–50 lean digital practices, which help in determining the digital strengths and weaknesses of an organization. Overall, there may be around 500 lean digital practices to assess the 12 business building blocks. Again, as per the industry segment, the weightage of lean digital practices varies. The weightage will be defined by the organization based on the industry context with support from internal and external sources.

Digital Assessment Method: Each business building block will have 40–50 lean digital practices, and many stakeholders could give their response (Delphi assessment method) on the level of adherence of the business building block to its respective lean digital practices.

For example, the CRM process could be assessed against 50 lean digital practices by the sales manager, marketing manager, customer and customer service representative.

Responses for each business building block by more than one stakeholder will nullify biased responses. Also, the standard deviation among the assessor's response gives clarity on which lean digital practice the stakeholder's response drastically differs. This helps in understanding multiple perspectives.

Following are the key considerations, when it comes to finding the LDQ of an organization:

1. The business building blocks, **lean digital thinking** principles, lean digital practices are the same for all industries.
2. The business anatomy has 12 business building blocks for all industries.
3. There are 12 **lean digital thinking** principles for all industries—only the weightage varies as per the industry context.
4. Lean digital practices to assess the organization's digital readiness for all industries are the same—include or exclude as per the industry context.

Lean Digital Quotient (LDQ): LDQ identifies the level of digital readiness of an organization in multiple dimensions. It all depends on the organization's choice and priorities.

For instance, a few organizations may want to find the LDQ only for the business building block, Process, which may have many instances. Some of them are mentioned here:

1. Instance 1: CRM
2. Instance 2: SCM
3. Instance 3: HCM

Similarly, for Product, there may be many instances, as mentioned below:

1. Instance 1: SUV Car IX100
2. Instance 2: Budget Sedan Car SD350
3. Instance 3: Two-Wheeler KP5i

LDQ could be determined at the instance level, business building block level or organization level as per the organization's requirement.

Business Building Block	Instance	LDQ
Business Process	CRM	Digital Opportunist
	SCM	Digital Beginner
	HCM	Digital Beginner
LDQ		**Digital Beginner**

Source: Author

Table 14.4: LDQ of Business Process Sample

S. No.	Business Building Block	LDQ
1	Business vision	Digital Cultivator
2	Business model	Digital Beginner
3	Business strategy	Digital Beginner
4	Business outcomes	Digital Beginner
5	Technology strategy	Digital Beginner
6	Technology governance	Digital Beginner
7	Technology architecture	Digital Beginner
8	Product	Digital Opportunist
9	Service	Digital Opportunist
10	Workplace	Digital Beginner
11	Business process	Digital Opportunist
12	Business operating model	Digital Beginner
Organization's LDQ		**Digital Beginner**

Source: Author

Table 14.5: LDQ of Organization Sample

The LDQ is determined using a predefined algorithm; the information in Table 14.4 and 14.5 is only for illustrative purpose.

Understanding Digital Strengths and Digital Weaknesses

For example, Figure 14.5 shows each business building block with its digital strengths and weaknesses. Each building block (BB1 to BB12) in the figure has 12 lean digital principles (P1 to P12), which are shown in rectangular boxes. The dark shade of the rectangular box reflects the digital weakness and white shade of the rectangular box reflects the digital strength.

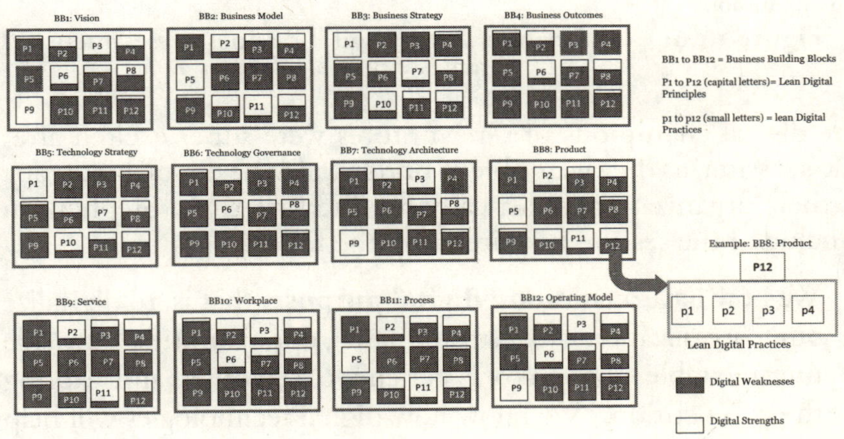

Source: Author

Figure 14.5: Current Digital Strengths and Weaknesses of an Organization

Rao summarized the digital assessment framework, with the help of Figure 14.6. Now, the core team had a complete understanding of the various elements of digital businesses and their relationship with each other. Stephenson and the team were now more confident than ever because of a better understanding of digital businesses.

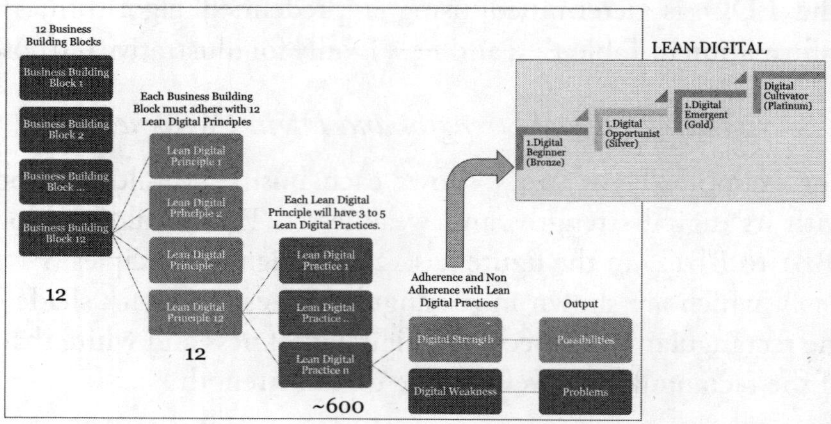

Source: Author

Figure 14.6: Summary of Lean Digital Business Assessment Framework

The digital champions of Green Stores were super excited after Rao's session, as they were able to understand the digital landscape of their organization holistically for the first time. Stephenson concluded the session by reiterating the following:

1. We now have a **common digital purpose,** that is, to digitalize our organization to create new opportunities, provide unforgettable moments of experience to customers and leapfrog the competition. We know how digital technologies will help us get there.
2. We all have a **common understanding of the key elements of digitalization** like business building blocks, instances of business building blocks, **lean digital thinking** principles, lean digital practices, digital strengths, digital weaknesses, possibilities, problems and LDQ. We will educate all our employees about these digital concepts.
3. We will use the **12 lean digital thinking principles as our digital language** during our reviews, discussions, debates and wherever possible. This will help in establishing a new culture.

Finally, Stephenson and team were happy with the outcome of the inspire phase as they got a great orientation and clarity on the whole digital business. Stephenson requested each digital champion to educate their teams about digitalization and ignite the spark to turn **digitalization into an organization-wide movement**.

Digitalization Roadmap Phase 2: Assess

Rao began the next session by stating that the assessment phase is highly critical when it comes to baselining the current digital state of an organization. Without knowing the current digital state, aiming at a better digital future puts too much pressure as the volume of work involved in moving from current state to future digital state is unknown.

Rao emphasized the importance of assessing the digital readiness and digital intensity of an organization or a building block. He continued to explain that 'digital intensity' means the extent to which each business building block has embraced the 12 **lean digital thinking** principles. For example, the extent to which the CRM process institutionalized all the 12 **lean digital thinking** principles reflects the digital intensity of the CRM process.

On the other hand, 'digital readiness' means the extent to which each **lean digital thinking** principle is institutionalized in the 12 business building blocks of an organization. For example, one of the 12 **lean digital thinking** principles is to design and deliver unforgettable moments of experience to stakeholders. Thus, to figure out the digital readiness, one would have to check to what extent does each building block embeds the principle of unforgettable moments of experience.

It may be overwhelming to assess and improve all the 12 business building blocks at the same time, which is also not the right approach. Rao asserted that it was important for Green Stores' leadership to prioritize their digital transformation journey and accordingly carry out the assessment sequence.

Rao instructed that one must always start with the assessment of the readiness of the digital business blueprint of the organization, irrespective of an incremental approach or big bang approach. This means asking questions like: What digital strengths are there? What digital weaknesses are there in the current business blueprint? How to finalize the digital business blueprint that becomes a reference and guiding document throughout the digital journey of Green Stores? Rao strongly recommended Green Stores to not start any digital initiatives without finalizing the **digital business blueprint (vision, business model, business strategy, business outcomes)**. The current state must be known for the future digital state to be finalized.

The elements of the digital business blueprint are given in Figure 14.7. These elements must be assessed to identify digital strengths and weaknesses using the LDQ finder.

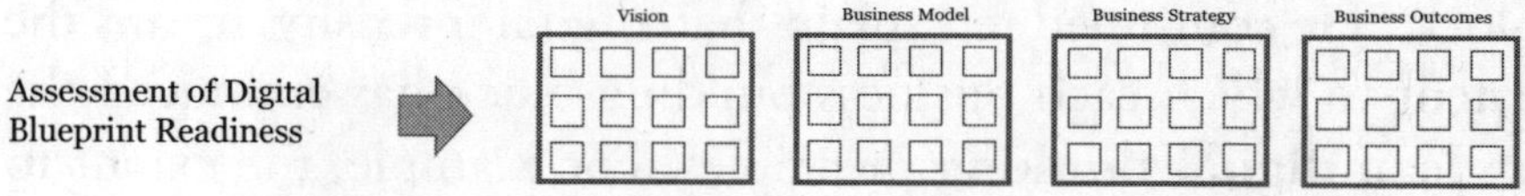

Assessment of Digital Blueprint Readiness

Vision | Business Model | Business Strategy | Business Outcomes

Source: Author

Figure 14.7: Digital Business Blueprint Cluster for Digital Readiness Assessment

After finalizing the digital business blueprint, based on the organization's priority, the digital readiness assessment could be done for other business building blocks.

Rao proposed that instead of assessing the business building blocks for digital readiness in an ad hoc manner, the following grouping of business building blocks could be done for digital readiness assessment. **There is no hard and fast rule to sequence the groups for digital readiness assessment.** It all depends on the organizations' strategy, priority and internal constraints as mentioned below (only illustrative):

1. Customer needs, wants and other focus areas
2. Market and competition trends
3. Internal skills availability
4. Internal resources availability

5. Budget availability

The grouping of business building blocks for digital readiness assessment helps in optimizing the effort and resources during the digital journey.

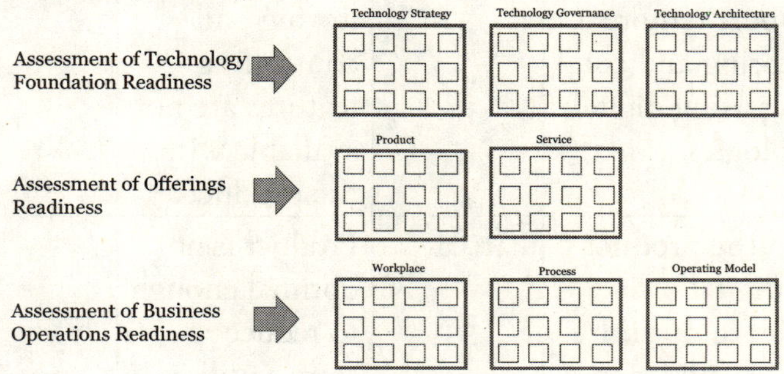

Source: Author

Figure 14.8: Business Digitalization Clusters for Digital Readiness Assessment

Digital Readiness Assessment

Next, Rao wanted to explain how the digital readiness assessment will be done. He chose the **Offerings** cluster (two business building blocks) for demonstration.

Source: Author

Figure 14.9: Digital Business Offerings Cluster for Digital Readiness Assessment

For Offerings (products and services) digital readiness assessment, there may be more than 80 lean digital practices, out of which 20 lean digital practices have been discussed here. A comprehensive list of more than 500 lean digital practices covering all 12 business building blocks is available in the LDQ finder. For the following examples, Green Stores is the manufacturer of washing machines and also sells apparels.

Example: Digital Readiness Assessment of Product (Washing Machine)

S. No.	Lean Digital Practice	Response	Digital Weakness	Digital Strength
1	Build the product to support remote monitoring and controlling using digital technologies	No	Remote monitoring and controlling features are not available with the product	—
2	Design the product using digital technologies to optimize resource utilization (people, machines, fuel, etc.)	Partial	Product is not optimal enough to reduce resource utilization (people, machines, fuel, etc.)	—
3	Design the product using digital technologies to reduce future maintenance costs	Yes	—	Digitalization has reduced the maintenance costs of the product
4	Conceptualize and integrate unforgettable moments of customer experiences while building the product	No	Moments of customer experiences are not integrated while building the product	—

5	Build the product with integrated cybersecurity features	No	The product does not have integrated cybersecurity features	—
6	Use simulation tools, built using digital technologies, to predict the performance of the product in different situations	No	Simulation tools are not used to predict the performance of the product in different situations	—
7	Work in close collaboration with the customers to innovate and add new digital features to the product	Partial	Not consistent in co-innovating digital features of the product with customers	—
8	Encourage a culture for employees to take risks while developing products suitable for digital business	Partial	Only a few employees take risks while developing products suitable for digital business	—
9	Design the product by leveraging digital technologies to have optimal consumption of power/energy during operation	Partial	The product is not efficient for the optimal consumption of power/energy during operation	—

Source: Author

Table 14.6: Example—Digital Readiness Assessment of Product (Washing Machine)

Example: Digital Readiness Assessment of Service (Apparel Selling Service)

S. No.	Lean Digital Practice	Response	Digital Weakness	Digital Strength
1	Majority of customer help desk activities are managed by virtual assistants/chatbots	No	Customer help desk activities are not yet managed by virtual assistants/chatbots	—
2	Use blockchain smart contracts (smart code) to minimize intermediary efforts while offering service	No	Blockchain smart contracts (smart code) are not leveraged to minimize intermediary efforts while offering service	—
3	Make sure that data collected in one channel can be accessed by other channels so that there is continuity in the service rather than the collection of the same information again (overprocessing)	Yes	—	Omnichannel service is available, which makes sure that the data collected in one channel can be accessed by many other channels so that there is continuity in the service rather than the collection of the same information again (overprocessing)

4	Design a robust digital supply chain system to avoid damage to goods during the transportation	No	There is no robust digital supply chain system to avoid damage to goods during the transportation	—
5	Create a customer journey map for each service/transaction and design experience at each critical interaction point in the customer journey	No	There is no customer journey map for the service/transaction to design experience at each interaction point	—
6	Design customer service in such a way that customer's safety, security and privacy are not compromised	Partial	Service has limited features for customer's safety, security and privacy	—
7	Field staff is offered real-time information/insight to facilitate faster troubleshooting and decision-making	No	The field staff is not offered real-time information/insight to facilitate faster troubleshooting and decision-making	—
8	Continuously monitor market changes to make the service digital ready	Yes	—	Market changes are continuously monitored to make the service digital-ready

9	Customer service representatives are trained to use digital tools and techniques to offer high-quality services	Partial	Customer service representatives are not trained to use all the relevant digital tools and techniques to offer high-quality services	—
10	Promote a culture of collaborative frugal innovation to generate uncommon ideas for designing a new service	Yes	—	There is a culture of collaborative frugal innovation to generate uncommon ideas for designing a new service
11	Design customer service in a way that they create no negative impact on the environment	No	Services designed have not factored in any features that reduce the negative impact on the environment	—

Source: Author

Table 14.7: Example—Digital Readiness Assessment of Service (Apparel Selling Service)

Consolidated List of Digital Weaknesses and Digital Strengths

Based on the digital readiness assessment of **Offerings**, the following are a few illustrative digital strengths and weaknesses captured under different categories. In real life, the assessment will be more comprehensive.

Digital Weaknesses

1. Customer experience
   a. Blockchain smart contracts (smart code) are not leveraged to minimize intermediary efforts while offering service.
   b. Customer help desk activities are not yet managed by virtual assistants/chatbots.
   c. Customer service representatives are not trained to use all the relevant digital tools and techniques to offer high-quality services.
   d. There is no customer journey map for the service/transaction to design experience at each interaction point.
2. Operational efficiency
   a. The field staff is not offered real-time information/insight to facilitate faster troubleshooting and decision-making.
   b. A robust digital supply chain system is not in place to avoid damage to goods during the transition.
3. Product design
   a. Digital twins (simulation tools) are not used to predict the performance of the product in different situations.
   b. Moments of customer experiences are not integrated while building the product.
   c. The product does not have integrated cybersecurity features.
4. Product management
   a. Not consistent in co-innovating digital features of the product with customers.
   b. Only a few employees take risks in developing products suitable for digital business.
5. Product operations
   a. Digitalization has not yet reduced the maintenance costs of the product.

b. Product is not optimal enough to reduce the resource utilization (people, machines, fuel, etc.).
c. Products are not efficient for optimal consumption of power/energy during operations.
d. Remote monitoring and controlling features are not available with the product.

6. Service design
   a. Service design has not factored any features that reduce the negative impact on the environment.
   b. Service has limited features for customer's safety, security and privacy.

Digital Strengths

1. Customer experience
   a. Omnichannel service is available, which makes sure that data captured in one channel can be accessed by many other channels so that there is continuity in the service rather than collecting the same information again (overprocessing).
2. Product operations
   a. Digitalization has reduced the maintenance costs of the product.
3. Service management
   a. Continuously monitor market changes to make the service digital-ready.
   b. There is a culture of collaborative frugal innovation to generate uncommon ideas for designing a new service.

In the above example, seven categories of digital weaknesses and three categories of digital strengths have been identified for two business building blocks Product and Service (Offerings). The other 10 business building blocks must be assessed similarly for digital readiness as per the business priorities.

Identify Problems, Possibilities and Future Digital State

CEO Stephenson and his team were now quite comfortable as they had gotten greater clarity on digital strengths and weaknesses of their offerings. They felt more confident about identifying various digital initiatives which would add tremendous value to Green Stores.

Next, Rao suggested that they derive the problems and possibilities from their digital weaknesses and digital strengths. Resolving the problems and taking advantage of possibilities using digital technologies would differentiate digital leaders from followers.

He provided quite a simple framework to capture the problems and possibilities based on digital weaknesses and digital strengths.

Business Problem Definition (Fix Digital Weaknesses)

In the current digital state, the digital weaknesses of Green Stores in customer experience area are the following:

1. Blockchain smart contracts (smart code) are not leveraged to minimize intermediary efforts while offering service.
2. Customer help desk activities are not yet managed by virtual assistants/chatbots.
3. Customer service representatives are not trained to use all the relevant digital tools and techniques to offer high-quality services.
4. There is no customer journey map for the service/transaction to design experience at each interaction point.

Based on digital weaknesses mentioned above, the business problem may be drafted as:

1. Customer journey maps are not available for our services/transactions. We have been failing to understand the customer pain points at each interaction/touchpoint. Hence, we are not able to provide unforgettable moments of experience to customers.

2. Our help desk is manually controlled. We do not have chatbots and robo coders to offer help to customers in real time. Our competition has been doing a great job in this area.
3. Our service representatives need to hone their skills in engaging and managing customers using various digital tools. In a few cases, our service representatives have to wait for senior management's approval to proceed which causes delays and customer dissatisfaction. We have not yet explored using blockchain smart contracts to apply business rules for automatic approvals.

The negative outcomes of the current situation were

1. customer attrition,
2. delays in providing complaints resolution and
3. high cost of acquiring a customer as compared to the competition.

Business Possibility Definition (Leverage Digital Strengths)

In the current digital state, the digital strength of Green Stores lay in the reduction of maintenance cost of products due to digitalization.

The various possibilities identified based on digital strength to further reduce the maintenance cost of the product are the following:

1. At present, a considerable amount of data came from their product through sensors, and by using a few AI algorithms they could predict the potential problems and proactively fix them to minimize the cost of maintenance.
2. They could build a digital twin for their product and enhance the ML algorithms to get automatic system recommendations (with no human interventions) for effective preventive maintenance.

3. By implementing digital technologies, performance could be monitored continuously and outliers could be identified proactively. The digital twin would provide a sophisticated product diagnosis system. However, remotely fixing the problem without a personnel visit was of critical importance.
4. It would be challenging to introduce remote troubleshooting for all troubles and problems. Hence, the remote troubleshooting of the product must be started by automatically upgrading the embedded software in the form of patches to fix related problems. In future, remote troubleshooting could be applied even for the mechanical components of the product.

Rao concluded the session on the **assessment** phase by sharing a pictorial representation of the assessment phase.

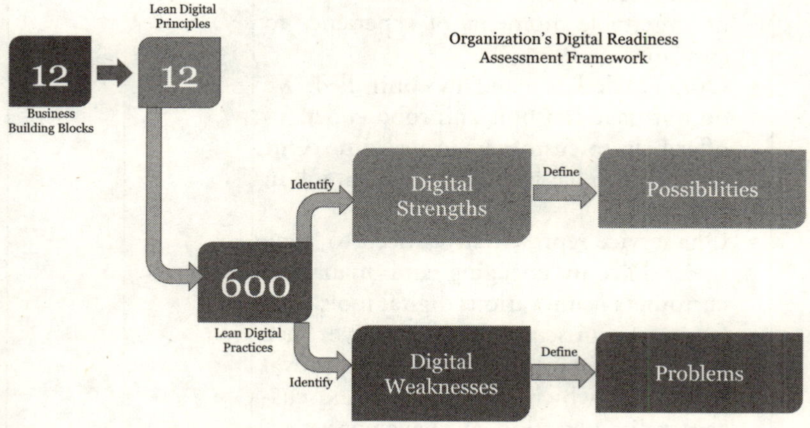

Source: Author

Figure 14.10: Summary of Digital Assessment Phase

The session with Rao on the identification of problems and possibilities had been enlightening for Stephenson and his team. They knew that after assessing Green Stores' digital strengths and weaknesses, they would be able to implement a comprehensive list of detailed lean digital practices for their organization. This would help them make Green Stores purely digital.

Problem-and-Possibility Catalogue and Prioritization

Next, Rao advised Stephenson and the team to categorize all the problems and possibilities as per the organization's strategy and priority. This is the problem-and-possibility (P&P) catalogue. The final step in the **Assess** phase is the prioritization of the problems and possibilities to design digital solutions (**future digital state**) in the next phase of lean digital execution (**Innovate**).

S. No.	Description	Category	Priority
1	Based on digital weaknesses, the problem definition may be drafted as: • Customer journey maps are not available for our services/transactions. We have been failing to understand the customer pain points at each interaction/touch-point. Hence, we are not able to provide unforgettable moments of experience to customers. • Our help desk is manually controlled. We do not have chatbots and robo coders to offer help to customers in real time. Our competition has been doing a great job in this area. • Our service representatives need to hone their skills in engaging and managing customers using various digital tools. In a few cases, our service representatives have to wait for senior management's approval to proceed which causes delays and customer dissatisfaction. We have not yet explored using blockchain smart contracts to apply business rules for automatic approvals. The negative outcomes of the current situation were • customer attrition, • delays in providing complaints resolution and • high cost of acquiring a customer as compared to the competition.	Problem	High

2	The various possibilities identified on the basis of digital strengths to further reduce the maintenance cost of the product are mentioned below: • At present, a considerable amount of data is coming from our product through sensors, and by using a few AI algorithms we can predict the potential problems and proactively fix the problems to minimize the cost of maintenance. • To get into the next level, we could build a digital twin for our product and enhance the ML algorithms to get automatic system recommendations (with no human interventions) for effective preventive maintenance. • We will implement digital technologies to monitor the performance continuously and identify outliers proactively. The digital twin provides a sophisticated product diagnosis system. However, remotely fixing the problem without a personnel visit is of critical importance. • It would be challenging to introduce remote troubleshooting for all troubles and problems. Hence, we will start remote troubleshooting of our product by automatically upgrading the embedded software in the form of patches to fix related problems. In future, remote troubleshooting could be applied even for mechanical components of our product.	Possibility	High
3		Problem	Low
4		Problem	Medium
5		Problem	Medium
100+		Possibility	Low

Source: Author

Table 14.8: P&P Catalogue and Prioritization

Digitalization Roadmap Phase 3: Innovate

Driving digital initiatives, without designing the future digital state is directionless and does not yield expected results. In the **innovation** phase, organizations use various ideation techniques like brainstorming, crowdsourcing and Six Thinking Hats to devise the best-in-class problem-and-possibility statement.

Business innovations focus on bringing changes or newness in business models, policies, processes, products, services, methods, structure and talent to provide exceptional value to customers and to bring operational excellence.

Digital innovations are different from traditional innovations in many ways. Some innovations in business are made possible only with digital technologies. For example, digital technologies have made the introduction of new customer channels, like social media, bot-driven help desk possible. We could add remote monitoring of the performance of a product only by using the internet of things (IoT) and analytics.

Hence, digital innovation means introducing changes and newness in the businesses building blocks using digital technologies that add tremendous value to stakeholders.

Rao used Figure 14.11 to explain the digital innovation life cycle.

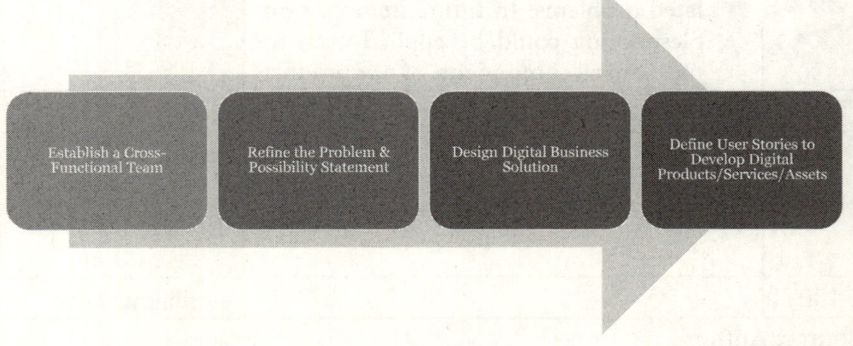

Source: Author

Figure 14.11: Digital Business Innovation Life Cycle

Establish a Cross-Functional Team

Rao asked Stephenson to form a cross-functional team to innovate solutions to resolve the problems and to leverage the possibilities. The cross-functional team may include specialists from various departments like sales, marketing, strategy, business process, HR, legal, anthropology, digital technology. The objective of forming this team was to understand the problems and possibilities from different perspectives to design best-in-class problem-and-possibility statement.

Refine the Problem-and-Possibility Statement

The problem-and-possibility statements derived from the digital readiness assessment could be further refined by directly meeting the customers and other stakeholders to understand their emotional needs better.

Stephenson's team spent quality time, participating in focused group discussions, virtual round table discussions, interviews and observation meetings to improve the quality of problem-and-possibility statements in terms of completeness, correctness, unambiguity and consistency.

A low-quality problem or possibility statement leads to inferior solutions. The cross-functional team reviews and approves the problem-and-possibility statements.

Design Problem-and-Possibility Statement

Stephenson and the team identified hundreds of problem-and-possibility statements for which Green Stores needed to design best-in-class problem-and-possibility statement. Rao provided them with an excellent framework to create high-quality solutions. He explained this framework for one of the problem statements. The first was to conduct a joint business value discovery workshop with the cross-functional team.

The cross-functional team collaboratively brainstormed the solution while also depending on open innovation platforms (crowdsourcing) to get uncommon solutions from external specialists. Rao guided the cross-functional team to capture many alternate solutions to one problem statement. The sample problem-and-possibility statement is explained here.

Problem Statement	Based on digital weaknesses, the problem definition may be drafted as: • Customer journey maps are not available for our services/transactions. We have been failing to understand the customer pain points at each interaction/touchpoint. Hence, we are not able to provide unforgettable moments of experience to customers. • Our help desk is manually controlled. We do not have chatbots and robo coders to offer help to customers in real time. Our competition has been doing a great job in this area. • Our service representatives need to hone their skills in engaging and managing customers using various digital tools. In a few cases, our service representatives have to wait for senior management's approval to proceed which causes delays and customer dissatisfaction. We have not yet explored using blockchain smart contracts to apply business rules for automatic approvals. The negative outcomes of the current situation were • customer attrition, • delays in providing complaints resolution and • high cost of acquiring a customer as compared to the competition.

Solution Design (what changes to be done in the business)	Business Model	Product	Service
	Identify new customer channels like, • social media, • intelligent website, • chatbots and • robo help desk.	IoT-enabled apparel could track heart rate and temperature.	The mannequin in the store automatically provides product details over the mobile phone using Bluetooth when the customer is near it. Robo coder to suggest attire as per current trends. AR and VR based virtual trials without dressing rooms.
	People	**Policies**	**Process**
	Enhance customer engagement skills. Learn virtual conversation etiquette	Encourage online purchasing and decrease the number of physical stores which reduce the cost of operations. Offer virtual superior buying experience to customers.	Create a customer journey map, to identify touchpoints and design unforgettable moments of experience at each interaction point using digital technologies.
	Structure	**Software Applications**	**Technology Infrastructure**
	Introduce a new role called virtual conversation specialist.	Design and develop software applications (mobile and web app or chatbot) using AI, AR, VR, analytics and blockchain to implement the proposed changes.	Size the hardware required and leverage AWS or MS Azure cloud infrastructure.

Future Digital State	• Unforgettable moments of experience to customer at each interaction point using digital technologies • Well-trained customer service representatives who use communication and collaboration tools effectively • A better and convenient buying experience for customers both online and at the stores using various digital channels and devices Following outcomes are expected to improve after solving this problem: • Customer retention • Real-time communication and engagement with the customer • Quicker customer complaint resolution • Reduction in the cost of acquiring customers
Complexity of Solution	Medium
Priority	High

Source: Author

Table 14.9: Problem-and-Possibility Statement Design Template

Define User Stories

Rao emphasised the importance of identifying user stories from already defined problem-and-possibility statement. User stories are small manageable chunks of features which will be implemented to build software applications, software products and platforms that are required to manage digital businesses. These user stories are significant inputs in the development of digital products, assets and services. He advised the Green Stores team to collate user stories and prioritize them as per the market and customer needs.

A few hundred user stories may be required to digitalize the whole organization.

User stories are grouped as per the business priorities. Selection of a few user stories for software development is called a sprint. Upon development of the user stories into software, which becomes useful as functional features that deliver business benefits. The duration of each sprint development is around 2–4 weeks. That means the software will be released every 2–4 weeks to support the dynamic changes in businesses and help the companies release new product features, services and refined business processes.

Unless the problem-and-possibility statement become working software applications, products and platforms, the real business value will not be unlocked. Rao proposed that the Green Stores' team use agile and Scaled Agile Framework (SAFe) methods to manage their user stories effectively and efficiently (see Table 14.10). Agile SAFe framework helps in implementing agile practices throughout the organization at team, programme and portfolio level. Agile SAFe framework is a body of knowledge that includes structured guidance on roles, responsibilities, and how to plan and manage the team, programmes and portfolio of projects using an agile methodology. It promotes collaboration and alignment among the multiple teams working on projects. A group of user stories is called epic and represents a core requirement of an organization. Table 14.10 has the list of user stories defined based on problem-and-possibility statement depicted in table 14.9.

Epic ID	Epic Name	User Story #	As a (Role)	I want (Something)	So That (Benefits)	Priority
Epic 001	Customer engagement	1	As a consumer	I want to get the job done using multiple digital channels.	So that, I get the services from anywhere, anytime using any device.	Phase 1
		2	As a consumer	I want to wear IoT-enabled apparel.	So that, I can track my heart rate and body temperature in real time.	Phase 1
		3	As a consumer	I want intelligent mannequins at the physical stores to automatically get product details on my mobile phone.	So that, I do not waste my time searching for the product and other details.	Phase 1
		4	As a consumer	I want to get recommendations using robo coder for the right attire.	So that, I can purchase the right attire according to my comfort and look.	Phase 2
		5	As a consumer	I want to use the virtual trials option using AR and VR.	So that, I can select the right attire in the store or online.	Phase 2
		6	As a consumer	I want to get unforgettable moments of experience at each interaction point.	So that, I enjoy my shopping.	Phase 1

Epic 002	Sales and marketing	7	As a sales manager	I want to motivate consumers using various digital tools to purchase the clothes online.	So that, we reduce the maintenance cost of physical stores.	Phase 1
		8	As a sales manager	I want to enhance customer engagement skills and virtual conversations etiquette using a digital learning platform.	So that, I can offer a superior experience to consumers	Phase 1

Source: Author

Table 14.10: Define Problem-and-Possibility Statement User Stories

Rao summarized the **innovation** phase with the help of Figure 14.12.

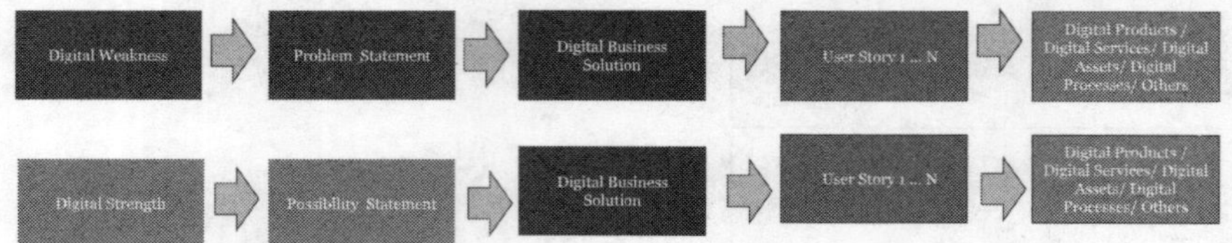

Source: Author

Figure 14.12: Outputs of Assessment and Innovation Phase

The user stories defined in Table 14.10 are for digital products, digital services and digital assets. But in practice, user stories will be defined to create new business models, new business processes, new workplaces, etc. In the rest of this chapter, reference will be given for digital products, digital services and digital assets for illustrative purpose. The assessment and innovation phases are very critical in the digital transformation journey. In Table 14.11, a summary of outputs of both the phases has been given for illustrative purposes. It includes digital strengths, digital weaknesses, problems, possibilities, problem-and-possibility statement and user stories. Digital transformation journey without effective assessment and innovation phases would fail.

Digital Readiness Assessment of Organization	Avg. Qty	Remarks
No. of digital strengths and digital weaknesses as per digital readiness assessment	500	Based on lean digital practices
No. of identified problems and possibilities derived from digital weaknesses and strengths	250	Group related problems and possibilities together
No. of problem-and-possibility statements designed	250	Brainstorming, crowdsourcing and other techniques
No. of user stories defined	**2000**	On an average, eight user stories for each problem-and-possibility statement

Source: Author

Table 14.11: Outputs of Digital Transformation Assessment and Innovation Phases

The critical challenge is how many of the 2,000 user stories will be implemented in each of the three years of digital transformation. How many user stories will be excluded as per business priorities? Stephenson and his team now had the complete picture of the complexity of the digital transformation programme, and they were convinced that prioritizing user stories would enable incremental digital value for all stakeholders of Green Stores.

Digitalization Roadmap Phase 4: Experiment

Stephenson was confident that the user stories identified by his team were relevant and aligned with the organization's digital mission. He felt that the digital readiness assessment done by using lean digital quotient (LDQ) finder tool had helped them in identifying the right user stories.

He knew that prioritizing these user stories was critical considering the budget and other resource constraints. He requested Rao to help Green Stores prioritize the user stories to develop digital software applications and experiment the same in selective markets.

Rao began by introducing the experimentation life cycle as given in Figure 14.13.

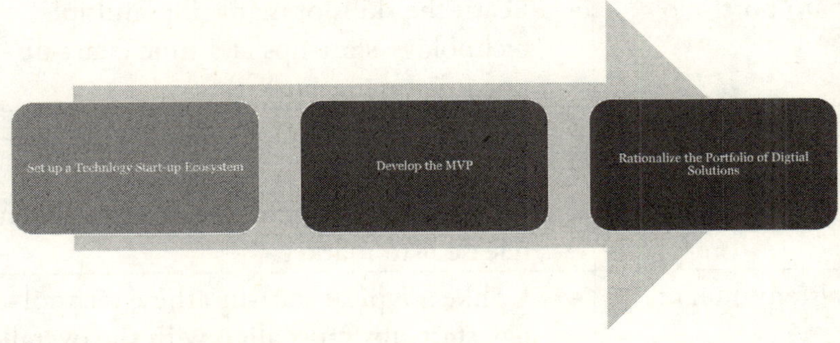

Source: Author

Figure 14.13: Digital Business Experimentation Life Cycle

Set up a Technology Start-up Ecosystem

Getting value from digital innovations at the speed of thought is essential and hence, it is imperative to set up a technology start-up ecosystem within the organization. Patrick, the senior technology architect, asked Rao what did he mean by setting up a technology start-up ecosystem. Rao replied that it was a new operating model, new mindset, uncertainty management, perseverance and hope. He explained how an organization could set up a technology start-up ecosystem, as mentioned in Table 14.12.

Elements of Technology Start-Up	Description
Funding	• Set up funding for digital innovations
Technology start-ups	• Establish a few technology start-ups within the organization to implement the prioritized user stories
Mindset	• Manage uncertainty with limited resources; leaders must acquire a mindset of "start small, fail fast and scale soon"
Operating model	• Business units act like angel investors or venture capitalists • Business units stop funding if the technology start-up is not performing well
Start-up portfolio	• Learn the skills of managing multiple technology start-ups at a time (start-up portfolio management)
Pivoting	• Pivoting is the critical tenet of any start-up; hence, each start-up must focus on the MVP, experiment, improve and then release it to market
Ecosystem integration	• Unlike a typical start-up, these technology start-ups must align with the overall organization's digital mission and build the solutions that could be seamlessly integrated with existing systems

Business focus	• Focus beyond product metrics like cost, time, quality; start-ups must focus on creating new revenue streams, acquiring new customers, expanding the current business to new markets and others

Source: Author

Table 14.12: Technology Start-up Ecosystem

Develop Minimum Viable Product

Considering the budget constraints and narrow profit margins, effective utilization of resources is very crucial. This means lean methods needed to be implemented to eliminate non-value-added activities. Rajan, the senior development manager, shared his experience of wasting a lot of money on software features development as the consumers do not use more than 30% of the features. This meant a lot of money was being wasted on unwanted features. Rao proposed to use the minimum viable product (MVP) method as explained in Table 14.13.

Benefits of MVP	Description
Manage uncertainty	• The MVP model is suitable for the dynamic nature of the digital age customer's needs, wants and expectations. They need continuous newness. • The MVP is a digital product/service/asset that is developed based on the user stories identified. • Developing digital solutions with too many features would be ineffective. Consequently, only **just enough features** must be built, in the form of **MVP** with constant feedback from the customer and multiple pivots to adapt to the changing needs of customers.

Treat customer as partner	• Customer development is carried out parallelly throughout the solution development life cycle. Managing the customer as a solution development partner minimizes a lot of risks. This means, at each phase of the digital solution development, customers must be involved to get timely feedback.
Leapfrog the competition	• With just enough features, it is possible to release the new product or service to the market ahead of the competition.
Test the market	• During MVP development, the involvement of a few selected customers helps in improving product quality and value. • After releasing the MVP to the market within the shortest possible time, getting comprehensive feedback from the representative customer segments helps in pivoting digital products and digital services that are highly valuable to consumers.

Source: Author

Table 14.13: Benefits of the MVP Model

Rationalize the Portfolio of Digital Solutions

Experimentation is an iterative process, and it is especially important to closely review the performance of each technology start-up within the organization. Each technology start-up works on digital products, services and assets (MVPs) in a very cohesive manner. Rao recommended that there must be stringent financial discipline in the technology start-ups. Regular reviews must be done to review the progress of each MVP using the following parameters given in Figure 14.14.

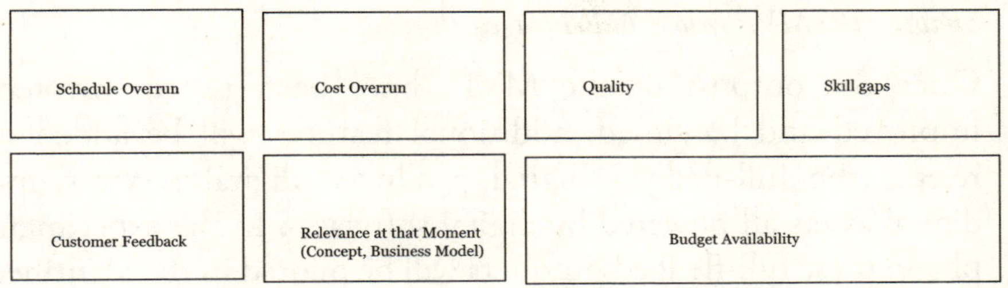

Source: Author

Figure 14.14: Rationalization Parameters for MVPs

In the journey of digital transformation, the organization may need to manage multiple technology start-ups at a time. Hence, it is essential to have insight into which start-up is doing well and which is not. A dashboard with all the key performance indicators (KPIs) of technology start-ups helps the senior management to make the right decisions at the right time to improve the technology start-up performance. A few examples are given in Table 14.14.

S. No.	Technology Start-up Name	Digital Products/ Services/Assets (MVP)	Current Perfor-mance	Decision
1	Connected store	• Virtual trials • Robo coders • In-store experience	**Red**	Continue
2	The companion	• Digital learning platform	**Green**	Continue
3	Intelligent apparel	• IoT-enabled clothes	**Red**	Suspend
4			**Amber**	Hold
5	Intelligent apparel		**Red**	Suspend
6	Intelligent apparel		**Red**	Continue
7	Intelligent apparel		**Red**	Hold

Source: Author

Table 14.14: Dashboard for Portfolio of Technology Start-ups

Shaping the MVP into a Full-Fledged Product

Once the outputs of the MVPs have been tested, pivoted, improved and baselined, additional features will be added to release the full-fledged digital products, digital services and digital assets all powered by digital software . In the experiment phase, these full-fledged products will be piloted in the identified areas/segments and prepared for institutionalization in the whole organization.

Digitalization Roadmap Phase 5: Institutionalize

Once the **experiment** phase is over, all the successful experimentations must move to operations. Operationalizing digital products, digital services and digital assets must be done in a very systematic manner to get the return on investments.

Introducing any new system requires considerable change in the culture, and hence managing change becomes a humongous task. Most of the products, services and assets, which are successful during the experiment phase, fail due to ineffective operationalization.

Stephenson asked Rao to share his experience on the reasons for the failure of operationalizing the digital products/services/assets. Table 14.15 lists the top five reasons shared by Rao.

Reasons for Failure	Description
Lack of shared vision	Employees do not understand the bigger picture; management fails to communicate the purpose of the new system and value to employees and other stakeholders
Ineffective execution	Lack of resources and a clear plan of action to operationalize the outputs of experiment phase (digital products, digital services, digital assets, etc.).

Operational overhead	Employees are busy with day-to-day tasks and do not have time to implement new products/services/assets; high manual work without automation creates new process overhead
Lack of recognition	No rewards and recognition for the additional effort being put in by employees
Talent gap	Skill gaps, lack of training and wrong understanding of the new system

Source: Author

Table 14.15: Reasons for Failure in Operationalization of Problem-and-Possibility Statement

Rao proposed a three-step process for the effective institutionalization of digital products, services and other outputs of MVPs like business models, business processes, etc. The three-step process, given in Figure 14.15, creates a positive culture, energy and common purpose for all the employees while the efforts put in by them are synergized.

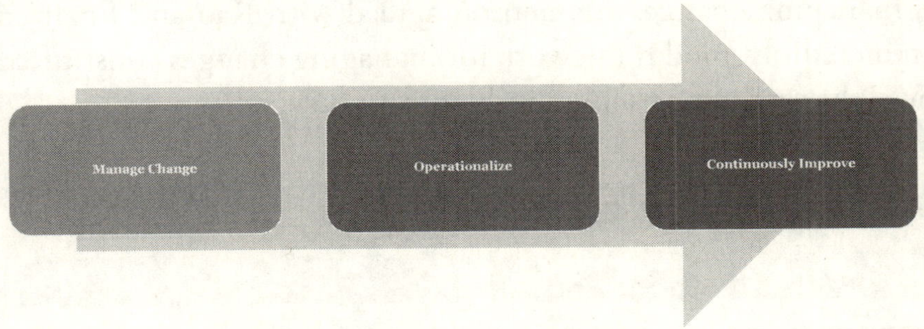

Source: Author

Figure 14.15: Three Steps of Institutionalization of Problem-and-Possibility Statement

Manage Change

Introducing a new product, service, process or method is not an easy task. Rao shared how most of the organizations fail to manage the change.

Usually, in any organization:

1. Some people embrace the change without much effort. They make up 20% of the people at the organization and do not hesitate to put 20% of additional effort apart from their regular work.
2. Another 20% never embrace the change, and they exist to find problems and loopholes.
3. The remaining 60% are opportunistic, and they swing their efforts based on the progress and outcomes of the change.

Most people hesitate to accept change due to the following reasons:

1. **Insecurity (job, position, etc.)**
2. **Resistance to coming out of their comfort zone**
3. **Lack of recognition or rewards for the additional effort being put in by them**
4. **Fear of failure in the new environment**
5. **Political reasons and conflicts**

The most challenging part of the digital transformation journey is managing change. Stephenson agreed with Rao and finalized an incredibly good framework for managing changes constituted by eight catalysts as shown in Figure 14.16.

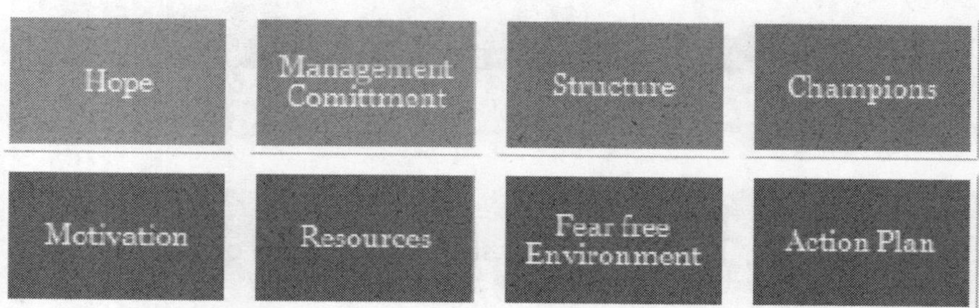

Source: Author

Figure 14.16: Change Management Catalysts

Rao explained the framework with the example of introducing virtual trials (digital service) which was very new to Green Stores.

Catalyst for Change	Description
Hope	Create hope for employees by sharing inspirational stories. "We will have more customers to buy from us due to virtual trials. Customers can try various apparels that suit their body and taste. Virtual trials can be done at home using a laptop or at our physical stores. This new service would improve our customer retention, revenue and profit. Hence, employees will get monetary benefits and career progression opportunities."
Management commitment	Get the management involved, their passionate commitment will create a positive work environment and motivate employees to embrace change.
Structure	For example, introduce a new role like the technical support executive to offer support to customers who use virtual trials.
Champions	Induct change management champions throughout the organization. These champions will train the employees for the new virtual trials system and will be available to provide regular mentoring and guidance. These champions will continuously motivate concerned employees.
Motivation	Introduce a transparent system that encourages employees to embrace the new system within a shorter time. For example, those employees who motivate customers to use virtual trials at home and minimize the visits to physical stores get high recognition. One size does not fit all. Personalized "motivation triggers" must be introduced for employees to be part of the change: • Provide an opportunity for the employee for self-expression within the organization (interviews and guest speeches). • Provide an opportunity for the employee to realize his/her career dream. • Provide an opportunity for the employee to take up a complex challenge. • Elevate the image of the employee. • Provide monetary benefits.

Resources	Provide the required budget, internet connectivity and video conferencing infrastructure to connect with customers, and a physical place for virtual trials in the stores.
Fear-free environment	Create a fear-free environment. Encourage employees to openly share feedback and reward the employees who provide feedback on the new system. Encourage employees to share risks and current issues without any hesitation and fear.
Action plan	Share an action plan to institutionalize the virtual trials system throughout the organization with clear milestones. This helps in getting the bigger picture to employees.

Source: Author

Table 14.16: Catalysts for Bringing Change

It would take Stephenson a few weeks to identify the change management team and create a detailed plan of action as per the framework proposed by Rao. Stephenson decided that he would give the go-ahead to operationalize the new digital products/services/assets only once he was confident that Green Stores was ready to embrace change.

Operationalize

While explaining the second step of institutionalization, Rao emphasized the difference between operations and projects. In the experiment phase, the work is executed in the form of time-bound projects. Operations involve nothing but performing recurring tasks effectively and efficiently. Projects are executed within finite timelines and once completed, the outputs of the projects are operationalized. Here are a few examples of project outputs of Green Stores derived during the experiment phase:

1. Virtual trials
2. Robo coder
3. In-store experience
4. Digital learning platform
5. IoT-enabled clothes

Operationalizing these project outputs was essential to get the desired benefits and Rao shared a simple framework, given in Figure 14.17, to do so.

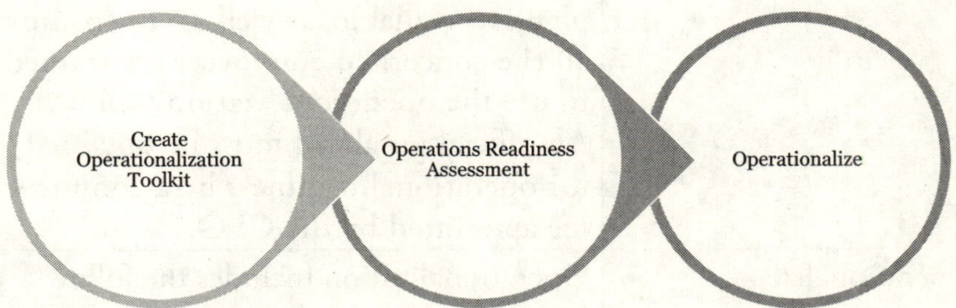

Source: Author

Figure 14.17: Operationalizing New Digital Products, Services and Assets

Rao explained the operationalization framework for the virtual trials project, (MVP) as described in Table 14.17, to give better clarity to Stephenson and the team.

Phase	Description
Create operationalization toolkit	The operationalization toolkit is prepared based on the experience gained from the many pilot projects done during the experiment phase. The operationalization kit for the virtual trials project included: • training material, • product demo, • product specifications, • forms, templates, • virtual trials process handbooks, • digital tools, • help manual, • troubleshooting knowledge base, • best practices and • lessons learned. The toolkit brings standardization and a shared understanding among the employees. It also provides ready-made solutions to the commonly encountered problems by employees.

Operations readiness assessment	An operations readiness assessment is mandatory to make sure that the required infrastructure, people, internet connectivity, budget, skills and action plan are available, as well as, to ensure all the concerned employees are trained to use the operationalization toolkit. An official go-ahead must be obtained for operational readiness by a committee appointed by the CEO.
Operationalize	Operationalization includes the following: • **Knowledge transfer:** Knowledge is transferred from the project team to the operations team using the operationalization toolkit and other training material. • **Secondary support:** The project team helps in driving the operations while the operations team play a support role. This is like on-job training for the operations team. • **Primary support:** The operations team plays a central role in the day-to-day operations and takes support from the project team as and when required. • **Hand over and takeover:** The project team exits from the operations and hands over all the artefacts to the operations team.

Source: Author

Table 14.17: Operationalization Framework for New Digital Initiatives

Continuously Improve

Rao further emphasized the point that it is not enough to just operationalize a product/service/asset. The performance of the same must be measured to verify whether the intended benefits have been achieved or not.

For the virtual trials service, Rao shared a few metrics, given in Table 14.18, which Green Stores could monitor on a regular basis.

Key Measurement	Intended Benefit	Actual Result	Root Cause
Customers using virtual trials	Increase	**Increase**	—
Revenue from the customers visiting physical stores	Increase	**Decrease**	After the virtual trial feature was introduced at the physical stores, there was an increase in customer footprint as customers enjoyed the experience. But there was no proportional increase in sales as all the visitors are not buying the clothes as their primary motive to visit the store is to enjoy virtual trials rather than buy the clothes.
Online purchase after virtual trials usage at home	Increase	**Increase**	—
Customers as brand ambassadors	Increase	**No Improvement**	The social media platform has not been utilized to its full potential to promote customers as brand ambassadors.

Customer retention	Increase	**Increase** ⬆	—
Repeat business	Increase	**Increase** ⬆	—

Source: Author

Table 14.18: Benefits Realization after Operationalization

Rao suggested that Green Stores establish a crowdsourcing platform to co-innovate new ideas and improvement actions to eliminate the root causes permanently. Active participation of customers and employees in the elimination of the root causes of poor performance was going to be the next big game changer for Green Stores.

Root Cause	Idea 1	Idea 2	Idea 3
After the virtual trial feature was introduced at the physical stores, there was an increase in customer footprint as customers enjoyed the experience. But there was no proportional increase in sales as all the visitors are not buying the clothes as their primary motive to visit the stores is to enjoy virtual trials rather than buy the clothes.	Exclusive discounts, who use virtual trials feature at home and visit physical store to purchase the selected cloth using virtual trials at home.	Introduce packaged clothes with good discounts so that on each visit, the customer purchases more items at the physical store.	Build many lean physical stores with less human intervention and more robot-based operations.

The social media platform has not been utilized to its full potential to promote customers as brand ambassadors.	Introduce innovative schemes to build thousands of customers as brand ambassadors on social media.	Provide a few clothes as gifts to customers who become the brand ambassador of Green Stores.	Introduce gamification technology that helps in motivating customers to become brand ambassadors.

Source: Author

Table 14.19: Improvement Actions for Operational Benefits

Rao cautioned that due to crowdsourcing, they might receive too many ideas. A few of them could be unique and cost-effective while others could be costly and yield results in the long term. Hence, it was important to establish a good prioritization engine and select the best ideas to eliminate the root causes of inadequate performance of the new digital products, services and assets permanently.

PART 5

Digital Business Solution Store

CHAPTER 15
Digital Business Solution Details

- Open Digital Innovation as a Service
- CX eWorker for Contact Center
- CX eWorker for Automated Order Processing
- Service Orchestration, Monitoring and Optimization of Distributed Frontline Workforce
- Real-Time Site and Asset Monitoring
- Real-Time Customer Experience, Engagement and Service Delivery Management
- HaaS: Happiness as a Service
- CQaaS: Construction Quality as a Service
- Lean Digital Quotient Finder
- Smart Vehicle Tracking System
- Connected E-Rickshaw
- Real-Time Tracking of Workforce
- Smarter, Resilient and Reliable Supply Chains
- OM: Order of Magnitude
- FoFo™: Folding Foropter™
- Meeting Wall
- ZeroCode
- GOKADDAL: Digital Solutions Exchange Cloud Platform
- SWAYAM: Anytime Health Monitoring
- VECAAS: Vertical Cloud as a Service

CHAPTER 15
Digital Business Solution Details

All solutions mentioned in this chapter are provided by the respective organizations. Author and publisher are not responsible for any IP/Copyright related issues or any other errors.

Open Digital Innovation as a Service

Problem Description
• Frequent changes in the needs and wants of digital age customers • Shorter product life cycles as customers need continuous newness • Outdated business models, products, services and operating models without leveraging technologies • Major gaps in digital skills and poor business innovations (closed innovation) • Wrong investment decisions and low ROI on digital initiatives

<table>
<tr><th colspan="2">Solution Description</th></tr>
<tr><td>Open digital innovation as a service (ODIaas) is a digital software platform, which enables crowdsourcing of digital business solutions from distributed solution providers for the various problems published by solution seekers.
<ul>
<li>Create a problem bank and curate business problems</li>
<li>Break the barriers and enrol globally distributed solution specialists (internal and external)</li>
<li>Matchmaking between solution seekers and solution providers</li>
<li>Receive many alternate solutions for one business problem</li>
<li>Voting and quantitative assessment of solutions using the Delphi method</li>
<li>Prioritize and approve or decline solutions</li>
</ul></td>
<td>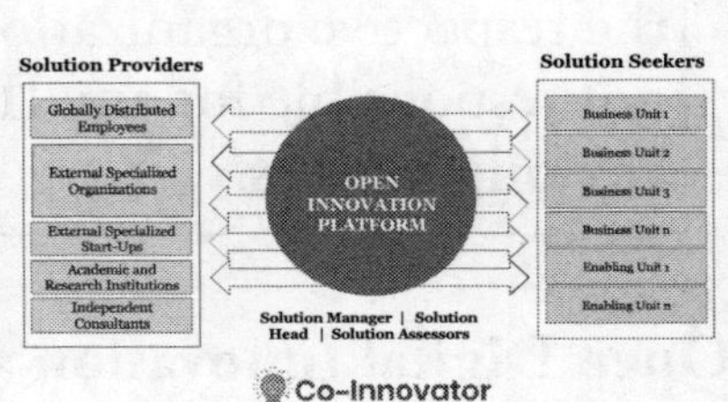

Open Digital Innovation Platform</td></tr>
</table>

<table>
<tr><td colspan="5">Digital Technologies Used for the Solution</td></tr>
<tr><td>Mobile</td><td>Social</td><td>Cloud</td><td>Analytics</td><td>AI</td></tr>
<tr><td colspan="5">Outcomes/Benefits of the Solution</td></tr>
<tr><td colspan="2">• One problem and many solutions from globally distributed solution experts (crowdsourcing)
• Investments in the right digital initiatives</td><td colspan="3">• Leapfrog competition with uncommon ideas
• Reduce 30% of the cost of designing digital solutions using open innovation</td></tr>
<tr><td colspan="2">Industry/Cross-Industry/ Horizontal</td><td colspan="3">Horizontal
Example: All industries</td></tr>
<tr><td colspan="2">Current Status of the Solution</td><td colspan="3">Production-ready</td></tr>
<tr><td colspan="2">No. of Case Studies Completed Using This Solution</td><td colspan="3">Ready to launch</td></tr>
<tr><td colspan="2"></td><td colspan="3"></td></tr>
<tr><td colspan="2">Name of the Company</td><td colspan="3">BT&BT and OSI Digital</td></tr>
<tr><td colspan="2">Website of the Company</td><td colspan="3">http://btbt.co.in/ and https://osidigital.com/</td></tr>
<tr><td colspan="2">Country of the Company</td><td colspan="3">USA and India</td></tr>
</table>

CX eWorker for Contact Center

<table>
<tr><th colspan="2">Problem Description</th></tr>
<tr><td colspan="2">Customer service teams across enterprises are staring at increased call volumes which means an additional workload on their current staff. The problem is further compounded because most teams are forced to work remotely and access to a lot of critical applications has been limited due to WFH.

Enterprises have to resort to reduced support hours to avoid agent burnout, or, in many cases, complete discontinuation of the access to support centres. All have resulted in the following issues:
• Increased handling time for customer service teams
• Increased average wait time per customer leaving them frustrated
• Average response time has gone up by over 90–95% in multiple cases
• Eventual loss of customer base due to dissatisfied customers, eroding brand loyalty</td></tr>
<tr><th colspan="2">Solution Description</th></tr>
<tr><td>An AI-powered contact centre solution for managing customer experience.
• Trained in six industries to perform inbound and outbound interactions
• Handles calls, emails, chats and social media
• Speaks 40+ languages
• Comes with 20+ enterprise CRM and help desk integrations and built-in robotic automation
• Cloud and on-premise options</td><td>CX eWorker Contact Centre</td></tr>
</table>

Digital Technologies Used for the Solution	
• NLP • RPA • APIs/digital integrations for multiple CRMs and helpdesk software • Voice-based automation across cloud telephony, Google contact centre AI and AWS	
Outcomes/Benefits of the Solution	
• Saves contact centre costs between 40–70% • Within a few months, improves the NPS and average handling times by upto 90%	• Reduced TAT from over 24 hours to under 5 minutes • Deflects over 65% of inbound calls to AI agent to improve agent productivity enabling them to focus on critical issues
Industry/Cross-Industry/Horizontal	Cross-industry Example: CPG, retail, banking, insurance, wealth management, airlines, hospitality, travel management, ecommerce, telecommunications, media and broadcasting
Current Status of the Solution	Production-ready
No. of Case Studies Completed Using This Solution	4 live customers across CPG, retail, non-profit agencies, government agencies and BFSI
Name of the Company	Techforce.ai
Website of the Company	https://www.techforce.ai
Country of the Company	USA and India

CX eWorker for Automated Order Processing

<table>
<tr><th colspan="2">Problem Description</th></tr>
<tr><td colspan="2">A large US manufacturing company with multiple plants across the US and B2B distribution model receives over 5,000 B2B orders via emails. Most of the sales orders include unstructured data, either as PDFs or scanned documents, making it difficult to pull information from them to be entered into ERP systems promptly.
The team has to rotate in eight-hour shifts to manually monitor, extract attachment and create order entry in ERP. This process is labour intensive and prone to human errors. This impacts cycle time and processing costs.</td></tr>
<tr><th colspan="2">Solution Description</th></tr>
<tr><td>eWorker leverages RPA to processes emails and extracts the purchase orders data and attachments. It processes the documents using an AI-powered computer vision by extracting unstructured data, including purchase order details. Then it sends this data to the low-code app that enables the business users to review any business exceptions using pre-configured business rules and workflow. Once approved by the business user (human-in-the-loop), the system again leverages RPA for automated data entry into the ERP system.</td><td>CX eWorker
Order Processing</td></tr>
</table>

Digital Technologies Used for the Solution	
• AI-powered computer vision • RPA • Integration with major ERP applications like SAP, Oracle's JD Edwards and Salesforce	
Outcomes/Benefits of the Solution	
• 90% faster straight-through processing of sales order • Over 70% reduction in manual effort leading to savings of over USD $200,000 in under 6 months	• Faster TAT for confirming orders leading to customer satisfaction
Industry/Cross-Industry/ Horizontal	Cross-industry Example: Manufacturing, CPG/retail, shipping, logistics, automotive and construction material
Current Status of the Solution	Production-ready
No. of Case Studies Completed Using This Solution	2 live customers across CPG and manufacturing
Name of the Company	Techforce.ai
Website of the Company	https://www.techforce.ai
Country of the Company	USA and India

Service Orchestration, Monitoring and Optimization of Distributed Frontline Workforce

Problem Description

- Opaque last mile people and processes
- No real-time information on task completion, productivity and performance
- Manual data flows and process flows
- Lack of accountability in managing distributed internal or third-party teams
- Lack of transparency and governance monitoring

Solution Description

- Unified platform to manage workforce and workflows
- Digitalization and automation of scheduled last mile processes and tasks
- Cognitive automation of dynamic issue management and resolution
- Expand real-time location system
- AI-enabled biometrics for attendance, presence, task completion and staff monitoring
- Real-time reporting and live dashboard for location, performance and issues
- GIS dashboard and reports; matrix and trends; time series and comparative analytics
- Comprehensive action and rules engine
- Multifactor AI algorithm-based service orchestration and optimization

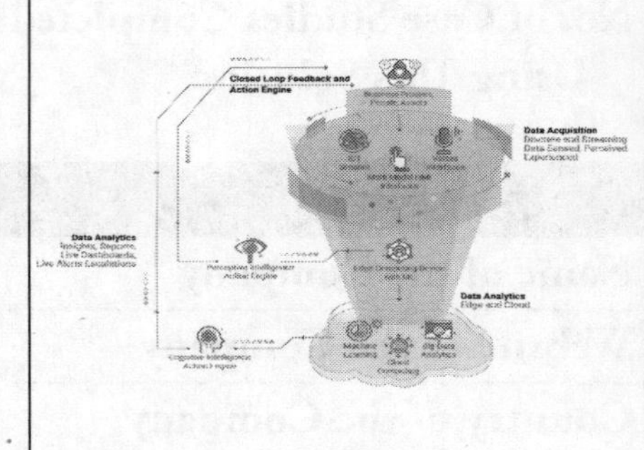

Digital Technologies Used for the Solution				
Mobile	IoT and RTLS	Cloud	Analytics	AI

Outcomes/Benefits of the Solution	
• Last mile challenge occurs across industries for managing last mile workforce • Unified platform enables end-to-end process automation	• Customizable to include client or industry-specific rules, processes and analytics requirements • Improve transparency and accountability • Improve 100% visibility • Improve 30% productivity (or losses due to penalties)
Industry/Cross-Industry/ Horizontal	Horizontal Example: Hospitals, government, transportation, logistics, hospitality, distribution, sales and service, facilities management and real estate
Current Status of the Solution	Production-ready; customizable for client
No. of Case Studies Completed Using This Solution	Ready to launch
Name of the Company	Gaia
Website of the Company	www.gaia.in
Country of the Company	India and USA

Real-Time Site and Asset Monitoring

<table>
<tr><th colspan="2">Problem Description</th></tr>
<tr><td colspan="2">
• Opaque last mile site performance
• No real-time information, onsite ambient condition or asset condition at distributed sites
• Manual data flows and process flows
• Lack of accountability in managing distributed internal or third-party teams
</td></tr>
<tr><th colspan="2">Solution Description</th></tr>
<tr><td>
• Unified platform to manage multiple assets, metrics and sites
• Digitalization and automation of last mile sensor and use-based data acquisition
• Edge and cloud analytics
• Real-time reporting and live dashboard for location, performance and issues
• GIS dashboard and reports; matrix and trends; time series and comparative analytics
• Comprehensive action and rules and alert-response management
• Use data-driven insights and alerts in real time to “close the loop” at the next level action, response or service orchestration by triggering automated workflows
</td><td>
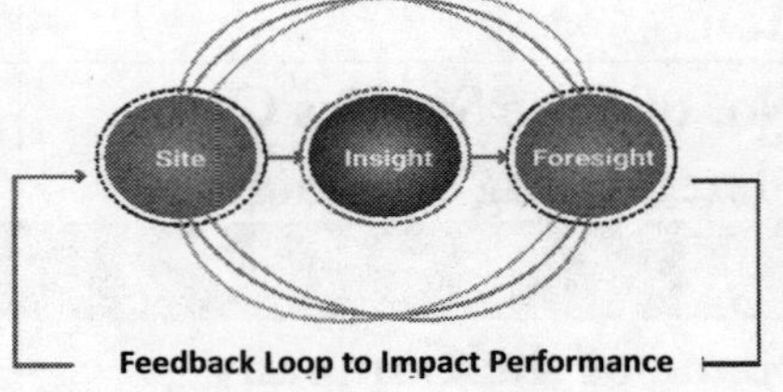

</td></tr>
</table>

Digital Technologies Used for the Solution				
Mobile	IoT	Cloud	Analyt-ics	AI/ML

Outcomes/Benefits of the Solution	
• Improve visibility into last mile sites and spaces • Unified platform enables end-to-end digitalization, remote monitoring and management of sites • Planned and preventive monitoring and optimization of assets and site ambience	• Customizable to include client or industry-specific rules, processes and analytics requirements • Improve transparency and accountability • Improve 30% efficiency (or cost optimization) • Improve 100% visibility into real-time site or asset level condition and status breaches; enable condition-based maintenance
Industry/Cross-Industry/ Horizontal	Horizontal Example: Warehouses and logistics, pharma and healthcare, real estate and retail
Current Status of the Solution	Production-ready; customizable for client
No. of Case Studies Completed Using This Solution	Ready to launch
Name of the Company	Gaia
Website of the Company	www.gaia.in
Country of the Company	India and USA

Real-Time Customer Experience, Engagement and Service Delivery Management

<table>
<tr><th colspan="2">Problem Description</th></tr>
<tr><td colspan="2">• Opaque last mile service delivery experience
• No real-time information on service delivery, customer experience and engagement at distributed sites
• Manual or periodic measurement of experience
• Delay in response to service level or exception issues
• Loss of accountability in managing distributed internal or third-party customer service teams
• Lack of transparency and governance monitoring</td></tr>
<tr><th colspan="2">Solution Description</th></tr>
<tr><td>• Unified platform to measure, monitor and manage customer experience
• For single or multiple experience metrics using IoT and digital tools
• Edge and cloud analytics
• Monitor customer experience/NPS/customer satisfaction/experience index
• Real-time reporting and live dashboard for location, performance and issues
• GIS dashboard and reports; matrix and trends; time series and comparative analytics
• Comprehensive action and rules and alert-response management</td><td>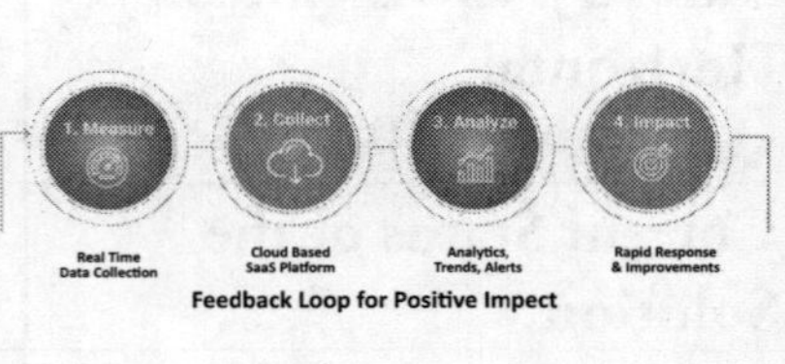
</td></tr>
</table>

Digital Technologies Used for the Solution				
Mobile	IoT	Cloud	Analytics	AI/ML

Outcomes/Benefits of the Solution	
• Last mile service delivery and service experience management • Unified platform enables end-to-end digitalization, remote monitoring and management of experience	• Customizable to include client or industry-specific rules, processes and analytics requirements • Improve transparency and accountability • Monitor and improve customer experience • Leverage customer experience to improve and motivate service teams
Industry/Cross-Industry/ Horizontal	Horizontal Example: Public spaces and infrastructure, transportation, airports and railway stations, hospitals, real estate and retail
Current Status of the Solution	Production-ready; customizable for client
No. of Case Studies Completed Using This Solution	Ready to launch
Name of the Company	Gaia
Website of the Company	www.gaia.in
Country of the Company	India and USA

HaaS: Happiness as a Service

<table>
<tr><th colspan="2">Problem Description</th></tr>
<tr><td colspan="2">
<ul>
<li>Lack of visible happiness in customers at a location, with a service or department while resolving complaints</li>
<li>Lack of real-time connections between people and service representatives regarding complaints and grievances</li>
<li>No collaborative platform for the customers to have a conversation among themselves regarding a complaint, which helps the organizations to know the real-time pulse or emotions of the customers about the complaint</li>
<li>No tracking of customers' emotions while resolving complaints</li>
<li>Poor communication and lack of information to customers about the various happenings in the organization</li>
<li>Scattered social media channels and swivel-chair effect</li>
</ul>
</td></tr>
<tr><th colspan="2">Solution Description</th></tr>
<tr><td>
Customers: Consumers, citizens, patients and employees
<ul>
<li>Multiple channels to raise complaints</li>
<li>GPS integration to know the location of complaints</li>
<li>Voice or text-based complaints capture</li>
<li>Voting and emotion capture by customers against each complaint</li>
<li>Real-time engagement with customers contextual to the complaint using chat, mobile phone</li>
<li>Collate the pulse of customer emotions and experience</li>
<li>Provide real-time information and communication (integrated YouTube, Twitter, podcast, blog, news, etc.)</li>
<li>Provides access to the event calendar of the organization</li>
<li>18-dimensional insight into customer complaints (analytics)</li>
<li>Know the mood swings of customers in real-time (happy, unhappy, angry)</li>
</ul>
</td><td>

</td></tr>
</table>

<table>
<tr><td colspan="4">Digital Technologies Used for the Solution</td></tr>
<tr><td>Mobile</td><td>Social</td><td>Cloud</td><td>Analytics</td></tr>
<tr><td colspan="4">Outcomes/Benefits of the Solution</td></tr>
<tr><td colspan="4">• Real-time happiness measurement of customers
• Faster response to complaints
• Real-time connect with customers
• Reduced cost of call centres
• Innovate solutions based on the actual pulse of customer experience</td></tr>
<tr><td colspan="2">Industry/Cross-Industry/Horizontal</td><td colspan="2">Horizontal
Example: All industries</td></tr>
<tr><td colspan="2">Current Status of the Solution</td><td colspan="2">Production-ready</td></tr>
<tr><td colspan="2">No. of Case Studies Completed Using This Solution</td><td colspan="2">2</td></tr>
<tr><td colspan="2"></td><td colspan="2"></td></tr>
<tr><td colspan="2">Name of the Company</td><td colspan="2">BT&BT and OSI Digital</td></tr>
<tr><td colspan="2">Website of the Company</td><td colspan="2">https://osidigital.com/ and http://btbt.co.in/</td></tr>
<tr><td colspan="2">Country of the Company</td><td colspan="2">USA and India</td></tr>
</table>

CQaaS: Construction Quality as a Service

<table>
<tr><th colspan="3">Problem Description</th></tr>
<tr><td colspan="3">• 3–5% of the project cost is spent on fixing the rework/quality issues</td></tr>
<tr>
<td>Builder/Developer
• High cash inflow from developers due to high rework and corresponding project delays
• Impact on the developer's brand due to low quality
• High maintenance cost due to low construction quality</td>
<td>General Contractor/ Contractors
• Low profitability for contractors due to high rework
• High penalties due to high rework and corresponding project delays
• Too much documentation and laborious effort for quality inspections
• Lack of comprehensive inspections due to non-availability of required quality specialists</td>
<td>End user
• End customer dissatisfaction due to low-quality construction</td>
</tr>
<tr><th colspan="3">Solution Description</th></tr>
<tr>
<td colspan="2">Customers: Real estate builders, general contractors, AEC companies and quality consultants
• Mobile/tablet-based inspections
• Pre-construction inspection of work items
• Inspection during construction
• Post-construction inspection of work items
• Life cycle snagging, handover snagging, off-line snagging
• Real-time visibility of the quality of construction
• Maintainability score and risk score of construction
• Readiness index to hand over the unit to end customer
• Configurable workflows and roles to suit the organization's specific requirements</td>
<td>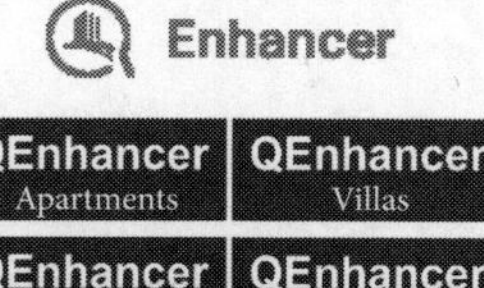
</td>
</tr>
</table>

Digital Technologies Used for the Solution			
Mobile	Social	Cloud	Analytics
Outcomes/Benefits of the Solution			
• Potential reduction of 3–5% of the project cost • Zero documentation • No need for highly qualified and skilled professionals to conduct the inspections; even low skilled persons will be empowered with system generated inspection checklists and recommendations for them to conduct the inspections easily • Reduced cost of maintenance due to preventive inspections • Enhance the happiness of customers with effective snagging			

Industry/Cross-Industry/Horizontal	Construction and real estate
Current Status of the Solution	Production-ready
No. of Case Studies Completed Using This Solution	3
Name of the Company	BT&BT
Website of the Company	http://btbt.co.in/
Country of the Company	India

Lean Digital Quotient Finder

<table>
<tr><th colspan="2">Problem Description</th></tr>
<tr><td colspan="2">
<ul>
<li>Organizations do not know the current digital readiness</li>
<li>Ambiguity about how, where and when to start the digital journey</li>
<li>Slow response of organizations to the digital age customer's needs</li>
<li>Fear of failure and resistance to change</li>
<li>Non-aligned digital initiatives and low ROI</li>
<li>Improper metrics to measure digital business performance</li>
</ul>
</td></tr>
<tr><th colspan="2">Solution Description</th></tr>
<tr><td>
<ul>
<li>Leverage the power of fusion of lean and digital</li>
<li>Institutionalize lean digital thinking</li>
<li>Assess the LDQ of each business building block of an organization</li>
<li>Identify digital strengths, weaknesses and create digital transformation problem statements and possibilities</li>
<li>Leverage open innovation digital platform for identifying digital business solutions</li>
</ul>
</td><td>

Lean Digital Quotient Finder
Prepare for Digital Future
</td></tr>
</table>

Assessment Method	
• 12 **lean digital thinking** principles • 12 business building blocks • 500 lean digital practices • Executive workshops on lean digital enterprises • Form core team (integrated business technology team) • Conduct interviews, discussions and workshops with the core team	• Complete lean digital assessment: Bronze, silver, gold and platinum • Release lean digital-strengths-and-weaknesses report and lean digital transformation roadmap • Introduce digital innovations by establishing project/programme management office • Force.com tool to find LDQ
Outcomes/Benefits of the Solution	
• 360-degree view of the digital readiness of business • High ROI from digital initiatives (new revenue streams, profit, customer delight) • Elimination of non-value-added activities from digital businesses • High synergy among different departments	
Industry/Cross-Industry/ Horizontal	Horizontal Example: All industries like travel, hospitality, manufacturing, automotive, public sector, retail, banking, healthcare and others
Current Status of the Solution	Production-ready
No. of Case Studies Completed Using This Solution	Ready to launch
Name of the Company	BT&BT
Website of the Company	http://btbt.co.in/
Country of the Company	India

Smart Vehicle Tracking System

<table>
<tr><th colspan="2">Problem Description</th></tr>
<tr><td colspan="2">• To reduce underwriting, control fraudulent claims and help processing the same faster for one of the leading insurance companies in India
• To assist in identifying and diagnosing the non-performing assets
• To optimize the process of repossession for a leading finance company</td></tr>
<tr><th colspan="2">Solution Description</th></tr>
<tr><td>• Investigate each case against the claims
• Vehicle trip report
• Crash/tow alerts
• Driving pattern at the time of the accident (hard braking, hard acceleration, overspeeding, sharp left turns, sharp right turns)
• Filter out vehicle owners who could be potential defaulters
• Business criterion defined for distance travelled by a commercial vehicle and number of days of usage within a month
• Live tracking of all vehicles which helps in locating non-performing assets and optimizing the process of repossession</td><td>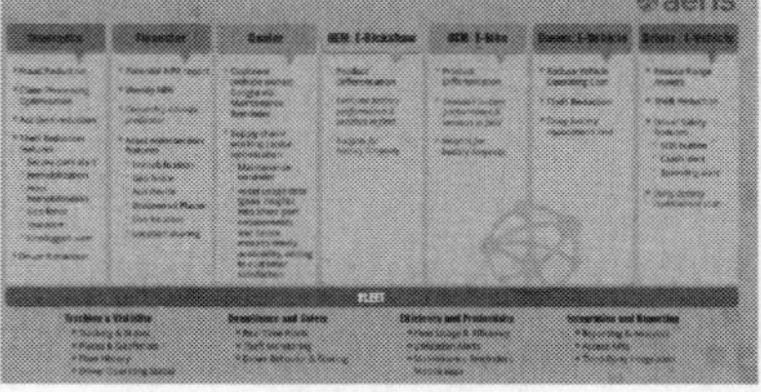
</td></tr>
</table>

Digital Technologies Used for the Solution				
Mobile	IoT	Cloud	Analytics	AI

Outcomes/Benefits of the Solution	
• Accurate information on driving pattern during accidents • Potential defaulters (vehicle owners) identification • Tracking of non performing assets • Faster recovery due to unplugged alert • Crash alert leading to a more informed reading of the accident scenario	
Industry/Cross-Industry/Horizontal	Cross-industry Example: Fleet, OEM, finance and insurance
Current Status of the Solution	Production-ready (in use)
No. of Case Studies Completed Using This Solution	10+
Name of the Company	Aeris Communications
Website of the Company	https://www.aeris.com/
Country of the Company	USA and India

Connected E-Rickshaw

Problem Description	
• Issues with last mile connectivity in rural areas • High maintenance cost due to the lack of real-time inspection of maintenance activities • Lack of real-time visibility on e-rickshaw health • Not able to prevent theft	
Solution Description	
• **Voltage** identifies catastrophic failures such as short-circuited cells and gives accurate visibility on battery performance • **Battery temperature measurement** helps analyze the health and performance of the battery • **Current monitoring** highlights potential problems within the battery string and allows early detection of issues and any significant leakage faults • **State of charge** describes the status of battery charge • **State of health** measures the long-term capability of the battery and indicates battery consumption • **Battery status** indicates whether the battery is charging, discharging or is in the standby mode	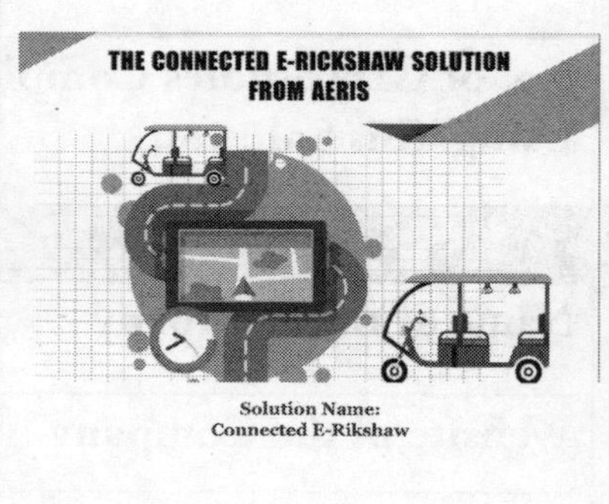 Solution Name: Connected E-Rikshaw

Digital Technologies Used for the Solution			
Mobile	IoT	Cloud	Analytics
Outcomes/Benefits of the Solution			
• B2B web portal access and mobile app • Single view dashboard • Map view of vehicles with status and live tracking • Complete trip data analytics	• Alerts on overspeeding, crash, theft, etc. • Easy navigation with route search, maps • Remote control with geofencing and immobilization • Report of vehicle details: Secure parking, immobilized vehicle, SOS and trips		
Industry/Cross-Industry/Horizontal	Transportation industry Example: Transportation		
Current Status of the Solution	Production-ready (in use)		
No. of Case Studies Completed Using This Solution	10+		
Name of the Company	Aeris Communications		
Website of the Company	https://www.aeris.com/		
Country of the Company	USA and India		

Real-Time Tracking of Workforce

<table>
<tr><th colspan="5">Problem Description</th></tr>
<tr><td colspan="5">

- Lack of real-time tracking of field personnel and their working locations (especially hazardous and unsafe locations)
- Safety and security incidents cause harm to filed personnel
- Lack of real-time alerts and notifications to all stakeholders during the emergency time to take appropriate actions

</td></tr>
<tr><th colspan="5">Solution Description</th></tr>
<tr><td colspan="3">

- Wearable devices and mobile app-based solution
- Configuration and monitoring of beat plan
- See the live location of staff anytime and anywhere in the world
- Single-press button on the device for SOS call during an emergency
- Supports SOS message to three pre-configured contact numbers
- Alerts/notice when the battery is running low
- Extremely portable solution—can be placed in a pocket or bag easily

</td><td colspan="2">

</td></tr>
<tr><th colspan="5">Digital Technologies Used for the Solution</th></tr>
<tr><td>Mobile</td><td>IoT</td><td>Cloud</td><td>Analytics</td><td>AI</td></tr>
<tr><th colspan="5">Outcomes/Benefits of the Solution</th></tr>
<tr><td colspan="3">

- Productivity improvement with real-time tracking of comprehensive view of the personnel geolocation and planned route
- Data-driven decisions

</td><td colspan="2">

- Regulatory compliance
- Portable and handy

</td></tr>
</table>

Industry/Cross-Industry/Horizontal	Horizontal Example: Railways
Current Status of the Solution	Production-ready (in use)
No. of Case Studies Completed Using This Solution	10+
Name of the Company	Aeris Communications
Weblink of the Company	https://www.aeris.com/
Country of the Company	USA and India

Smarter, Resilient and Reliable Supply Chains

<table>
<tr><th colspan="2">Problem Description</th></tr>
<tr><td colspan="2">
• Quality and compliance: Need for real-time and granular monitoring and actionability to regulate temperature compliance, spoilage control and physical damage
• End-to-end traceability: Farm-to-fork view of the product's movement, including sub-components; intelligent management of traceability and recall
• Operational efficiency: Optimization of yard management, FIFO, scheduling, staff productivity, and digital operational enablement and tracking
</td></tr>
<tr><th colspan="2">Solution Description</th></tr>
<tr><td>
• Tag products and assets with BLE-powered wireless networks at any desired granular level
• Blend IoT data from the client's ERP and public data to enable control and analytics
• Focus on root-cause analysis, predictive insights and AI-based action enablement
• Automated quality and operational traceability and visibility
• Contextualize and customize the product to fit into the customer's operating environment
</td><td>
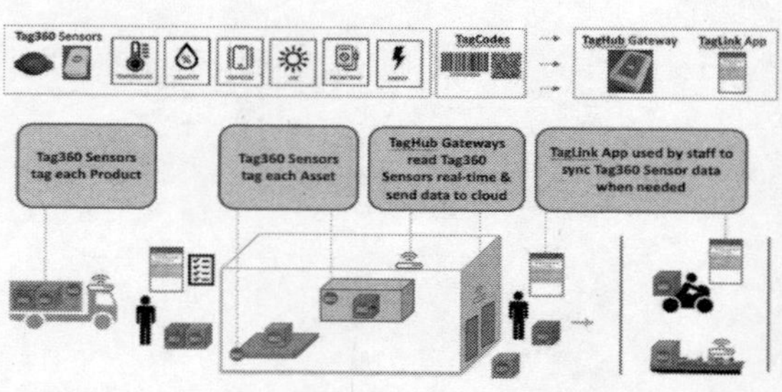

Smarter Supply Chain Solution
</td></tr>
</table>

Digital Technologies Used for the Solution				
IoT	ERP	Cloud	Analytics	AI

Outcomes/Benefits of the Solution		
F&B (e-commerce, cloud kitchens, FMCG) • Two times improvement in e-com delivery temperature compliance • 80% standard operating procedure (SOP) compliance in cloud kitchens • 70% inventory visibility improvement, 30% asset loss reduction	Pharma, diagnostics, and e-pharma • 95% product/sample traceability • 20-30% Phlebotomist and runner productivity improvement • Automated QC and CAPA • 50% reduction in overseas release time	High-tech, manufacturing • 40% reduction in product damage in transit • 90% pallet, bin traceability • 60% improvement in SOP compliance

Industry/Cross-Industry/Horizontal	Cross-industry Example: Logistics, transportation and supply chain
Current Status of the Solution	Production-ready (in use)
No. of Case Studies Completed Using This Solution	10+ live customers with more than 10,000 devices on the field
Name of the Company	TagBox
Website of the Company	https://www.tagbox.co
Country of the Company	India, Singapore, APAC and US

OM: Order of Magnitude

<table>
<tr><th colspan="2">Problem Description</th></tr>
<tr><td colspan="2">

- Blindness and visual impairment is disproportionately prevalent in developing countries, 60% of which is avoidable
- High infrastructure, cost of ownership and dependency on trained professionals limit the adoption of necessary diagnostic solutions at the vision screening level

</td></tr>
<tr><th colspan="2">Solution Description</th></tr>
<tr><td>

- OM is an intelligent vision assessment system that can be used to screen more than 60% of the causes of avoidable blindness
- A software solution that can convert any off-the-shelf VR headset into a clinical-grade vision screening system
- In-built automated clinical decision support reduces the dependency on trained professionals by allowing for easy understanding of the results and next steps
- Cloud-based design makes it suitable for telemedicine

</td><td>

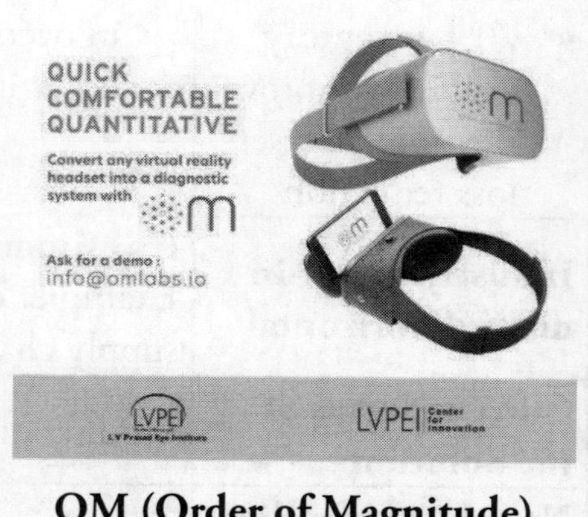

OM (Order of Magnitude)

</td></tr>
</table>

<table>
<tr><td colspan="5">Digital Technologies Used for the Solution</td></tr>
<tr><td>Mobile</td><td>Social</td><td>Cloud</td><td>Analytics</td><td>AI</td></tr>
<tr><td colspan="5">Outcomes/Benefits of the Solution</td></tr>
<tr><td colspan="2">• Cost-effective alternative to conventional diagnostics in rural and remote settings
• Empowering healthcare workers with more quantitative information at the vision screening level</td><td colspan="3">• Software solution built on a vendor-agnostic VR platform allows for easy adoption
• Quick and comfortable test process enables a better care experience for patients compared to diagnostic counterparts</td></tr>
<tr><td colspan="2">Industry/Cross-industry/ Horizontal</td><td colspan="3">Industry: Medical diagnostics (ophthalmology, neurology)
Cross-industry: Patient education</td></tr>
<tr><td colspan="2">Current Status of the Solution</td><td colspan="3">Production-ready</td></tr>
<tr><td colspan="2">No. of Case Studies Completed Using This Solution</td><td colspan="3">• Application in rural eye care at 25 centres of LV Prasad Eye Institute network
• Application in private practices at Shinde Eye Care Centre</td></tr>
<tr><td colspan="5"></td></tr>
<tr><td colspan="2">Name of the Company</td><td colspan="3">OM Labs</td></tr>
<tr><td colspan="2">Website of the Company</td><td colspan="3">https://omlabs.io</td></tr>
<tr><td colspan="2">Country of the Company</td><td colspan="3">India</td></tr>
</table>

FoFo™: Folding Foropter™

<table>
<tr><th colspan="2">Problem Description</th></tr>
<tr><td colspan="2">
<ul>
<li>Lack of awareness and access to refractive error and vision-related problems and solutions in low resource areas, resulting in poor academic performance, dropouts and poor lifestyle, due to low vision and blindness</li>
<li>Three billion people in the world have some form of refractive error; of whom 153 million are visually impaired due to uncorrected refractive error; around 90% of them live in low resource areas (Source: WHO)</li>
<li>US$269 billion in productivity lost and nearly 158 million people are vision-impaired because of uncorrected refractive error - Kevin Frick</li>
<li>Lack of trained healthcare professionals in low resource areas to provide services for eye check-ups</li>
<li>Lack of affordable and easy-to-use devices for refractive error screening</li>
</ul>
</td></tr>
<tr><th colspan="2">Solution Description</th></tr>
<tr><td>
<ul>
<li>An affordable and shareable refractive error screening device, made of paper enveloping lens optics supported by a mobile app</li>
<li>Simple to use for the target age group (12–40 y/o) who needs intervention</li>
<li>Easy deployment of spherical equivalent glasses using FoFo device by community healthcare workers with an optimal accuracy</li>
<li>An app as an assistive tool to register check-up cycles, error assessment and geographical coverage</li>
</ul>
</td><td>

Folding Foropter™
</td></tr>
</table>

<table>
<tr><td colspan="4">Approaches Used for the Solution</td></tr>
<tr><td>Social</td><td>Digital assistive</td><td>Eco-friendly material and packaging</td><td>First principles (the fundamental concepts or assumptions on which theory, system or method is based)</td></tr>
<tr><td colspan="4">Outcomes/Benefits of the Solution</td></tr>
<tr><td colspan="2">• Easy-to-use screening tool for healthcare workers
• Faster patient checkouts
• Easy onboarding and short learning curve</td><td colspan="2">• Low-cost set-up and high impact (one device can screen one remote village)
• Low overall carbon footprint
• Better quality of life and awareness about vision-related issues</td></tr>
<tr><td colspan="2">Industry/Cross-industry/ Horizontal</td><td colspan="2">Example: Healthcare (eye care)</td></tr>
<tr><td colspan="2">Current Status of the Solution</td><td colspan="2">• Officially launched in 2019 and in use (around 18 countries)
• App in validation phase</td></tr>
<tr><td colspan="2">No. of Case Studies Completed Using This Solution</td><td colspan="2">• 3,000+ screening data to prove the accuracy of the device
• Ongoing NCD screening in villages of South India by Bare Foot Network (four-year programme)</td></tr>
<tr><td colspan="2"></td><td colspan="2"></td></tr>
<tr><td colspan="2">Name of the Company</td><td colspan="2">LVPEI Center for Innovation and LV Prasad Eye Institute</td></tr>
<tr><td colspan="2">Website of the Company</td><td colspan="2">www.fofo-project.com, www.lvpei.tech and www.lvpei.org</td></tr>
<tr><td colspan="2">Country of the Company</td><td colspan="2">India</td></tr>
</table>

Meeting Wall

<table>
<tr><th colspan="2">Problem Description</th></tr>
<tr><td colspan="2">• U$37 billion loss every year because of inefficient meetings
• Low focus on productivity of meetings
• Digital divide in the meeting rooms
• Lack of meetings standardization
• Improper meeting management</td></tr>
<tr><th colspan="2">Solution Description</th></tr>
<tr><td>• Meeting Wall is a digital platform for meetings of all sizes and purposes
• Large interactive touch screen unified platform
• Meeting science, standardized meeting management software
• Content management and analytics for meetings
• Ink-to-text annotation
• Multi-user, multi-device and multi-touch interaction
• Integrated video conferencing
• Group productivity tools
• Webinar management
• Gesture shortcuts</td><td>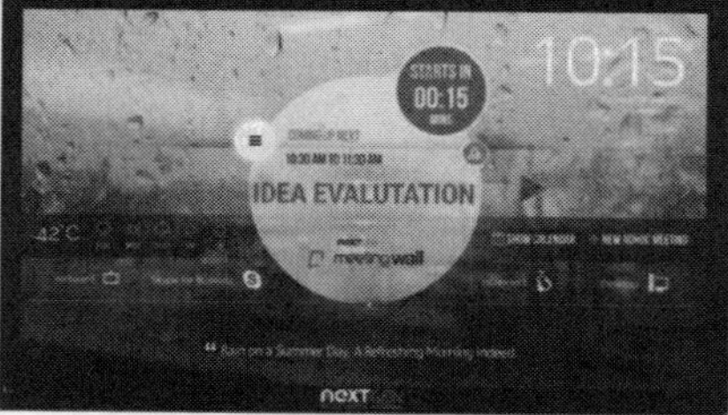
</td></tr>
</table>

<table>
<tr><td colspan="5">Digital Technologies Used for the Solution</td></tr>
<tr><td>Desktop</td><td>Mobile</td><td>Cloud</td><td>Analytics</td><td>AI</td></tr>
<tr><td colspan="5">Outcomes/Benefits of the Solution</td></tr>
<tr><td colspan="2">• Suits any type of modern-day meetings, be it remote, in-room or blended
• Perpetual engagement with zero distraction meetings
• No app installation required for end users</td><td colspan="3">• End-to-end meeting solution that has meeting schedulers, meeting management, multi-VC integration, device casting, content management and content distribution
• An inclusive approach that consolidates all the existing meeting room investments</td></tr>
<tr><td colspan="2">Industry/Cross-industry/Horizontal</td><td colspan="3">Horizonal
Example: All industries</td></tr>
<tr><td colspan="2">Current Status of the Solution</td><td colspan="3">Production-ready</td></tr>
<tr><td colspan="2">No. of Case Studies Completed Using This Solution</td><td colspan="3">25</td></tr>
<tr><td colspan="2"></td><td colspan="3"></td></tr>
<tr><td colspan="2">Name of the Company</td><td colspan="3">NextGen Multitouch Pvt Ltd</td></tr>
<tr><td colspan="2">Website of the Company</td><td colspan="3">http://themeetingwall.com/</td></tr>
<tr><td colspan="2">Country of the Company</td><td colspan="3">India</td></tr>
</table>

ZeroCode

<table>
<tr><th colspan="2">Problem Description</th></tr>
<tr><td colspan="2">• Faster changes in customer needs and demand for continuous newness
• Conventional and slow development methods for software applications
• Significant investments in custom-built software development
• Non-availability of high skilled people to develop software applications</td></tr>
<tr><th colspan="2">Solution Description</th></tr>
<tr><td>ZeroCode is like playing a LEGO game with a visual drag and drop of ready-made coding blocks. Anyone can be a coder irrespective of their computer background. ZeroCode works easily with browser and internet connectivity.
ZeroCode can be used to build
• web, mobile or enterprise-grade solutions,
• an end-to-end solution,
• a backend application exposing API to IoT device and
• a mashup of third-party applications.</td><td>

Highlights:
• Instant mobility
• In-built security
• Integrated project management
• Auto documentation
• Single-click deployment
• Automated testing

Coming up:
• Code conversion
• Conversation to code
• UnZero code
• DeepTech ZeroCode</td></tr>
</table>

<table>
<tr><td colspan="5">Digital Technologies Used for the Solution</td></tr>
<tr><td>Web</td><td>Mobile</td><td>Cloud</td><td>Analytics</td><td>AI</td></tr>
<tr><td colspan="5">Outcomes / Benefits of the Solution</td></tr>
<tr><td colspan="3">• Coding for anyone and for everything
• A true on-cloud developer platform where anyone can code from anywhere and juxtapose to develop a software
• Business continuity is assured along with guaranteed deliveries</td><td colspan="2">• Faster development by 8X
• Reduction in time to market by 50%
• 65% overall cost saving
• No resource headaches</td></tr>
<tr><td colspan="3">Industry/Cross-industry/ Horizontal</td><td colspan="2">Horizontal
Example: All industries</td></tr>
<tr><td colspan="3">Current Status of the Solution</td><td colspan="2">Production-ready</td></tr>
<tr><td colspan="3">No. of Case Studies Completed Using This Solution</td><td colspan="2">50+</td></tr>
<tr><td colspan="3"></td><td colspan="2"></td></tr>
<tr><td colspan="3">Name of the Company</td><td colspan="2">ZeroCode Innovations Private Limited</td></tr>
<tr><td colspan="3">Website of the Company</td><td colspan="2">http://zeroco.de</td></tr>
<tr><td colspan="3">Country of the Company</td><td colspan="2">India</td></tr>
</table>

GOKADDAL: Digital Solutions Exchange Cloud Platform

<table>
<tr><th colspan="3">Problem Description</th></tr>
<tr><td colspan="3">• Global adoption for digital solutions and technologies is at an all-time high with the next 5-year market size pegged at US$1 trillion (Source: IDC/Forbes)
• There is no common medium or interface for solution providers, seekers and integrators to connect</td></tr>
<tr><td>Provider problems
• High cost of entering new markets and reaching out to global clients
• No feasible method to position and sell solutions online (unlike Amazon for products)
• Limited awareness outside their base market
• High cost of delivery</td><td>Seeker problems
• No access to innovative solutions globally, only local and close circuit vendor access available
• Cost optimization not feasible with limited solutions
• No repository or one-stop-shop to search, compare and buy/deliver digital solutions
• High cost of adoption</td><td>Integrator problems
• No common medium available to connect to solution providers and seekers
• No access to global innovative solutions outside their work zone or region
• High cost of adoption/delivery</td></tr>
<tr><th colspan="3">Solution Description</th></tr>
<tr><td colspan="3">GOKADDAL, a cloud-based solutions exchange platform, acts as solutions aggregator and marketplace. It is created to help solution seekers (customers) realize their digital transformation vision, with the right solutions using the right tools and skills.</td></tr>
</table>

<table>
<tr>
<td>

Providers
- Global focus
- Solution showcase to global customers
- Cloud-enabled
- Searches buyers
- Alerts for new buyers
- Timely marketing promotions
- Bidding/reverse bidding
- New models creation for SaaS

Seekers
- Cloud-enabled
- Solution comparison
- Bidding/reverse bidding
- TCO/ROI calculators
- One-on-one dialogue between solution seeker and providers
- EOI/RFP circulation
- SaaS model adoption

Integrators
- One-stop for all solution components
- Meet solution buyers and sellers
- Get solutions for projects
- Uncover new projects
- Solution design, deployment and management opportunities
- Pick bids and RFPs
- Help providers with new markets and opportunities

</td>
<td>

4A Focus

Artificial Intelligence

Automation

Analytics

Augmentative

</td>
</tr>
</table>

<table>
<tr><td colspan="3">Digital Technologies Used for the Solution</td></tr>
<tr><td>Cloud</td><td>Analytics</td><td>AI</td></tr>
<tr><td colspan="3">Outcomes/Benefits of the Solution</td></tr>
<tr><td colspan="3">• Reduced TCO for seekers and integrators
• Low-cost global outreach and business development
• Variety of solutions to choose from
• Access to new and global markets
• Less time is taken to sell and adopt solutions</td></tr>
<tr><td colspan="2">Industry/Cross-industry/Horizontal</td><td>Horizontal
Example: All industries</td></tr>
<tr><td colspan="2">Current Status of the Solution</td><td>Version 1.0 is in the market</td></tr>
<tr><td colspan="2">No. of Case Studies Completed Using This Solution</td><td>23</td></tr>
<tr><td colspan="3"></td></tr>
<tr><td colspan="2">Name of the Company</td><td>Merkado RHA Technology Pvt. Ltd</td></tr>
<tr><td colspan="2">Website of the Company</td><td>www.gokaddal.com</td></tr>
<tr><td colspan="2">Country of the Company</td><td>India and UAE</td></tr>
</table>

SWAYAM: Anytime Health Monitoring

<table>
<tr><th colspan="3">Problem Description</th></tr>
<tr><td colspan="3">• Millions of people lack access to quality healthcare
• A significant burden of diseases on developing countries
• Health issues due to delay in diagnosis
• Failure to take preventive measures to safeguard the health of the individual</td></tr>
<tr><td>Healthcare provider problems
• Cost of providing quality healthcare access at doorsteps is high
• Cost of setting up testing centre/diagnostic clinics is high
• Getting clients to come to the centre is not feasible for all</td><td>Government problems
• Providing quality mass level diagnostic services is not highly feasible
• Cost of mass diagnostic services is high
• No budget for local/neighbourhood clinics
• Access for masses is a challenge</td><td>End user problems
• Access to quality diagnostic services in the neighbourhood is low
• Cost is high
• Too much time is taken for testing and report</td></tr>
<tr><th colspan="3">Solution Description</th></tr>
<tr><td colspan="3">SWAYAM, India's first AHM kiosk, is an IoT- and AI-enabled system. It provides more than 60 diagnostic tests round the clock, at any convenient place: shopping mall, society, community centre, rail/bus station, etc. It is a health ATM which provides instant test reports and the option for video conferencing with doctors.</td></tr>
</table>

<table>
<tr>
<td>

Healthcare providers
- Use SWAYAM as a last mile diagnostic system
- Access to a large client base
- Provide low-cost service at the doorstep
- 24-hour service option

Government
- Use in rural/semi-urban services
- Use in PHCs or village clinic for diagnostic tests
- Use at events (e.g., Kumbh Mela), large gatherings and community spots
- Data security and cloud access

End user
- Access to economical diagnostic test service at the doorstep
- Round-the-clock access
- Immediate reports
- Video conference with doctors
- Data security and cloud access

</td>
<td>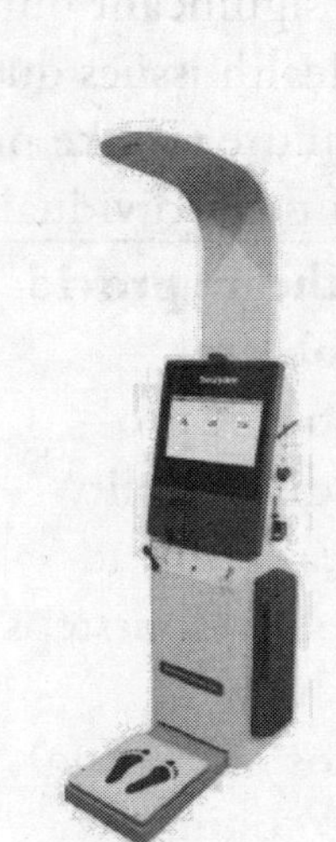</td>
</tr>
</table>

<table>
<tr><td colspan="4">Digital Technologies Used for the Solution</td></tr>
<tr><td>Kiosk</td><td>Cloud</td><td>Security</td><td>Network</td></tr>
<tr><td colspan="4">Outcomes/Benefits of the Solution</td></tr>
<tr><td colspan="4">• Reduced TCO for providers and the government
• Low-cost health access to users
• Cloud-based access to reports
• Mass health access for government/users
• Healthy society</td></tr>
<tr><td colspan="2">Industry/Cross-Industry/Horizontal</td><td colspan="2">Industry
Example: Healthcare</td></tr>
<tr><td colspan="2">Current Status of the Solution</td><td colspan="2">In the market</td></tr>
<tr><td colspan="2">No. of Case Studies Completed Using This Solution</td><td colspan="2">62</td></tr>
<tr><td colspan="2"></td><td colspan="2"></td></tr>
<tr><td colspan="2">Name of the Company</td><td colspan="2">Merkado RHA Technology Pvt. Ltd</td></tr>
<tr><td colspan="2">Website of the Company</td><td colspan="2">www.merkadorha.com</td></tr>
<tr><td colspan="2">Country of the Company</td><td colspan="2">India and UAE</td></tr>
</table>

VECAAS: Vertical Cloud as a Service

Problem Description		
• Cloud is the new default for any technology application and solution • Cloud as a service is available but vertical-focused cloud is still a dream • Market for vertical solutions is huge, e.g., smart city applications is a $200 billion market but a focused and dedicated cloud offering for smart cities is not available		
Solution provider problems • Multiple levels and intricacies when approaching vertical client base • Creating cloud offering and infrastructure is very costly • Cloud delivery of solutions needs new capabilities	**Cloud provider problems** • Getting vertical applications is costly • Creating focused service needs technical capabilities • Market branding needs time and money	**End user problems** • No such focused application and service cloud available • Cloud offerings are sparse • Private cloud is costly and public cloud has security concerns

<table>
<tr><td colspan="3">Solution Description</td></tr>
<tr><td colspan="3">VECAAS (pronounced vikas) is a vertical-focused cloud. It is a set of solutions focused on one vertical and sold to any vertical and hence VECAAS is a cross-industry solution.</td></tr>
<tr><td colspan="3">Solution providers</td></tr>
<tr><td colspan="2">• Build and offer cloud-based services—XaaS/SaaS
• One-stop for all vertical solutions—opportunity to cross-connect
Cloud providers
• Offer one-stop for all vertical-focused applications and services
• New revenue streams
• Access to wider and target customer market
End user
• Option to get all vertical services from one-stop-shop
• Focused applications and services
• Low TCO and high ROI</td><td>FEW FOCUS AREAS
Smart City Cloud
Healthcare Cloud
Manufacturing Cloud
Education Cloud</td></tr>
<tr><td colspan="3">Digital Technologies Used for the Solution</td></tr>
<tr><td>Mobile</td><td>Cloud</td><td>AI</td></tr>
<tr><td colspan="3">Outcomes/Benefits of the Solution</td></tr>
<tr><td colspan="3">• Moving to cloud offering is easy for solution providers
• Creating vertical-focused application and services is easy for cloud providers
• Access to one-stop-shop for all vertical solutions and services</td></tr>
<tr><td>Industry/Cross-Industry/Horizontal</td><td colspan="2">Cross-Industry
Example: Cities, government, healthcare</td></tr>
<tr><td>Current Status of the Solution</td><td colspan="2">Design stage</td></tr>
</table>

No. of Case Studies Completed Using This Solution	–
Name of the Company	Merkado RHA Technology Pvt. Ltd
Website of the Company	www.merkadorha.com
Country of the Company	India

PART 6

COVID-19: Digital Interventions

<table>
<tr><td></td><td>CHAPTER 16
COVID-19: The New Normal
• A Global Concern
• Business New Normal
• Government New Normal
• Society New Normal
• Lifestyle New Normal</td></tr>
<tr><td>1010
1010</td><td>CHAPTER 17
Digital Interventions in the New Normal That Emerged with COVID-19
• Impact of the New Normal
• Possible Digital Interventions in the Business New Normal
• Possible Digital Interventions in the Government New Normal
• Possible Digital Interventions in the Society New Normal
• Possible Digital Interventions in the Lifestyle New Normal</td></tr>
</table>

CHAPTER 18

COVID-19: Employee Safety and Health

- Challenges in the Time of COVID-19
- Uncertain Future and Low Employee Morale
- Employee Journey Map and Potential Virus Transmission Points
- Digital Interventions for Employee Health and Safety
- Preparing Employees for the New Normal

CHAPTER 16
COVID-19: The New Normal

A Global Concern

The secretary-general of global nations, Nelson, in his video conference with all the nation heads expressed his concern about the impact of COVID-19. There was pin-drop silence and most of the country heads were very anxious and worried about the future.

Nelson shared that millions of people had lost their lives, and they had lost the battle to an invisible enemy. He saluted the doctors, paramedical staff, police, farmers and many other front-end staff, emphasizing that they were the real soldiers in this war.

The COVID-19 disaster has been an entirely different experience as compared to other disasters, after which it usually takes a few days/weeks to come back to normal. But COVID-19 will be with us for a long time and would severely impact the lives of everyone.

Nelson pointed out that the repeated lockdowns in each country had been impacting public life, health and economy. Almost all the countries were expected to experience negative growth of GDP that year as there was no significant activity in the industry and government except fighting the deadly virus.

Nelson praised the unified effort being put in by all the countries in these tough times. For the first time in the last hundred years since 1920, humans in more than 200 countries have been affected by a virus. "While the virus wins the battle today," Nelson confidently said, "we will win the final war. But we need to fight many battles for at least the next 12–18 months until we find a vaccine or cure."

Individuals, society, industries and governments have been experiencing a new environment and new challenges, which are different from the past. COVID-19 has been creating a new normal, in which people have different needs, wants and expectations.
In this new era, people have been acquiring new habits and new attitudes, working remotely and experiencing virtual socialization, virtual entertainment and expect to get most of the services over the internet.
This means industries, governments and society must prepare for this new normal. New policies, processes, systems and infrastructure must be built by these entities to be successful in this new normal.

Nelson was worried that a continuous lockdown may lead to economic collapse and they may even fail to feed people, which will lead to social unrest. Hence, immediate measures had to be taken to improve the economy with caution.

Once the lockdown is selectively lifted,

1. children will go to schools,
2. patients will go to hospitals for seasonal and chronic diseases,
3. people will go out to purchase essentials,
4. workers will go to sites and
5. professionals will go to offices.

This would be the most significant risk in a pandemic. This means that everyone might need to wear personal protection equipment (PPE) based on their profession (light, moderate, full-fledged). It will be no surprise if we all look like astronauts walking on the roads.

People's social life has been disturbed due to the fear of the virus. Tourism, entertainment, hospitality industries have been severely impacted. Manufacturing industries are facing serious supply chain issues, which has affected production.

Nelson urged each country head to understand the **new normal** and start implementing the right measures to bounce back. He requested them to collaborate and exchange knowledge and best practices to speed up the discovery of a cure or vaccine.

He requested his executive assistant, Tejas, to present the definition of the **new normal** and the problems, possibilities and opportunities that arise out of it. All the country heads concurred with Nelson and agreed to work on reviving the economy while taking care of their people's health and safety. Nelson emphasized that each country must explore to best leverage technologies to improve the economy and public life.

Tejas and his team presented an incredibly detailed presentation on the four dimensions of the **new normal**, as given in Figure 16.1.

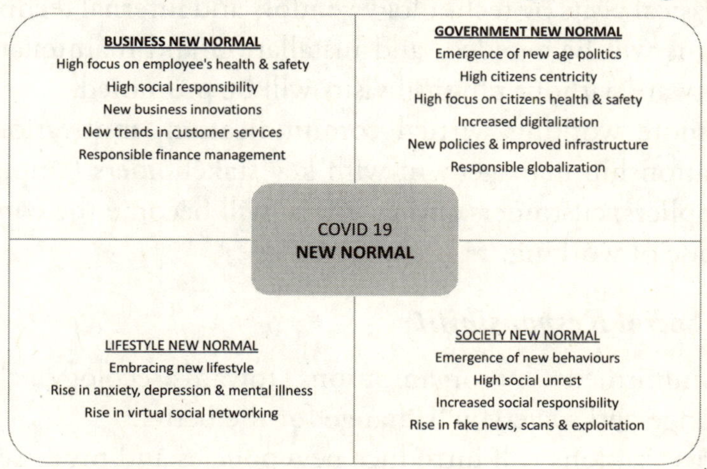

Source: Author

Figure 16.1: Four Dimensions of the New Normal

Business New Normal

Many changes can be expected in the way business is conducted in the future. Some of them have been listed in the following sections.

Heavy Focus on Employee Health and Safety

1. Biosecurity of employees will be safeguarded by new policies, guidelines, digital initiatives, physical distance measures and PPEs.
2. Drones, robots and chatbots will be used to reduce the risk of employees getting infected (e.g., camera-/robot-based security and surveillance, delivery of grocery using drones, intelligent help desks without human intervention).
3. Low-touch/no-touch solutions, devices and services at the office to avoid infections
4. No or limited business travels and more virtual meetings will become the norm. Physical meetings will be limited (only on need basis).
5. Physical visits by technology vendors and internal technology teams will be avoided, and installation and maintenance of software without physical visits will be promoted.
6. Remote working, virtual communication, motivation and relationship management with key stakeholders (employees, suppliers, customers and partners) will become the common mode of working.

High Social Responsibility

1. Commitment by organizations towards ecology, climate change and society will change for the better.
2. Organizations will introduce new policies and processes
    a. to safeguard the animals, insects, forests and maintain biodiversity,

b. to reduce the pollution of air, water and soil due to business transactions and
c. to contribute to the country's economy and resolve social problems while getting revenue and profits.

New Business Innovations

1. External specialists—not the permanent staff—will work on identified deliverables within given timelines (gig economy).
2. New incentives and policies will be introduced by the government to support businesses.
3. New policies, guidelines and performance monitoring mechanisms by organizations will make the work from home (WFH) model successful.
4. Open innovation, crowdsourcing and virtual collaboration among globally distributed specialists will become the norm.
5. The emergence of new manufacturing hubs (domestic supply chain) in various countries will minimize the dependency on a few select countries.
6. Working from home is the new norm but many homes were not built with this feature in mind. Organizations must accept a few distractions at home. Also, scheduled time for activities like cooking, usage of noisy home appliances at home might be introduced to avoid disturbance to the professionals working at home.

New Trends in Customer Services

1. Customer services with no-touch and low-touch modes will be encouraged.
2. E-commerce ecosystem (more delivery executives, drones) will increase the volume of delivery (three times more) to meet the demand of substantial online orders.
3. Organizations will be pressurized to reduce the price of products/services due to less consumption power of people. This is mainly due to job loss and salary cuts.

Responsible Financial Management

1. A flexible cost structure will help embrace changes and manage uncertainty.
2. Frugal and intelligent investments by business leaders will be required to sustain the current business.
3. Constraints in clearing loans, because of low revenue and cash flow issues will lead to high financing costs.
4. The focus will be on survival, operational excellence, cash flow and fixed cost reduction.
5. One must be prepared for hiring freeze, retrenchment of employees and paid leaves, salary cuts and no bonuses for some time.

Government New Normal

The pandemic has also affected the functioning of governments across the globe. In light of the changing world and rapid digitalization, the expectations of citizens are also changing. Some of these changes and expectations have been discussed in this section.

Citizen-Centric Approach

1. Bureaucrats are expected to be more empathetic and citizen-centric. With the emergence of social media, there will be a lot of pressure on government officers.
2. Governments collect a lot of data about individuals covering spending pattern, income, travel history, electronic health records, social media and criminal records. All this information will be linked to a national identification number. This data will be used by the government as and when required as per pre-defined policies and guidelines.
3. Governments may introduce smart cards for each citizen. This smart card could be used for storing health data, travel data, payments history, transportation and government benefits.

Data is an important asset with which the quality of services, responsiveness to citizens can be improved significantly.

4. People expect their government to spend on the infrastructure that takes care of migrant workers and people at the bottom of the social pyramid. Many loopholes in government policies, processes, systems and infrastructure in the face of natural calamities, like non-availability of required shelters, isolation zones and the weak supply chain of essentials, was exposed by the pandemic.
5. There will be pressure from people for a shift in expenses and focus on family care (elders, pregnant women, children)—public health will become a significant part of election manifestos.

Emergence of New Age Politics

1. People expect electoral reforms to avoid corruption, to encourage virtual voting and stringent policies to minimize criminals getting into politics. People have realized that a strong political will is required to safeguard them, especially during a crisis/pandemic period.
2. People will call for the induction of specialists and experienced leaders for nominated positions. This would help in managing crises better.
3. People will prefer to elect experienced citizen-centric political leaders with ethics over inexperienced, unstable, irresponsible and short-term focused leaders.
4. One will see the emergence of a new age politics focused on humanity and people centricity. Online tools and platforms will be used to connect with citizens and cadre in real time.
5. There will be an increase in respect for doctors and paramedical staff just like the defence personnel.

Heavy Focus on Citizen Health and Safety

1. Airports will be modernized by integrating rapid blood tests and biosecurity scanners. Infected travellers will be sent to hospitals.
2. Governments will promote public and private partnerships to establish millions of digital primary health centres (DPHCs), which will implement telemedicine and any time medicine machines (ATMMs). DPHCs will reduce crowding of patients and help in maintaining physical distance among them.
3. Governments will promote the development of innovative diagnostic tools and technologies to baseline the disease at a faster pace by complementing the doctor's effort.
4. High expectations from people for the establishment of a long-term sustainable public healthcare system by governments (doctors, paramedical staff, resources, infrastructure, leadership, etc.) will lead to significant changes.
5. Local bodies will be strengthened with the required infrastructure to automatically spray disinfecting chemicals using robots at identified places regularly.
6. People may demand a reduction in the military/defence budget and focus on public health, drugs innovation and citizen happiness initiatives.

Increased Digitalization

1. Governments will introduce digital literacy programmes, promote vernacular languages and voice-based mobile apps to enable most of the population to embrace the digital lifestyle.
2. Most of the schools will upload content online for students to access and participate in online classes, quizzes and online examinations due to the uncertain future.
3. People will demand seamless and straight-through processes from the Federal/Central government, state governments and local bodies that help in improving the citizen's quality of life.

4. There will be more government apps that regularly provide proactive information about risky areas in the province or country (security, health, safety) and provide recommendations to the public about taking precautions.

New Policies and Improved Infrastructure

1. Governments will strengthen the institutions in a few specific areas (education, health, agriculture) with respect to investments, infrastructure and leaders.
2. Governments will focus more on new policies, infrastructure and resources to take care of the informal/disorganized sector's workforce during the crisis period.
3. Governments will revisit privatization policies for healthcare, pharma and education. For private entities, the focal point is profit, while they also focus on public health. Hence, governments will spend a good percentage of GDP on medical innovations and public health. We will also see high collaboration among different nations in medicine and public health innovations.
4. New drone policies will allow drones to provide various announcements and authenticated news/information to people at different locations in a country.
5. New policies and regulations will be introduced to regularly disinfect the premises and to reconfigure the physical infrastructure of hospitals to avoid crowding of patients. Private hospitals may increase their fees.
6. There will be a lot of pressure for changing the regulatory rules, policies, laws and acts to reduce imports and high dependency on other countries.

Responsible Globalization

1. Governments may impose sanctions on countries which promote the eating of wild animals, and become climate change or pandemic hotspots.

2. There will be an increased pressure to rationalize globalization initiatives. For example, for a component manufacturing company, shutdown in one country halts the production in another country. Local initiatives will be encouraged and sought to reduce dependency on other countries.

Society New Normal

Significant changes in various spheres of society, as well as social practices, are already visible and a few positive and negative changes have been discussed here.

Emergence of New Behaviours in School Students

1. Children are forced to stay at home due to the fear of COVID-19. Home environment will play a vital role in impacting children's mindset.
2. Parents no longer allow their children to freely play with their friends due to the fear of COVID-19. This will continue until their children are vaccinated.
3. Students will experience hybrid schooling—physical and virtual schooling. There will be changes in seating arrangements in school buses, classrooms to maintain the required physical distance.
4. There will be a serious impact on government school education in developing countries, as most of the parents may hesitate to send their children to government schools due to a lack of confidence in the government's preparedness to provide a healthy, safe and secure environment at schools.

High Social Unrest

1. Layoffs, lack of social life, staying at home for a long time and low or no income will lead to mental illness like anxiety and depression.

2. There will be a major impact on the economy and its severe impact on the people at the bottom of the social pyramid will make it difficult for them to meet their basic needs like food, clothes and shelter. This may lead to an increase in crimes.

Increased Social Responsibility

1. Public communities will become more vigilant and become whistle-blowers who raise the issues and risks associated with the pandemic, security, crime, pollution, plastic usage and other diseases.
2. Spiritual gurus will conduct various advocacy programmes to educate the common people on consciousness, yoga, meditation, hygiene, good health, immunity-boosting by referring to the multiple scriptures in the holy books.
3. There will be a rise in volunteerism for the betterment of society. Governments and political leaders will trigger patriotism and nationalism to get more support from society.
4. There will be a significant increase in civic awareness and commitment to the environment, biodiversity, ecology and public health.
5. There will be an emergence of the digital nation—millions of global communities work for the common good. Committed global virtual communities fight against poverty, corruption, dictatorship and climate change.

Rise in Fake News, Scams and Exploitation

1. Self-claimed gods will promote superstitions and false propaganda by claiming to predict future pandemics.
2. There will be an increase in fake news factories and intentional attempts to create gaps between different races, religions and regions by attributing COVID-19 to particular sections of the society.

3. There will be an increase in the number of scammers and cheaters (physical and digital), who want to gain monetary and other benefits from the fear of the pandemic.
4. There will be many debates on privacy vs safety of the people. Some governments and a few private organizations may misuse public data.

Lifestyle New Normal

The pandemic has also affected the lifestyle of people across the globe. In light of the changing world and rapid digitalization, the expectations of citizens are also changing. Some of these changes and expectations have been discussed in this section.

Embracing a New Lifestyle

1. A digital lifestyle will be embraced by everyone, as they get connected to the internet using the web, mobile devices, wearable devices, consoles, TVs. Digital life means most of the activities are done over the internet without physical visits. Digital also means sensor-enabled things like locks, cars, doors, switches which can be operated without touching them. People will prefer low-/no-touch digital activities.
2. Due to the fear of infection from hospital visits, telemedicine will become the norm for non-critical health issues. Hence, there will not be much travel or waiting time at the hospital and no risk of contagious diseases.
3. Job loss and salary cuts will lead to lower consumption of goods and services, as spending will be prioritized according to needs and wants.
4. New behaviours, new culture, new customs, new traditions and new etiquette will emerge with a digital lifestyle.

5. People will embrace a new lifestyle, avoid travel and avoid going to densely populated places (public events, restaurants, tourist locations, theatres). People will prefer to interact with each other over digital media. They will have to wear PPEs at work as per their profession.
6. People will also increase their focus on spirituality, humanity, meditation, yoga, natural herbal medicines (scientific Ayurveda) to improve immunity.
7. The emergence of cloud kitchens exclusively focused on cooking and home delivery with no dining option. Customers do online ordering, online payment and maintain a direct virtual connection with the chef to share their food preferences.
8. There will be an increase in the digitalization of homes. For example, no-touch or low-touch switches and doorknobs. Also, robot maids may become a norm to minimize the number of outsiders working at home.

Rise in Anxiety, Depression and Other Mental Illnesses

1. Alcoholics and addicts may be affected by mental illness due to regular lockdowns and rationed supply.
2. There will be a serious impact on the lifestyle of labourers and migrant workers as they will not get work and income due to the impact of COVID-19 on industries.
3. There will be an increase in incidents of domestic violence due to layoffs, lack of social life, staying at home for a long time and low or no income.

Rise in Virtual Networking

1. A decrease in physical visits to the homes of friends and relatives accompanied by an increase in virtual social networking and virtual happy hours using various collaboration tools.
2. Friends, relatives, visitors, guests, employees will avoid greeting each other with physical touch. The whole world

may embrace touchless greetings (Namaskar or bowing down).

3. People get habituated to join virtual concerts using web conferencing tools like Zoom, GoToWebinar, Webex, MS Teams and others. Virtual event management companies build unique tools, events to offer a personalized experience to individuals or communities. A few examples could be a virtual music concert, virtual magic show and others.
4. Virtual reality (VR) and augmented reality (AR) create immersive experiences that simulate the physical environment. With the fear of going outside due to pandemic, more and more VR- and AR-based virtual environments are emerging. Hence, people get into virtual socialization, virtual happy hours, virtual entertainment, virtual mental wellness sessions and many more.

At the end of the presentation, the country heads thanked Tejas and his team for their excellent homework and for sharing the details of the new normal (business new normal, government new normal, society new normal and lifestyle new normal). They were confident that this presentation had given them the direction for the next steps to tackle the current crisis. Nelson concluded the meeting by thanking all the participants.

CHAPTER 17
Digital Interventions in the New Normal That Emerged with COVID-19

Impact of the New Normal

The new normal, which is emerging during the COVID-19 pandemic, has been bringing changes in business, government, society and lifestyle. While a few of the changes may be short-lived, most of the changes that help organizations gain a competitive advantage, design business models, products and services will be absorbed by them.

For example, remote working is the new normal and it will continue even post the pandemic. Biosecurity checks at airports may continue even after the pandemic is over. Therefore, the digital interventions given here are not limited to the COVID-19 pandemic, they will help organizations accelerate their digital journey for at least 3–5 years.

Possible Digital Interventions in the Business New Normal

The minister of finance, minister of industries and minister of HRD of one of the fastest-growing countries convened a meeting, chaired by the prime minister, with all the major industrialists and start-ups to discuss the new normal. As per the minister of the council's direction, a lot of emphasis was placed on the best leveraging digital technologies to simplify policies, rules and regulations and establishing a hyperautomated system to deliver faster, better and steadier services to business stakeholders.

Minister of Finance James was not happy. He was nervous that the country's GDP will be negatively impacted and that will have a severe impact on the social and economic situation of the country. Due to COVID-19, most of the industries were facing significant challenges, and it was no surprise that a few companies may even shut down due to severe cash flow issues. He urged the industry leaders to come together and address the challenge with the vision of one nation-one goal.

James explained how each industry was facing a grave situation due to the COVID-19 situation in their country. He classified the industries into three categories:

1. Negatively impacted
2. Moderately impacted
3. Positively impacted (new opportunities will be available)

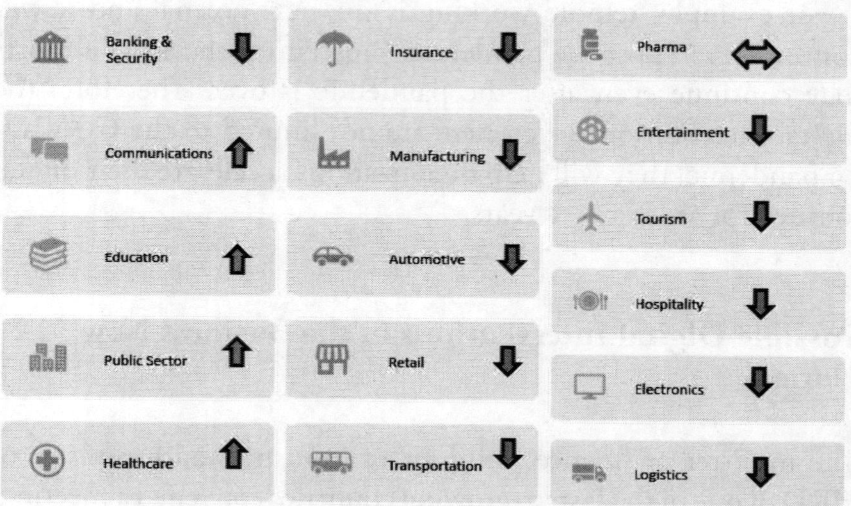

Legends

Source: Author

Figure 17.1: Impact of the COVID-19 Pandemic on Various Industries

James formed a task force with representatives from each ministry, industry and start-up. They were entrusted with the task of sharing possible digital interventions, which would create new opportunities for the country to improve the economy and operational excellence. He requested the task force to present the list of digital interventions within two weeks.

The task force spent two weeks on focused efforts and came up with a few innovative suggestions. The chair of the task force, Malcolm, presented the predictions on possible digital interventions that would create a positive impact in the new business normal. Malcolm added that a few of the digital interventions might become a permanent part of business activities and a few others would be short-term measures.

Business Value

1. Immediate business opportunities, new business models powered by digital technologies must be identified in the new normal to generate new revenue streams. Optimization using digital technologies must become the priority instead of transformation in the immediate future.
2. Organizations must change their products, services and value chains using digital technologies to reduce the impact on society and the environment. For example, home appliances will be designed with the internet of things (IoT) and artificial intelligence (AI) technologies that will reduce a lot of electricity consumption with automatic switch-off feature and reduce carbon emissions.

3. Organizations must start leveraging crowd innovation digital platforms to get new ideas and concepts from globally distributed specialists to resolve various problems effectively and efficiently within a shorter period.
4. Identifying the right initiatives and investments, with the help of AI/machine learning (ML) algorithm-based prioritization engines, would save money and add business value immediately. Rescoping of the initiatives must be done as well.

Customer Engagement

1. Consumers may want to subscribe to the best cars, home appliances and electronic devices and pay as they use. Consumers may show less interest in one-time capital investments.
2. In light of physical distancing and fear of travel due to COVID-19, organizations must offer virtual tourism services using augmented reality (AR) and virtual reality (VR) technologies. Customers will be able to visit any tourist location of their choice virtually and enjoy the immersive experience.
3. New business models will emerge in the entertainment space that cover virtual entertainment events like virtual happy hour, virtual live concerts, personalized live concerts and virtual kitty parties. New organizations will emerge to professionally manage virtual entertainment services.
4. Organizations will implement NFC-/Bluetooth-based applications with beacons while offering services to customers to minimize contact with objects and ensure the safety of the customers from the virus at retail malls and other locations.
5. Organizations will implement robot-based customer help desks that optimize human interventions and provide personalized services to customers.

Business Performance

1. AI/ML algorithms must be built for effective demand and supply mapping that eliminates waste. Establish IoT, unmanned vehicles and blockchain-based solutions for effective logistics and goods tracking from source to destination. These solutions help in reducing the cost and ensure the delivery of quality goods in a timely manner.
2. Organizations must establish smart shop floors powered by the industrial IoT to improve production capacity, machine utilization, workers productivity and safety.
3. Organizations must promote virtual socialization events, virtual relationship with stakeholders, virtual education, virtual events, virtual negotiations, etc. Telepresence, video conferencing and chat tools need to be used extensively.
4. Organizations must spend on digital supply chain solutions to optimize the lead time to deliver the products to customers. Digital supply chain solutions also help in reducing the transportation cost, tracking the health of perishable goods from farm to fork and provide the real-time status of the products in transit to manage the customer expectations in case of any delay.
5. Account receivables will be delayed, which would create serious cashflow issues. MSMEs (micro, small and medium enterprises) will face serious liquidity issues. Organizations could build a daily cashflow tracker app that collates data from different systems and provides automatic recommendations using AI/ML algorithms to manage cash flow effectively.

Employee Engagement and Performance

Digital employee engagement and productivity systems will be developed using AI/ML algorithms and gamification technologies. These digital systems will bring a lot of transparency in employee performance, workload and the potential of the employee for future roles. Based on the data available about the employee from various systems in the organization, the digital system will automatically propose recommendations for retrenchment, salary cuts, etc. Transparency and objectivity will become the norm.

Safety and Security

1. A few companies will invest in manufacturing integrated masks, protection coats, sanitizers and smart face shields. This integrated kit will be easy to use and remove, and will be washable. We may see people in attires similar to that of astronauts on the roads for a few months.
2. Just like security equipment, thermal scanners will be established at each mall, restaurant, movie theatre and other crowded places. Central air conditioning system will be used only on a need basis.
3. Many innovations will emerge from the digital age start-ups to produce affordable and smart antimicrobial PPEs (personal protection equipment) using digital technologies. Not limited to producing PPE, these organizations must introduce free orientation sessions using mobile apps to take precautionary steps while moving from "home to work to home" and "home to school to home" to avoid infections. Games will be introduced to make people aware of protection measures against COVID-19 or any other epidemics/pandemics.

Sales and Marketing

1. Organizations must reduce travel overheads and manual interventions through the digitalization of services. The sales process must be completely digitalized. Internal IoT initiatives would reduce the cost of operations. All these initiatives will reduce the cost of acquiring a new customer which would help in reducing the cost of the goods.
2. Organizations will reduce spending on marketing and costly advertisements. Crowdsourcing will involve customers in the process of advertising products. For example, Coca-Cola

introduced a global challenge for innovative video ads. Out of all, the best ads were broadcasted by the organization on social media. This reduced the cost of ad production and broadcasting.

Technology Management

Today, much fake news is being published on social media. Abusive and offensive language is being used as well. Deepfake technology uses AI/ML algorithms to morph lip movement and gestures in a video and suppress the original content to create and deliver fake content. For example, if a person passionately praises his country in a video, deepfake technology can morph the video in such a way that the person appears to be talking against the country. These sort of morphed videos from influencers spread like wildfire on social media and create negative emotions and perceptions.

1. Organizations must invest in building AI/ML algorithms to identify abusive/offensive posts and delete them as appropriate.
2. Organizations must introduce new tools and policies to control deepfake technology issues in the new normal. Deepfake technology has been creating an unpleasant experience for celebrities and common people. During a crisis period, anti-social people act smart and use such technology to create a negative impact on businesses and society.
3. Organizations must introduce new policies, digital devices and tools to enhance the biosecurity of the employees, customers, partners and suppliers.
4. The technology departments must enable remote software and infrastructure management using various digital tools that minimize travel and physical visits.

Work From Home

1. Organizations must focus on establishing a sustainable work from home (WFH) system and modify their applications and infrastructure in such a way that the home network is secured. Application and infrastructure monitoring with integrated analytics must be introduced.
2. Organizations must introduce innovative WFH virtual induction programmes and provide WFH digital toolkit to empower their employees to deliver faster, better and steadier services with high productivity.

Possible Digital Interventions in the Government New Normal

Prime Minister Subhash convened another meeting with the council of ministers to discuss the impact of COVID-19. Minister of Finance James shared updates from the meetings he and his colleagues had had with the industry leaders. James distributed a booklet that consisted of the list of possible digital interventions in various industry sectors, which would help in generating new revenue streams, optimizing costs while safeguarding the security and safety of the multiple stakeholders against the COVID-19 virus.

Subhash lauded their effort and advised the minister of finance to release a similar booklet on possible digital interventions by the government to implement real-time digital governance. Subhash emphasized the importance of leveraging technology to improve the economy and quality of life of citizens during the COVID-19 pandemic. James and his team spent the next two months with other ministers and different government agencies, analyst firms and came up with a list of interventions. He shared the list with the council in the next meeting.

Digital Governance

1. Both the government and various organizations must implement the General Data Protection Regulation (GDPR) Act. This means all software applications and infrastructure will be updated in such a way that the data privacy of people is safeguarded. Each country must refine its cyber laws, data protection policies and laws to safeguard citizen data privacy. The government must include a few special provisions on data privacy, especially concerning disasters and pandemics.
2. Citizen centricity and empathy must be inculcated in public servants as essential qualities. Self-centric and non-empathetic civil servants will be taken for a ride on social media by citizens.
3. Governments must introduce drone fly zones with proper air traffic management system and protocols. The drones can be used for surveillance, announcements and disinfection. Private organizations could use drones to deliver goods and as flying taxis.
4. Most countries are experiencing zero or negative GDP growth due to COVID-19 which has put much pressure on governments to create short-term bailout plans. For long-term growth, the country must start working on various digital products and services to increase their exports to create new markets and revenue streams to contribute to GDP growth.
5. Most of the critical citizen services must be provided using digital kiosks powered by voice technologies that support vernacular languages. Internet Protocol television (IPTV) at home will become the norm with the support of fiber grid. Using IPTV, citizens will be able to directly interact with government officers and get their service. The government must accelerate the digitalization of court cases, immigration services, passport services and other citizen services.
6. The federal governments must introduce digital platforms that would integrate the systems of provinces/states and local

bodies. This would help in straight-through processing, faster delivery of services and improvement in transparency. The focus must be on removing disintegrated and isolated systems. For a few technology initiatives, national-level integration must become the norm using digital platforms approach.

7. The government must create a national animal consumption index and enforce stringent measures to regulate the consumption of animals which spread or cause deadly diseases.
8. The government must emphasize on digital governance, under which, citizens will have the provision to share their concerns, suggestions, ideas and solutions to shape the nation. Digital governance will enable inclusive governance. Citizens will be able to actively participate in the formation of government policies.
9. The government must introduce an integrated Citizen Alert System (CAS). This system will provide real-time information to citizens and tourists about the locations which might be risky to visit due to an epidemic, security issues or law and order issues.
10. The government must mandate bank accounts for citizens and introduce direct digital payments to the people who are at the bottom of the social pyramid and are in poverty. This will help remove intermediaries and delays. For example, India has successfully introduced the Pradhan Mantri Jan Dhan Yojana (PMJDY), which is a national mission for financial inclusion that ensures banking services are made available to every citizen. PMJDY also ensures direct benefits transfer of various government schemes without any middlemen.
11. The government must slowly move away from the old-fashioned governance systems. Public servants should not be limited only to government officers. More and more leaders from the private sector must be given responsibilities to help in shaping the nation. A district administrator could be a proven industry leader, who could be appointed for

five years to work under the governance framework with comprehensive training.

12. The performance of public servants will be driven by citizens with no political affiliations using digital systems.
13. The rise in the usage of digital technologies will lead to many cyberattacks and cybercrimes. This will disturb day-to-day lives and lead to financial burdens. The government and professional organizations must conduct digital literacy and cybersecurity knowledge enhancement programmes to prevent people from falling into the trap of cyber rogues.
14. Virtual walls with QR codes could be installed in bus stations, railway stations, theatres and airports. Citizens could scan the code with their mobile phones and order, say, groceries. Once they have made the payment, their order will be delivered to their homes, by retail stores (private or public). In this way, crowds could be minimized in retail malls or stores.

Public Distribution System

1. The government may also establish millions of digital distribution centres like vending machines at various locations and encourage citizens to purchase items with a smart card.
2. The government must put considerable focus and money into digitalizing the public distribution system. COVID-19 has exposed many loopholes in the system. Citizens should be able to order essential commodities from home without visiting the government stores or shops. The government must use digital technologies to improve the supply chain system and based on the requests put in by the citizens through the mobile app or web applications, commodities could be delivered to citizens within a shorter period. Government delivery executives must be provided with protective gear to safeguard them from pollution, heat and viruses.
3. The government must estimate the demand and supply at different locations and plan the distribution of essentials as

per pre-defined schedule every month. In case the government does not have the required resources for home delivery, it may set up mini-warehouses at identified locations. Each mini-warehouse may have a fixed number of smart boxes (IoT-enabled) for each residential address in the area. The resident visits mini-warehouse and opens the lock of the smart box using an NFC-enabled mobile phone. The smart boxes will send automatic alerts to the citizens when it is ready to be picked up or delivered. Citizens will scan the code on the box to send the delivery report to the government.

4. The government must focus on re-engineering and modernizing the home shelters used during natural calamities. These home shelters will be rebuilt like colonies with all necessary amenities and services supplied by a government public distribution system using vending machines and digital payments. The shelters will be designed to minimize the effects of the pandemic. The home shelters could be used as budget hotels when there are no calamities, like revenue generation hubs.

Public Health

1. Discounts will be provided to doctors and other healthcare professionals on various products and services as a way of respecting their work. Digital profiling of doctors on the basis of public voting via a mobile app will help identify the real heroes.
2. Airports will become micro DPHCs (digital primary health centres) and travellers will be expected to go through these DPHCs to ensure that they are not carriers of dangerous viruses. Self-tests will become the norm at airports. Government investments in rapid test kits and affordable diagnostic tools must be increased.
3. Hospitals must install surveillance cameras, which will help in spot crowding of patients waiting for meeting the doctor in

the hospital. Based on this information, the hospital mobile app, which is installed by the patients, sends alert messages to patients to sit in the designated waiting halls with their estimated time to see the doctor. Waiting halls must be designed as per the requirements for maintenance of physical distance.

4. Pharmacies at hospitals must be replaced with medicine vending machines. Patients will use mobile phones with NFC/Bluetooth features for digital payments and get medicines from the vending machines.
5. The government must invest a high percentage of GDP in improving public healthcare systems using digital technologies. DPHCs to enable telemedicine from remote locations and any time medicine machines (ATMMs) which will require a smart card or passcode to get prescribed medicines.
6. Super speciality government hospitals in urban areas must be connected with DPHCs in remote locations using the internet in the hub and spoke model. Telemedicine, robotics-driven isolation wards will become the norm. Before entering hospital premises, mandatory tests/verification must be done to separate healthy and infected patients. A strict appointment system will be implemented to avoid crowding of patients in the hospitals.
7. The government needs to spend considerable money on biotechnology innovations and new medicines without leaving this part completely to private organizations. Global collaborative initiatives for drugs innovation among different nations will become the norm.
8. While the government must encourage private businesses, it must also spend a considerable amount of money on public health, education and agriculture. Establishing many research centres and technology innovation centres will make the country a hub of education, healthcare, medicines and

food production. The country must become self-sustainable in these areas.

Education

1. Establish reskilling centres to prepare those who have lost jobs for their next assignment. More digital learning platforms will emerge with gamification techniques and personalized learning modules.
2. The government must focus on introducing virtual education and must make sure that even the less privileged students get world-class education sitting at home. IPTV will become an instrument for virtual home education.

Manufacturing

1. By leveraging the power of Industry 4.0, the government must focus on manufacturing essential items like PPEs, medical devices, medicines, etc. Reverse innovation must be promoted.
2. Political leaders will leverage the COVID-19 situation and raise the nationalism and protectionism sentiment. Due to COVID-19, the global supply chain has been disturbed and production has been stopped by many manufacturing companies. Hence, most of the countries have started working on various initiatives to strengthen their local manufacturing facilities for components, spare parts and other products.
3. The government must encourage industrial and academic collaborative innovations that promote indigenous technologies for the manufacturing of products. Reverse innovation will become the norm and developing nations will become manufacturing hubs for developed countries.

New Age Politics

1. Digitalized electoral systems will emerge out of IoT, blockchain smart contracts and strong cybersecurity

measures. This would help avoid fraud in elections. Anywhere, anytime and any device-based voting will become the norm. After the pilot stage, it will take a few years to institutionalize digitalized electoral systems. Cyberattacks and hacking will become a significant challenge, and the government must invest in tackling the same.

2. Political science universities and institutions must be established to train the next generation of political leaders on new age politics. More and more digital tools will be available to connect with citizens in real-time and to provide personalized interaction and support.
3. The government must also introduce new policies and norms to share the profile of the leaders for the nominated posts. Digital profiles of the leaders will be shared with the public to improve transparency.
4. There will be a surge of electoral candidate profiling apps. Digital profile of the candidates will be created by collating public information available about them like education, criminal records and attitude based on social media and other news items. Citizens will get an excellent picture of the type of candidate they are going to elect. This will ensure candidates are careful about their public image and work.

Technology Management

1. The government must introduce vernacular voice-based digital apps/bots to offer services and prepare the citizens for digital life.
2. The government must invest in digitalizing villages, towns and cities. Digitalization generates a lot of data from various things like cars, bins, water pipes, surveillance cameras, energy distribution feeders, social media, geospatial and other systems of government agencies. The command control centre in the city will mine the data received from various

sources and apply AI algorithms to predict crimes, women safety, health issues, pollution and other natural calamities. Automatic alerts and recommendations will be sent to the citizens and concerned authorities to take necessary action.

3. Erstwhile Andhra Pradesh Chief Minister N. Chandrababu Naidu established a sophisticated Real-Time Governance Centre (RTGC). RTGC was expected to become a state-level predictive and preventive centre to safeguard human lives, infrastructure and other physical assets against natural calamities with no or minimal destruction.
4. The government must issue a smart card with an identification number for each citizen. This smart card could be used to purchase items, travel, hospital visits, avail government services and withdraw cash. "One Nation, One Card" for a smarter life.

Possible Digital Interventions in the Society New Normal

While the COVID-19 virus has created a major disruption in human life, it has also reminded everyone to be socially responsible. Humans have destroyed the planet in the last 50–60 years and have polluted the air and water, led to the extinction of various species and encroached forests. Humans must coexist with nature, but unfortunately, we humans have started dictating the terms of existence to nature. Now nature has reminded us that we are only visitors and must honour its rules.

Rakesh Jain, a famous social activist, had organized a workshop which was attended by many social activists and technology specialists. They discussed how technologies can play a very vital role in safeguarding the environment and biodiversity. They identified a few important digital interventions that could enhance social responsibility and empower people to start adopting nature-friendly activities.

Social Responsibility

1. The COVID-19 pandemic has allowed most people to reflect upon themselves and to realize the importance of accomplishing sustainable development goals (SDGs).
2. Global virtual communities were formed by like-minded people who have taken up the social responsibility to address issues in the areas of ecology, environment, climate change, public health, poverty, hunger, etc. These virtual communities use digital crowdsourcing platforms to identify affordable digital solutions for social issues.
3. With these virtual communities and crowdsourcing initiatives, the real global citizens will emerge, working for a common purpose. Not-for-profit organizations will utilize digital technologies to motivate and inspire global volunteers to work for sustainable development goals.
4. Citizens use social media and other mobile apps to share information about hotspots of diseases and crimes. The data gathered by citizens can be shown on Google Maps based on the risks, like red would be high risk, medium risk in case of amber and green as low risk. In this way, citizens could become whistle-blowers.
5. Citizens have become more socially responsible. They follow a few rituals like monthly observance of "No Vehicle Day", "Self-home Locking Day", "Zero Electricity Day". These rituals help in reducing pollution, CO_2 emissions to some extent. Those citizens who voluntarily participate in these days must be rewarded a few digital tokens and reward points by the government. These digital tokens and reward points could be used by citizens to get discounts at different stores and special privileges to avail government services.

Education

1. Schools and colleges must invest in rearranging their physical infrastructure in such a way that physical distance is maintained at schools and in school buses.

2. Schools must mandate lightweight PPE for each student to avoid the risk of viral infections.
3. Schools and colleges must introduce virtual classrooms with AR/VR immersive experience and gamification techniques to encourage students to participate in classes virtually from their homes.
4. Personal computers, personal mobile devices must be secured using cybersecurity software.
5. Social media etiquette must be evolved to encourage the safe use of social media and virtual games. A cybersecurity course could be introduced in the school curriculum for students from eighth grade onwards to help them best leverage the digital world.

Animal Rights

Every year worldwide, a few million animals are killed or used for various scientific experiments. Social activists have been and will continue to protest and resist the practice of animal trials. Hence, there will be more focus on creating virtual organs using 3D modelling and other digital technologies, which are like physical organs and will be used for clinical trials. In a few cases, the virtual organs can be printed using 3D printing technology, and clinical trials could be conducted on those 3D printed organs.

Possible Digital Interventions in the Lifestyle New Normal

The COVID-19 virus has changed the lifestyle of millions of people in the world. In a way, the freedom of social networking has been taken away by the pandemic. An element of fear has been created in the minds of people.

Sourabh, a software engineer from Hyderabad, had to make many changes in his lifestyle because of the COVID-19 pandemic. He has started using the digital apps available to him to shape his new lifestyle, which include yoga and meditation. He has learned to actively participate in virtual social events to connect with friends and relatives regularly. Working from home has taught him how to work in a distracting work environment.

He decided to share his experience with his colleagues. They were curious to listen to the possible digital interventions that could enhance their lifestyle.

Digital Lifestyle

1. Due to new technology innovations, people have acquired new behaviours and attitude. They
   a. live together with intelligent machines and things,
   b. do less physical work and use natural motor skills less frequently,
   c. remotely manage their work and
   d. have become less tolerant.
2. There has been a reduction in physical movement as people have gotten accustomed to a digital lifestyle, in which anything—fans, switches, cars—can be operated remotely. Information and services can be obtained using the internet and mobile devices without going out. This means day by day, people use significantly less physical motor skills and become prone to health issues like neck pain, back pain, head pain, hearing loss and obesity.
3. Human eyes are not designed to read the tiny font/text on small screens with too much brightness. Hence, there has been a decline in the quality and quantity of sleep.
4. In the digital era, most people use the internet, mobile phones, mobile tablets, laptops, wearable devices, smart TVs and many more. Some are even addicted to these devices and continuously look for information, content and communicate with colleagues and friends constantly. They feel impatient if they are away from these devices. This is

called a heavy digital diet, which may cause many health issues like back pain, neck pain and other non-communicable diseases too. Hence, the government and businesses will have to introduce digital diet programmes and policies to make sure that people use new age digital technologies and devices carefully and judiciously.

5. People have started practising yoga, meditation and physical exercises to reduce the negative impact of the digital lifestyle.

Entertainment

1. Virtual celebration of festivals, family functions and meeting with friends will become the norm. Physical visits will be minimized. VR, AR, unified communications and social networking technologies will enable emotional connection among people.
2. Event management companies will transform themselves to arrange innovative, entertaining and emotionally engaging virtual events.

Food

1. Cloud kitchens will maintain biosecurity throughout the cooking life cycle (cleaning of vegetables to delivery at home). They do not have any dining facility. Customers can place the order using a mobile app. They can directly chat/speak with the chef about their order. Bio-secured chefs and cooks who wear personal protection kits while working will be the new norm.
2. COVID-19 has created much fear in the minds of people about going to restaurants for dining. The government must introduce stringent policies to certify cloud kitchens which follow the standard hygiene practices from preparation to delivery of food.
3. For food delivery, mandatory safety assurance barcodes will be pasted on the delivery boxes, which would display the

body temperature of the cook and the delivery personnel and the safety precautions they have taken while cooking the food.

Personal Finance

There will be an increase in the use of personal finance apps, which gather information on income and expenditure to suggest optimal spending plan for the user. These apps will send alerts if users cross their monthly spending budget proposed by the app. They will also bring financial discipline into the lives of the users. By integrating all the bank accounts of an individual, these apps will be able to give a 360-degree view of income and expenditure.

Wellness

1. Brain-computer interface technologies help in preventing smoking, alcoholism and motion sickness. They also help in detecting brain tumours, brain disorders, sleep disorders and psychological disorders in advance.
2. Homes will be digitalized with digital locks, wireless switches, fans, IPTV, IP cameras. This would increase security and comfort by ensuring that physical touch is minimized to avoid viral infections.
3. More than 300 million people in the world suffer from depression today. Digital therapeutics behavioural interventional technologies will be used by people to avail psychotherapy using video conferencing. Simulation technologies create a virtual environment in which the activities performed by people would ease their anxiety and depression.
4. Digital social communities are emerging which allow people to connect with others who have been suffering from mental illness. These virtual communities also include counsellors,

mental health specialists and mentors. These communities help depressed people find the support they need.

5. VR- and AR-based apps, powered by AI, released by spiritual organizations will encourage millions of people to practice yoga, meditation and physical exercise. These apps can help motivate even lazy people to do physical exercise and yoga regularly. Motivational speeches by spiritual gurus telecasted through mobile apps will enhance the humanity of people.
6. Wearable devices play a significant role in taking care of the seriously suffering people from mental illness. These wearable devices share the details of vital health indicators in real time with concerned doctors.

Cybercrimes

People will be exposed to new types of cybercrimes. Cybercriminals use malware and ransomware and shut down the servers, individual devices and demand heavy ransom. Data stored without encryption on the cloud infrastructure become an easy source of hacking. During a pandemic or similar emergencies, criminals will use different techniques to create fake websites and collect private data in the guise of offering solutions to ailments. Law and order officers will have to introduce new policies, processes and technology solutions to deal with cybercrimes.

CHAPTER 18 COVID-19: Employee Safety and Health

Challenges in the Time of COVID-19

The CEO of Srini Corporation, Koushik, was sitting anxiously in his office. He had already consumed three cups of coffee within the past 45 minutes. It had been 90 days since Srini Corp had been shaken up by an invisible enemy called COVID-19. Srini Corp manufactures washing machines.

In the last 90 days, employee morale had gone down, so did 50% of their revenue. They had also been experiencing major cash flow issues because of a 20% decrease in orders. In the current situation, the cost of operations was higher than the revenue.

This invisible enemy had challenged humanity by wholly disrupting the nation, people's lives and businesses. CEO Koushik had been facing a situation like this for the first time in his life. The business continuity strategies designed by Srini Corp proved to be ineffective against the COVID-19 pandemic.

The new normal of Srini Corp was still uncertain, and Koushik knew it may evolve in the coming months. He was very worried about his employees and their future at Srini Corp. Srini Corp was well connected with its employees and their families. The leadership considered employees as one of the critical assets of Srini Corp. The CEO spoke with Raj, the chief human resources officer (CHRO) of Srini Corp, to immediately start working on employee safety and health.

The CHRO along with other senior leaders identified the following focus areas concerning employee health and safety in the new normal:

1. Manage uncertain future and low employee morale.
2. Create employee journey map and identify potential virus transmission points.
3. Identify digital interventions for employee health and safety.
4. Even if the pharma companies develop a vaccine for COVID-19, it may take 12–18 months for everyone in the country to get vaccinated. Hence, it is vital to focus on employee safety and health.

Uncertain Future and Low Employee Morale

One of Koushik's employees called him up to inform him that he has been suffering from severe depression and anxiety. The lockdown and working from home had disturbed him. He was concerned whether he will have a job in the future or not.

CHRO Raj shared with Koushik that many such concerns had been raised by their employees. One of them was pregnant and she was under extreme psychological stress as coming to the office these days made her worry about her yet to be born baby.

Even after the decline in the number of COVID-19 cases, most of the shop floor workers were concerned about working on the shop floor without protection measures. Workers felt that the COVID-19 impact was not yet over, it may attack again anytime.

Raj further shared that the families of their employees had also been sending messages to him. They are very concerned about their spouses coming to the office. They wanted to ensure that Srini Corp had established the necessary protection measures at the office. Another employee had insisted on coming back to the office because of the pressure she was under due to domestic violence.

While there were many challenges concerning the supply chain, production shut-down, customer order management and cash flow, Koushik felt that immediately addressing his employees' issues was the most critical task. After that, with the support of employees, they could streamline other functions.

Koushik asked Raj to arrange a leadership council meeting immediately to discuss employee health and safety at Srini Corp in the context of the COVID-19 pandemic.

Raj was asked to form a team, analyze employee issues and come up with possible solutions, before the meeting. Koushik also emphasized the importance of digital technologies in the enhancement of employee safety and health. Contactless operation, which was crucial in the current situation, was made possible by digital technologies.

Employee Journey Map and Potential Virus Transmission Points

The coronavirus that causes COVID-19 is an invisible, deadly and contagious virus that spreads from human to human through droplets dispersed when an infected person coughs. If the infected person touches any doorknobs, elevator buttons, table, gates and others exposed surfaces, the virus may be transmitted to the healthy person who touches those contaminated objects/ surfaces.

Raj shared the lifespan of the virus on different kinds of surfaces, as given in Table 18.1, with Koushik and other leaders. They were surprised by the information. Raj warned that the COVID-19 virus is 1,000 times more infectious than other viruses. Koushik took a deep breath and said, "I think, we are getting into a severe crisis period, which we have never experienced in our professional careers before."

Surface	Examples	Lifespan of Coronavirus*
Metal	Doorknobs, jewellery, silverware	5 days
Wood	Furniture, decking	4 days
Plastics	Milk containers, detergent bottles, subway and bus seats, backpacks, elevator buttons	24 hours
Stainless steel	Refrigerators, pots and pans, sinks, water bottles	2–3 days
Cardboard	Shipping boxes	24 hours
Copper	Pennies, tea kettles, cookware	4 hours
Aluminium	Soda cans, tin foil, water bottles	2–8 hours
Glass	Drinking glasses, measuring cups, mirrors, windows	5 days
Ceramics	Dishes, pottery, mugs	5 days
Paper	Mail, newspaper	Some strains of coronavirus live for only a few minutes on paper, while others live for up to 5 days

Food	Takeout, produce	Does not appear to spread
Water	Drinking water	Coronavirus has not been found in drinking water
Fabrics	Clothes, linens	Not as long as it does on hard surfaces
Shoes	—	Ambiguous
Skin and hair	—	No research yet

Source: https://www.webmd.com/lung/how-long-covid-19-lives-on-surfaces

Note: *The data provided above is only for reference and the exact details must be obtained from the subject matter experts for any decision-making or action.

Table 18.1: Lifespan of the COVID-19 Virus on Different Surfaces

Sirisha from HR shared the employees' journey from **home to office to home** and the 10 interaction points and 23 activities which might enable virus transmission, as mentioned in Table 18.2.

S. No.	Touchpoint/ Interaction Point	Activity	Potential Virus Transmission
1	Home	• Preparing to go to office	• Hugging family members, possibly virus-infected family members • Contaminated objects/surfaces/items at home • Delivery executives of grocery, vegetables, medicines and milk at home
2	Pick-up point	• Moving from home to the pick-up point of the company bus	• Infected persons less than 1m physical distance
3	Vehicle	• Getting into the company bus	• Infected persons less than 1m away while getting into the bus or within the bus
		• Getting into the car	• Virus contaminated car door handles • Carpooling, no physical distance from other employees
4	Stopping point	• Getting off the bus at the bus stop near the office	• Infected persons less than 1m physical distance
		• Parking the car in the parking spot	• Infected persons less than 1m away
5	Walking path	• Walking from the bus stop to the company reception	• Infected persons less than 1m away
		• Walking from the parking spot to the company reception	• Infected persons less than 1m away

6	Attendance point	• Scanning the ID card for attendance	• Infected persons less than 1m away
7	Workplace	•Operating computer	•Computer, water bottle, surfaces, telephone and other contaminated items at the workstation
		•Operating machines/equipment	•Virus contaminated machines/equipment •Infected persons less than 1m away
		•Visiting in the organization	•Infected persons less than 1m away
		•Carrying files/documents	•Contaminated files/documents •Infected persons less than 1m away
		•Maintaining facilities	•Infected persons less than 1m away •Contaminated objects/surfaces
		•Attending physical meetings/ training	•Infected persons less than 1m away •Contaminated objects/surfaces
		•Attending conference calls	•Infected persons less than 1m away, sitting in a conference room to attend video or audio conference calls •Contaminated objects/surfaces
		•Meeting visitors	•Infected persons less than 1m away
		•Addressing the team	•Infected persons less than 1m away •Contaminated objects/surfaces
		•Visiting the restroom	•Infected persons less than 1m away •Contaminated objects/surfaces

8	Cafeteria	•Having snacks/coffee/lunch/inner	•Infected persons less than 1m away •Contaminated objects/surfaces
9	Office	•Preparing to go back home	•Infected persons less than 1m away on the walking path, at the bust stop, within the bus, at the car parking place •Contaminated objects/surfaces on the way back home
10	Home	•Parking the car at home or walking from the bus stop to home •Remove shoes, masks, socks and enter home	•Hugging family, possible virus-infected family members •Contaminated objects/surfaces at home

Source: Author

Table 18.2: COVID-19 Virus Transmission Points

Digital Interventions for Employee Health and Safety

Koushik pointed out that with the help of the preliminary analysis done by Raj and his team, they had identified ten interaction points of potential COVID-19 virus transmission throughout the employee journey. Koushik reiterated that as per their organization's philosophy, employees were their assets and they must safeguard them.

CFO Satya raised a concern that to address all the virus transmission points with the required health and safety infrastructure and resources, they would need a considerable budget. Arranging such a budget at this juncture did not seem like a good idea to him. Satya remarked, "In fact, we need to lay off a few employees to manage our cash flow."

Koushik stopped Satya and requested him to focus on employee safety and health against COVID-19. Koushik agreed that Srini Corp faced severe cash flow challenges and they must come up with innovative ideas to prepare their organization for the new normal. But he also stated that without employee support, they would not be successful in this **new normal** era.

CHRO Raj, CIO Rob and Corporate Admin Officer Steve presented their game plan to implement digital solutions for improving the employee safety and health and minimize/avoid COVID-19 infections. The six digital initiatives, which they wanted Srini Corp to implement are mentioned in Figure 18.1

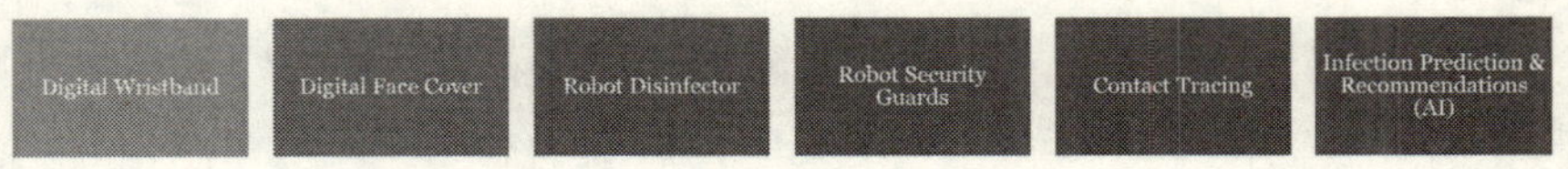

Figure 18.1: Digital Interventions for Employee Health and Safety

Koushik was curious about how these initiatives would minimize the risk of employees getting infected with the COVID-19 virus. Rob explained each initiative as given in Table 18.3.

S. No.	Digital Initiative	Details
1	Digital wristband	Issue digital wristbands to every employee. The digital wristband will have the following features: • Temperature reading of employee at home, bus pick-up points, attendance entry location and cafeteria • Attendance tracking with digital wristbands and NFC • Digital payments at cafeteria using mobile phones • Alerts to wash the hands or use the sanitizer • Alerts about physical distance o Alerts when an employee maintains the physical distance lesser than prescribed o Alerts when the chairs are placed at a physical distance lower than prescribed • Alerts when hands are too close to the face
2	Digital face shield	Issue AI-enabled digital face shield for each worker on the shop floor. Sensors embedded face shield covers the whole face of the workers. With the ML algorithm, the face shield reads the body temperature of the workers with the help of sensors. It alerts the workers if any worker in the shop floor has body temperature beyond the prescribed limit.
3	Robot disinfector	Introduce robots of various sizes and capacities to disinfect the office premises, objects, surfaces, vehicles, restrooms, elevators and cafeteria at regular intervals.

4	Robot security guards	Reduce the burden of the security staff and allow them to monitor the robot security guards in the office premises.
5	Contact tracing	Usage of contact tracing apps may become a norm by organizations. These apps in the mobile phone logs the details about other phones which come in contact using Bluetooth. The information captured use random numerical ID codes without personal information, and after a designated number of days, the information will be deleted automatically. That means the information about the employees who come nearer to each other will be stored in the form of anonymous keys on their mobile phones locally. Contact tracing apps may also log the length of the time a person has been in contact with another person and how far away using Bluetooth. If any employee in this contact network gets infected with COVID-19, the information is updated by the employee in the contact tracing app, the anonymous keys or random numeric ID codes of the employees whom the infected employee contacted in the last 15 days will be uploaded to the central server. Then all the employees in the infected employee's contact network will be informed about the infected employee (anonymously) and recommend getting tested.

6	Infection prediction and recommendations (AI)	Based on the physical distance maintained by the employee, body temperature and contact tracing app information, it will predict the potential virus-infected employees using a software application. The software application will provide automatic recommendations to employees to act appropriately (isolation, quarantine, testing and others). Employees who do not respond to the alerts sent by the digital wristband or face cover will be given level 1 and level 2 warning messages before disciplinary action is taken against them. Employees who follow the recommendations and alerts given by digital solutions will be rewarded.

Source: Author

Table 18.3: Digital Initiatives for Employee Health and Safety

Rob expressed candidly that the digital initiatives mentioned above required considerable investment, culture change and leadership commitment. Most of the employees may even feel that the organization is compromising the privacy of the employees. Koushik interrupted Rob and said that they will prioritize employees safety and health over privacy until things get streamlined. Rob added that once the impact of the pandemic is nullified and businesses come back to normal, the digital devices, solutions and apps developed by Srini Corp will be tailored/customized to enhance the health, safety and happiness of the employees by safeguarding the privacy of the employees.

Koushik asked CFO Satya to work with Raj, Rob and Steve to finalize phase-wise investments. Koushik also suggested that in order to provide additional support and safeguard employee and visitors, they must be categorized as follows:

1. **High exposure to people:** Provide personal protection equipment (PPE) as required (eg., security guards, cafeteria workforce, etc.)
2. **Medium exposure to people:** Provide face shields and face masks (e.g., shop floor workforce, delivery executives, etc.)
3. **Low exposure to people:** Provide face masks (e.g., employees working in the office)

Preparing Employees for the New Normal

Raj pointed out that most of the employees may feel that taking precautions at various interaction points is an overhead. In a few cases, there may be apprehension in employees about data privacy compromise.

Hence, Raj suggested that a series of programmes could be held to educate the employee on their health and safety and various initiatives, which Srini Corp has taken. There must be a change management task force, which is available to resolve any complaints and issues raised by employees.

Koushik and other senior leaders concurred with Raj. They emphasized the importance of identifying passionate volunteers from employees who will also become the brand ambassadors of Srini Corps' employee health and safety digital initiatives.

Employee Health and Safety Communication

Raj shared the initiatives taken by the HR team as mentioned below:

1. Posters and slogans at different places in the office
2. Daily internal news broadcast at 06:00 PM, summarizing the pandemic situation
3. A weekly newsletter on COVID-19 trends and future
4. Monthly open house with the CEO on employee safety and health
5. Family Connect—weekly webinar with family members
6. Employee safety and health communication champions for each unit/floor (as applicable)
7. Prediction and recommendations based on the AI algorithms to safeguard the safety and health of employees in the context of COVID-19

Employee Health and Safety: Training and Certification

Sirisha from HR shared different training programmes planned by them:

1. Introduction to the 23 (refer to Table 18.2) possible virus transmission points from home to office to home
2. Orientation on the digital initiatives introduced by Srini Corp and the Srini Corp COVID-19 Prevention App
3. Mandatory certification in "COVID-19 101" for every employee

Employee Health and Safety Champions Award

Raj emphasized the importance of promoting a positive culture and shared a reward and recognition mechanism:

1. Track the employee's performance and adherence to the organization's policies and practices based on the data received from various sources (digital wristband, digital face cover, etc.).
2. Educate the employees who have not been adhering to COVID-19 policies and conduct Family Connect sessions with them.
3. Reward the employee health and safety champions, who consistently adhere to COVID-19 policies and practices.

Koushik was now confident that with all the initiatives identified by Srini Corp, they could take care of employee's safety and health in the COVID-19 environment. He wished his team the best for their plans to return to a healthy work environment at the earliest. But he also cautioned that the new normal will be different from the past and they needed to be ready to handle the uncertainties and unknows with courage and confidence.

Weblinks

3D printing

https://amfg.ai/2018/05/29/3d-printing-transforming-spare-parts-industry/

Bimodal IT

https://www.cio.com/article/2875803/what-gartner-s-bimodal-it-model-means-to-enterprise-cios.html

Business ecosystem

https://www2.deloitte.com/cn/en/pages/about-deloitte/articles/pr-business-ecosystem-governance.html

Business model

https://www.intemarketing.org/marketing-information/marketing-models/osterwalders-business-model-canvas

Business operating model canvas

https://operatingmodelcanvas.com/

Business process

https://www.businessmapping.com/blog/business-processes/

https://www.sweetprocess.com/what-are-the-differences-between-a-policy-a-process-and-a-procedure-why-knowing-this-is-the-key-to-scaling-and-automating-your-business/

Citizen IT management

https://www.gartner.com/smarterwithgartner/the-mse-cios-guide-to-citizen-it/
https://blog.appsheet.com/5-key-practices-to-enable-citizen-developers

COVID-19

https://www.thestar.com.my/opinion/columnists/2020/04/12/life-after-covid-19-what-will-our-new-normal-be-like

Cybersecurity

https://insights.sei.cmu.edu/insider-threat/2019/07/cybersecurity-governance-part-1-5-fundamental-challenges.html

Data governance

https://www.pwc.in/consulting/technology/data-and-analytics/govern-your-data/insights/global-and-industry-frameworks-for-data-governance.html

Digital dexterity

https://emtemp.gcom.cloud/ngw/globalassets/en/insights/executive-guidance/documents/insights/executive-guidance-2018q3-digital-dexterity-at-work.pdf

Future of work

https://www.cognizant.com/whitepapers/from-to-everything-you-wanted-to-know-about-the-future-of-your-work-but-were-afraid-to-ask-codex4799.pdf

IT service continuity

https://invenioit.com/continuity/4-real-life-business-continuity-examples/
https://www.forbes.com/sites/cognitiveworld/2019/03/15/how-ai-can-and-will-predict-disasters/#c065761 5be26
https://www.mdsny.com/5-top-machine-learning-use-cases-for-security/

List of Tables

List of Figures

List of Abbreviations

AEC	Architecture, engineering and construction
AHM	Anytime Health Monitoring
AI	Artificial intelligence
AISP	Account information service provider
API	Application programming interface
AR	Augmented reality
ATMM	Any time medicine machine
AWS	Amazon Web Services
BFSI	Banking, financial service and insurance
BLE	Bluetooth Low Energy
BOK	Body of knowledge
BPaaS	Business process as a service
CAS	Citizen Alert System
CCM	Continuous change management
CEO	Chief executive officer
CeFi	Centralized Finance
CFO	Chief financial officer
CHRO	Chief human resources officer
CIO	Chief information officer
CIoT	Consumer internet of things
CoE	Centre of excellence
COO	Chief operating officer
CPG	Consumer packaged goods
CQaaS	Construction Quality as a Service
CRM	Customer relationship management
CSR	Corporate social responsibility
CSO	Chief strategy officer
CTO	Chief technology officer

CXO	C-level officer
DAO	Decentralized autonomous organization
DCAM	Data Management Capability Assessment Model
DDoS	Distributed denial-of-service
DeFi	Decentralized Finance
DoS	Denial-of-service
DL	Deep learning
DPHC	Digital primary health centre
ELM	Experience lifecycle management
EOI	Expression of interest
ERP	Enterprise resource planning
F&B	Food and beverages
FIFO	First in, first out
FMCG	Fast-moving consumer goods
GCP	Google Cloud Platform
GDP	Gross domestic product
GDPR	General Data Protection Regulation
GIS	Geographic information system
GPS	Global Positioning System
HaaS	Happiness as a Service
HCI	Hyperconverged infrastructure
HCM	Human capital management
HDLI	Heavy data, light intelligence
IaaS	Infrastructure as a Service
I&I	Intelligent and Immersive
IaC	Infrastructure as code
IoT	Internet of things
IPTV	Internet Protocol television
KPI	Key performance indicator
LAN	Local area network
LDQ	Lean digital quotient
LiDAR	Light detection and ranging
M&A	Mergers and acquisitions
ML	Machine learning
MR	Mixed reality
MSME	Micro, small and medium enterprises
MVP	Minimum Viable Product

NC	Non-conformance
NCD	Non-communicable disease
NFC	Near-field communication
NLP	Natural language processing
NPS	Net Promoter Score
OEM	Original equipment manufacturer
OTA	Over-the-air
P&P	Problem and Possibility
PaaS	Product as a Service
PCI DSS	Payment Card Industry Data Security Standard
PHC	Primary health centre
PLM	Product lifecycle management
POS	Point of sale
PPE	Personal protection equipment
PSP	Payment service provider
PWA	Progressive web application
R&D	Research and development
RACI	Responsible, accountable, consulted and informed matrix
RFP	Request for proposal
RFID	Radio-frequency identification
RTLS	Real-Time Location System
ROI	Return on investment
RPA	Robotic process automation
RPO	Recovery point objective
RTGS	Real Time Governance Society
RTO	Recovery time objective
SaaS	Software as a Service
SAFe	Scaled Agile Framework
SCM	Supply chain management
SDG	Sustainable development goals
SDK	Software development kit
SLA	Service-level agreement
SOP	Standard operating procedures
SPA	Single page application
SRM	Supplier relationship management

SVP	Senior vice president
TAT	Turn around time
TCO	Total cost of ownership
VECAAS	Vertical Cloud as a Service
VR	Virtual reality
VTM	Video teller machine
WEF	World Economic Forum
WFH	Work from home
WHO	World Health Organization
XaaS	Anything as a Service

ABOUT THE AUTHOR

V. Srinivasa Rao, or VSR as he is known in his circle, is a senior technology business executive, digital business leader and technology entrepreneur. He has a unique blend of experience of over 25 years in managing P&L of large-scale conventional, digital technology businesses and technology start-ups. He is one of the pioneers of digital business in the Indian IT industry. He worked as the de facto chief digital officer at Tech Mahindra, one of the top five Indian IT organizations, and prepared 39 business units and practice units for the Fourth Industrial Revolution and the digital economy. VSR has worked with major Indian IT companies, like TCS, Infosys and Satyam Computers. His name was featured in the *Fortune India* magazine as "Rubber Soul" for his role in the Satyam Computers' turnaround story, pulling it out of one of the biggest scams in the world.

Email ID: vsr@btbt.co.in

Podcast: https://anchor.fm/VSR

YouTube: https://bit.ly/3jvH09c

LinkedIn: http://bit.ly/2kPl8LF

LinkedIn Pulse: https://bit.ly/2IJs5uv

SlideShare: http://www.slideshare.net/vsr0001

Blog: http://vsr-blog.blogspot.com

VSR is the chairman of Smart Cities Working Group, Institute of Engineering and Technology (IET, India). In this position, he has been working on recommending policy changes and technology innovations to shape up Indian cities. He has authored a book on driving cities with the fusion of technology, titled *Software Defined Smart Cities*. It was published by IET, India.

An investor and digital business visionary, VSR has pioneered digital business solution innovations in his capacity as the chairman and managing director of BT&BT. He has conceptualized and built digital platforms like Platform Weaver, Co-Innovator; digital software products like myVoiceToday and QEnhancer and the Lean Digital Quotient (LDQ) Finder by leveraging mobile, cloud, analytics, artificial intelligence and social technologies in a leaner way.

VSR has advised and mentored CXOs as well as government officers on digital strategies, business models, policies, processes and governance. He was instrumental in establishing the International Institute of Digital Technologies (IIDT) in Andhra Pradesh, India. He also created a business blueprint and execution model for a fibre grid organization to acquire 5 million subscribers and enterprise business. VSR is credited with designing digital platforms for media house and smart energy meters for organization. He has created a blockchain strategy for one of the government agencies. He has provided direction to a few consultants of global big five consulting firms while working in a few customer engagements.

VSR has also been a panel member and keynote speaker at events held by dynamicCIO, HYSEA, Scrum Conference, NASSCOM, Jawaharlal Nehru Technological University (JNTU), Hyderabad, Confederation of Indian Industry (CII), Institution of Engineers (IEI), Summenter Pro, All India Management Association (AIMA), FTAPCCI, Intel and others. His interviews and articles have been published by NASSCOM, *Business Line*, *The Week*, *The Economic Times*, *Computerworld*, *The Financial Express*, *PCQuest* and others.

As a techno-social reformer, VSR has made many contributions to society in the following capacities:

1. IT advisor for FTCCI (https://www.ftcci.in/)
2. Convener of Bureau of Indian Standards (Blockchain Smart Contracts ISO Standard) (https://bis.gov.in/)

3. Former chairman of Volunteer Organization, Project Management Institute, Team India (PMI Team India, a conglomeration of 7 PMI Chapters in India with more than 6000 members) (https://www.pmi.org)
4. Former charter member and mentor, TiE, Hyderabad (https://tie.org/)
5. Evangelist of Leadership Development for students of age group 12–22 years (trained 2,000+)